THE CHANGING FLORA AND FAUNA OF BRITAIN

THE SYSTEMATICS ASSOCIATION PUBLICATIONS

1. BIBLIOGRAPHY OF KEY WORKS FOR THE IDENTIFICATION OF THE BRITISH FAUNA AND FLORA
3rd edition (1967)
Edited by G. J. KERRICH, R. D. MEIKLE *and* NORMAN TEBBLE

2. THE SPECIES CONCEPT IN PALAEONTOLOGY (1956)
Edited by P. C. SYLVESTER-BRADLEY, B.Sc., F.G.S.

3. FUNCTION AND TAXONOMIC IMPORTANCE (1959)
Edited by A. J. CAIN, M.A., D.Phil., F.L.S.

4. TAXONOMY AND GEOGRAPHY (1962)
Edited by DAVID NICHOLS, M.A., D.Phil.

5. SPECIATION IN THE SEA (1963)
Edited by J. P. HARDING *and* NORMAN TEBBLE

6. PHENETIC AND PHYLOGENETIC CLASSIFICATION (1964)
Edited by V. H. HEYWOOD, Ph.D., D.Sc. *and* J. McNEILL, B.Sc., Ph.D.

7. ASPECTS OF TETHYAN BIOGEOGRAPHY (1967)
Edited by C. G. ADAMS *and* D. V. AGER

8. THE SOIL ECOSYSTEM (1969) *Edited by* J. G. SHEALS

LONDON. Published by the Association

SYSTEMATICS ASSOCIATION SPECIAL VOLUMES

1. THE NEW SYSTEMATICS (1940)
Edited by JULIAN HUXLEY (Reprinted 1971)

2. CHEMOTAXONOMY AND SEROTAXONOMY (1968)*
Edited by J. G. HAWKES

3. DATA PROCESSING IN BIOLOGY AND GEOLOGY (1971)*
Edited by J. L. CUTBILL

4. SCANNING ELECTRON MICROSCOPY (1971)*
Edited by V. H. HEYWOOD

5. TAXONOMY AND ECOLOGY (1973)*
Edited by V. H. HEYWOOD

* Published by Academic Press for the Systematics Association

THE SYSTEMATICS ASSOCIATION
SPECIAL VOLUME No. 6

THE CHANGING FLORA AND FAUNA OF BRITAIN

*Proceedings of a Symposium held at the
University of Leicester
11-13 April, 1973*

Edited by

D. L. HAWKSWORTH

Commonwealth Mycological Institute, Kew, Surrey, England

1974

Published for the

SYSTEMATICS ASSOCIATION

by

ACADEMIC PRESS · LONDON · NEW YORK

ACADEMIC PRESS INC. (LONDON) LTD.
24-28 Oval Road
London NW1 7DX

U.S. Edition published by
ACADEMIC PRESS INC.
111 Fifth Avenue
New York, New York 10003

Library of Congress Catalog Card Number: 73-19008
ISBN: 0 12 333450 0

PRINTED IN GREAT BRITAIN BY
T. & A. CONSTABLE LTD., EDINBURGH

Contributors and Participants

Contributors are indicated by an asterisk (*)

BEST, J. A. *Department of Geography, University of Leicester, Leicester.*

*BOOTH, Dr C., *Commonwealth Mycological Institute, Ferry Lane, Kew, Richmond, Surrey.*

BUCKLAND, P. C., *York Archaeological Trust, 47 Aldwark, York.*

CHATER, A. O., *Botanical Laboratories, University of Leicester, Leicester.*

CONOLLY, Miss A. P., *Botanical Laboratories, University of Leicester, Leicester.*

*COOKE, Dr A. S., *Monks Wood Experimental Station, The Nature Conservancy†, Abbots Ripton, Huntingdon.*

*COPPINS, B. J., *King's College Field Centre, University of London, Rogate, nr Petersfield, Hampshire.* [Present Address: *Royal Botanic Garden, Edinburgh.*]

*CORBET, Dr G. B., *Department of Zoology, British Museum (Natural History), Cromwell Road, London SW7.*

*CORBETT, K. F., *British Food Manufacturing Industry Research Association, Randles Road, Leatherhead, Surrey.*

CRITTENDEN, P. D., *Department of Botany, University of Sheffield, Sheffield.*

CUMMING, Miss R. M., *18 Orleans Road, Twickenham, Middlesex.*

CURTESS, Miss G., *Natural Environment Research Council, 27-33 Charing Cross Road, London WC2.*

DANIEL, R. J., *Department of Environmental Sciences, The Polytechnic, Plymouth.*

DAVEY, S. R., *Hampshire County Museum Service, Chilcomb House, Winchester.*

DAWSON, Miss J. E., *Leicester Museum, New Walk, Leicester.*

DEADMAN, A. J., *Estate Office, Twyford, Barrow upon Trent, Derbyshire.*

*DUFFEY, E. A. G., *Monks Wood Experimental Station, The Nature Conservancy†, Abbots Ripton, Huntingdon.*

EASTER, T. D., *Botany Department, University College of Wales, Aberystwyth.*

EVANS, I. M., *Leicester Museum, New Walk, Leicester.*

*FELTON, J. C., *Woodstock Laboratory, Shell Research Ltd, Sittingbourne, Kent.*

*FITTER, R. S. R., *The Fauna Preservation Society, c/o Zoological Society of London, Regent's Park, London NW1.*

GADSON, Miss A., *18 Wellington Road, Enfield, Middlesex.*

GEARING, Mrs M., *53 Shanklin Drive, Leicester.*

GILBERT, Dr O. L., *Department of Landscape Architecture, University of Sheffield.*

GROVES, E.W., *Department of Botany, British Museum (Natural History), Cromwell Road, London SW7.*

HALL, Miss J., *Commonwealth Mycological Institute Ferry Lane, Kew, Richmond, Surrey.*

*HAMMOND, P. M., *Department of Entomology, British Museum (Natural History), Cromwell Road, London SW7.*

*HAWKSWORTH, Dr. D. L., *Commonwealth Mycological Institute, Ferry Lane, Kew, Richmond, Surrey.*

*HEATH, J., *Monks Wood Experimental Station, The Nature Conservancy†, Abbots Ripton, Huntingdon.*

HILL, R. N., *6 Kingsway East, Newcastle-under-Lyme, Staffordshire.*

*JONES, Dr W. E., *Marine Science Laboratories, University College of North Wales, Menai Bridge, Anglesey.*

KENWARD, H. K., *Department of Entomology, British Museum (Natural History), Cromwell Road, London SW7.*

LANSBURY, I., *Hope Department, University Museum, Oxford.*

LAWLEY, M., *29 Chamberlain Road, King's Heath, Birmingham.*

LEGG, C., *Botany Department, University College of Wales, Aberystwyth.*

LINDLEY, D. J., *44 Dovedale Rise, Allestree, Derby.*

LOADER, R. J., *31 Knowle Road, Knowle, Bristol.*

*MACAN, Dr T. T., *Freshwater Biological Association, Windermere Laboratory, Ambleside, Westmorland.*

McLAREN, P., *308 Victoria Park Road, Leicester.*

*MARSHALL, Mrs J. A., *Department of Entomology, British Museum (Natural History), Cromwell Road, London SW7.*

MASON, Dr J. L., *South Regional Office, The Nature Conservancy, Foxhold House, Crookham Common, nr Newbury, Berkshire.*

*MELLANBY, Professor K., *Monks Wood Experimental Station, The Nature Conservancy†, Abbots Ripton, Huntingdon.*

MESSENGER, K. G., *27 South View, Uppingham, Rutland.*

MOON, Professor H. P., *formerly of the School of Biology, University of Leicester Leicester.*

MORTON, J. A., *Department of Adult Education, University of Leicester, Leicester.*

NICHOLSON, P., *East Midlands Sub-Regional Office, The Nature Conservancy, The Lodge, Oakham, Rutland.*

MULDER, Dr J. L., *Commonwealth Mycological Institute, Ferry Lane, Kew, Richmond, Surrey.*

OWEN, Professor D. F., *University of Lund, Lund, Sweden.*

PARKER, Dr P. F., *Botanical Laboratories, University of Leicester, Leicester.*

PENDLEBURY, J. B., *Peak District Sub-Regional Office, The Nature Conservancy, Aldern House, Bakewell, Derbyshire.*

*PERRING, Dr F. H., *Monks Wood Experimental Station, The Nature Conservancy†, Abbots Ripton, Huntingdon.*

POTE, R. F., *85 Coleridge Drive, Enderby, Leicestershire.*

*PRESTT, I., *Central Unit on Environmental Pollution, Department of the Environment, London. [Present address: The Nature Conservancy, 19 Belgrave Square, London SW1.]*

RATCLIFFE, Dr D., *Botanical Laboratories, University of Leicester, Leicester.*

*Reid, Dr D. A., *The Herbarium, Royal Botanic Gardens, Kew, Richmond, Surrey.*
Ritchie, A., *City Museum, Albert Square, Dundee.*
*Rose, Dr. F., *Department of Geography, King's College, University of London, Strand, London WC2.*
*Sharrock, Dr J. T. R., *Ornithological Atlas, British Trust for Ornithology, 59 Curlew Crescent, Bedford.*
*Smith, K. G. V., *Department of Entomology, British Museum (Natural History), Cromwell Road, London SW7.*
Smith, Mrs K. G. V., *c/o* Mr K. G. V. Smith, *Department of Entomology, British Museum (Natural History), Cromwell Road, London SW7.*
Sneath, Professor P. H. A., *Medical Research Council Microbial Systematics Unit, University of Leicester, Leicester.*
*South, Dr A., *Department of Biological Sciences, City of London Polytechnic, London.*
Swindells, R. J., *12 Hawthorn Road, Wallington, Surrey.*
Thomson, A. G., *Director's Research Laboratory, The Nature Conservancy Headquarters for Wales, Bangor, Caerns.*
Tulloch, Mrs M., *Commonwealth Mycological Institute, Ferry Lane, Kew, Richmond, Surrey.*
*Wallace, E. C., *2 Strathearn Road, Sutton, Surrey.*
*Wheeler, A., *Department of Zoology, British Museum (Natural History), Cromwell Road, London SW7.*
*Whitton, Dr B. A., *Department of Botany, University of Durham, Durham.*
*Williamson, K., *British Trust for Ornithology, Beech Grove, Tring, Hertfordshire.*

† Now the *Institute of Terrestrial Ecology.*

Preface

In recent years there has been an increasing interest in changes which are occurring in our environment and the effects they are having on the flora and fauna of particular countries and regions. In most groups of organisms it is only the specialist systematists who have precise information on present distributions and changes which have taken (or are taking)place in them. Broad generalizations about "declines" and "threatened" species are often made in the Press and semi-popular publications, but scientists concerned with conservation need accurate authorative information on which to base their opinions. Specialist taxonomists and ecologists are naturally aware of the changes occurring in the groups with which they work, but tend to know relatively little of those in others. There is consequently always the danger that a specialist will recommend a conservation measure for a member of his group which will be harmful to that of another.

One of the main aims of The Systematics Association is to promote co-operation between workers in different branches of systematic biology, and the principal purpose of the present Symposium on *The Changing Flora and Fauna of Britain* was to increase the awareness of specialist taxonomists dealing with British species of changes occurring in groups of organisms other than their own, and the factors responsible for these changes.

More information is probably available on the flora and fauna of Britain than that on any other area of the same size in the world, and so a consideration of what has occurred in Britain is also of some interest to systematists and conservationists working in areas whose biota are less well known.

The last decade has seen a number of symposia, books and papers dealing with changes in particular groups or the effects of particular environmental factors, but no attempt to approach the subject of change in all groups from the standpoint of the systematist has been made in recent years. The last comparable meeting to have been held in Britain appears to have been that organized by the Linnean Society of London on 5th December, 1935, at which taxonomists presented papers on mammals, birds, insects, flowering plants, marine algae and fungi under the general title of *Changes in the British fauna and flora during the past fifty years* (Proc. Linn. Soc. Lond. **148**: 33-52, 1935). The present Symposium, held at the University of Leicester on 11-13th April, 1973, includes 23 papers, however, which between them deal with changes in most of the major groups of plants and animals in Britain since systematic recording in them began. As the scientific organizer of this Symposium I am only too conscious of groups which were omitted because of limitations of space and time, or the lack of suitable speakers. The main gaps are the bacteria, protozoans, and many groups of small invertebrates, but in these too few data are often available on which to base any accurate assessment of change. Relatively little information is included on littoral and marine groups (with the exception of marine algae) because these will be treated in detail in a forthcoming meeting of the Association.

Many of the contributions presented here include a great deal of hitherto unpublished information and are the first attempts to provide accurate synopses of changes which have occurred in their groups. The present volume is consequently of interest to a wide spectrum of both amateur and professional conservationists, ecologists, naturalists and taxonomists, presenting a reasonably comprehensive account of the present state of the flora and fauna of Britain and the changes occurring in it for the first time. The wide interest in this subject is reflected in the fact that this is one of the few recent meetings of the Association to have featured in the Press (*Nature, Lond.* **243**: 9, 4 May, 1973; *The Times* 7th May, 1973, p. 4). I also hope, however, that this volume may prove a useful base-line from which changes in the flora and fauna of Britain in the future may be considered.

The Leicester Symposium approached the problem of change from the systematic viewpoint. The effects of particular environmental factors on the various groups considered can be traced through the Subject Index. It is to be hoped that some of the threads which recur in various chapters (see Ch. 23) may be taken up as the themes for future symposia organized by other bodies.

The Systematics Association is very grateful to Mr R. S. R. Fitter and Professor K. Mellanby for chairing the sessions on 12th and 13th April, respectively; and to Dr P. F. Parker for dealing with the local organization of the meeting. As editor of the Symposium volume I wish to thank the Biological Records Centre of the Institute of Terrestrial Ecology for their assistance in preparing maps included in many of the Chapters and Mrs B. Renvoize of Academic Press for her invaluable help in the preparation of this volume.

D. L. HAWKSWORTH

January, 1974

Contents

1 | The Changing Environment

K. MELLANBY

Monks Wood Experimental Station, The Nature Conservancy,
Abbots Ripton, Huntingdon

Abstract: The environment in Britain is constantly changing. There are climatic changes, probably unaffected by man's activities. The more mobile and affluent population asks for greater pressure on land and wildlife, though suburban development and gravel exploitation produce habitats suitable for some organisms. Air pollution still affects some plants, but it is decreasing at least in urban areas. Inland waters are more seriously polluted, but even here notable improvements have recently been recorded. Forestry is producing new habitats, particularly where exotic conifers are planted, but recent trends towards restoring native hardwoods in England are welcomed. Agriculture, removing hedges and becoming increasingly intensified, is probably the greatest danger to our flora and fauna.

This volume is concerned with the changes which are occurring in the flora and fauna of Britain. My purpose is to describe how the environment itself is changing, and to try to show how these changes may affect animals and plants. Many of the other contributors will deal in greater detail with particular factors in relation to their chosen groups, and my purpose is to attempt to give an overall picture of the whole environment.

The climate of any region is perhaps the most important environmental factor. We can easily distinguish the flora and fauna of the tropics from that of a temperate region, and changes in the climate of any region will obviously have important ecological effects. We know that the climate of Britain has differed at different periods. We had the last ice age only some 12 000 years ago. Since then we have had warmer and cooler periods, and the flora and fauna has been affected. We know that for the first 50 years of this century Britain became slightly warmer, and that for the past 20 years this trend had been

Systematics Association Special Volume No. 6, "The Changing Flora and Fauna of Britain", edited by D. L. Hawksworth, 1974, pp. 1–6. Academic Press, London and New York.

reversed. Some people think that these changes may have been caused or exacerbated by man—the rise of temperature is parallel with a rise in atmospheric carbon dioxide and may possibly be a manifestation of the "greenhouse effect"; the subsequent fall may be associated with a rise in the global levels of smoke and other pollution. However, it seems most probable that any fluctuations in climate in the past have been mostly, if not entirely, "natural" in origin, and that man's contribution was negligible in comparison except very locally, e.g. in large conurbations where temperatures are a degree or two higher than those in the surrounding country. Fortunately the whole question of climatic change in Britain was authoritatively covered by H. H. Lamb and others in 1969 at the Botanical Society of the British Isles conference on "The Flora of a Changing Britain" (Perring, 1970) and so needs no further consideration here.

In Britain, as in most other countries, we are experiencing an increase in the human population, increasing affluence and mobility of this larger population, industrial growth and a farming revolution. All these factors are having profound environmental effects.

The increasing population and the demand for better housing is causing widespread "suburbanization". Some 50 000 areas of farmland—often the most productive farmland—is built on annually. The population of our actual city areas is not increasing, it is falling dramatically in London and most other cities, and central areas are often becoming semi-derelict or the sites of unoccupied monstrosities like the notorious Centre Point building. The bomb-damaged sites of the last war with their rich flora of exotic and native species, and with their corresponding and perhaps unexpected fauna, are gradually being transformed into dull concrete jungles of one kind or another, but there are still many fascinating ecological problems in our central urban areas.

It is the suburbs and new towns—modern "garden cities"—where growth continues. A great many biologically interesting sites are being lost on the outskirts of existing and growing towns, but the new habitats are not without value. Provided it is not overrun by that most dangerous predator, the domestic cat, the suburban garden may contain many more woodland birds than a similar area of natural deciduous hardwoods, and the fox is probably more numerous in some London suburbs than it is in the hunting shires adjacent to the city of Leicester. Foxes appear to find the ample contents of dustbins, where conspicuous waste is clearly a symptom of affluence, a more reliable food supply than the poultry now immured in battery cages and so almost completely invulnerable to predation.

The motor car, in one way or another, is having considerable ecological effects. Cars kill many hedgehogs, toads, rats and many other animals including, in early summer, even a proportion of young, migrating moles. We do not know whether this slaughter is affecting the national or the local population of any of the affected species. Better roads, and particularly motorways, are covering huge areas of the country, for the roadside verges alone exceed in extent the National Nature Reserves. The verges of motorways and freeways are almost totally protected from unauthorized interference, and so should become wildlife sanctuaries. Some work to improve their aesthetic and botanical interest is in progress, though the best policy may be to leave them to develop naturally. This has happened on some railway embankments, which are now becoming the largest areas of self-sown, semi-natural woodland in the country. The developing oakwoods along many mainline railway lines are a major gain to our countryside.

Cars bring people into the country, and this increases the pressure especially within the short distance most town dwellers are willing to walk. Fortunately this is not far. Cliffs popular with climbers, the bits of mountain adjacent to chairlifts and the surroundings of car parks on the shore or inland are being rapidly destroyed, but the majority of the areas which can only be reached on foot by walking more than a mile are actually under less pressure than 50 years ago. A proper planning policy should prevent too much damage, except perhaps from over-enthusiastic naturalists or even scientific research workers, who still need a good deal of education. These topics were discussed at a Monks Wood Symposium on "The Biotic Effects of Public Pressures on the Environment" in 1967 (Duffey, 1967). In many places litter is a serious problem, and tins and milk bottles may kill so many small mammals and other creatures that population reduction is not impossible.

Air pollution is known to affect many plants, particularly lichens and bryophytes, and we have papers on these groups. It is perhaps encouraging to be able to report that, on the whole, the air in Britain's cities is less polluted today than it was 20 years ago. Today central London enjoys as much sunshine as rural areas at a corresponding latitude, and the smog and "pea-souper" fog of the Victorian era has now disappeared. Better dispersion has even lowered the levels of sulphur dioxide in our cities, though there is some evidence that there may have been a slight rise in some parts of the countryside. These levels of sulphur dioxide are almost certainly harmless to man, but they still affect susceptible plants, and so their ecological effect is still far from negligible. Trends in sulphur dioxide air pollution and emissions in Britain have recently been reviewed by Saunders and Wood (1973). Local concentrations of other pollutants, such as fluorides near smelters and brickworks, still continue, though in almost all cases, except where industrial accidents occur, the situation is generally one of steady improvement. Cars undoubtedly pollute the air with many substances which, in sufficient quantities, are harmful to life, but there is little evidence that wildlife suffers appreciably. Lead levels in grass verges adjacent to heavy traffic are increased, and voles living on the grass have lead levels higher than normal in their tissues, but there is no evidence to suggest that the animals are seriously incommoded.

The majority of our larger rivers are seriously affected by man-made pollution, from sewage and from industry (see Mellanby, 1972). Here again the picture is not entirely discouraging. During the last 10 years almost every industrial river has become somewhat less dirty, and we can expect further improvements as more of the water is needed for human consumption. Unfortunately man is able to drink water (after it has been treated) which would not support trout or many invertebrate animals, so we may not get the improvement that would be technically possible. While most rivers are becoming less dirty, some of our estuaries have deteriorated, but even here the future outlook is moderately optimistic. These changes, for better or for worse, in our inland waters, are clearly having important ecological effects.

Industry, particularly the electrical industry, is warming many of our rivers, and many fear widespread "thermal pollution". This could be serious, because when water is warmed the solubility of oxygen decreases, though the rate of metabolism of aquatic animals (and so their need for oxygen) increases. Fortunately, in Britain, no serious thermal pollution has occurred, though the moderate warming of some rivers is likely to have ecological effects. Thus the Thames did not freeze in early 1963, though during previous centuries such temperatures enabled fairs to be held on the ice. The Trent is warmed by its power stations, but the levels of oxygen are actually raised, for the incoming water is so polluted. If the river is really cleaned up the heating may be more detrimental.

The need for water is causing more reservoirs to be built, and we may soon see large areas of impoundment on some of our estuaries. Building and road-making increase the extraction of sand and gravel, with the destruction of farmland and other habitats. However, the area may not always be lost, for new wetland habitats are produced, and these have attracted many birds and increased the breeding populations of some.

Pollution from industry is being increasingly controlled, but industrial products are sold and may then affect our environment. Thus sewage effluent containing detergents formerly caused serious foaming of our rivers, a problem partially solved by a change in the nature of the chemical used. Detergents still cause other pollution problems, particularly because they may be rich in phosphates and so add to the growing eutrophication of inland waters. Attempts to reduce phosphate levels have given rise to greater risks from other pollutants. Recently we have found other new chemicals becoming widespread, and we have so far developed no satisfactory mechanism to prevent this happening. Thus large amounts of polychlor biphenyls ("PCBs") have been found in birds and elsewhere, following the widespread industrial use of these substances (Holdgate, 1971). Responsible voluntary action by industry in withdrawing the PCBs from the more dangerous uses probably saved the situation, but the incident should have forewarned us that unexpected and harmful side effects of new chemicals must always be guarded against.

Man suffers from man-made noise. Sometimes wildlife may be damaged, for birds have been said to be driven from their nests at critical periods by sonic booms, with harmful effects on breeding. But on the whole most animals seem remarkably adaptable to increased levels of noise, particularly when this is more or less constant in level.

We have no evidence that radiation levels have arisen so as to endanger any species, but a great increase in the number of atomic power stations would make the possibility of accidents causing local rises to dangerous levels more likely, and some day the disposal of enormous amounts of radioactive waste may become a serious environmental hazard.

Agriculture and forestry continue to change the whole face of the country-side. Many areas are now covered with exotic species of conifers, which have replaced upland areas of moor, or in southern Britain, the native hardwoods. Ecologists now welcome the change in emphasis in the policies of the Forestry Commission, and we are beginning to see evidence of new hardwood planting and of increasing diversity in all areas. The effects on wildlife in the future will be immense. For the present, we still need far more studies of the effects of different types of forestry practice on the whole ecosystem.

The agricultural revolution since 1945 has had the greatest effect on lowland Britain. Hedge removal, particularly in Eastern England, has altered the whole appearance of the countryside (Hooper and Holdgate, 1970). Fewer animals graze out of doors, more are kept indoors, where their excrement instead of becoming valuable manure becomes a potential source of water pollution. Herbicides and fertilizers are destroying large areas of species-rich old grassland which we cannot replace. Chemical fertilizers make some contribution to the growing eutrophication of inland water. Insecticides have killed many seed-eating birds and reduced the population of some predators (Mellanby, 1967). However we now have better methods of controlling pesticides, and their danger in Britain at least is decreasing. Nevertheless

K. Mellanby

agricultural intensification, certain to increase in our more fertile areas with our entry to the Common Market, is likely to have the greatest effect on our native flora and fauna as a whole.

REFERENCES

DUFFEY, E. (ed.) (1967). "Biotic effects of Public Pressures on the Environment." Monks Wood Symposium No. 3, The Nature Conservancy, Huntingdon.

HOLDGATE, M. W. (ed.) (1971). "The Seabird Wreck of 1969 in the Irish Sea." Natural Environment Research Council, London.

HOOPER, M. D. and HOLDGATE, M. W. (eds) (1970). "Hedges and Hedgerow Trees." Monks Wood Symposium No. 4, The Nature Conservancy, Huntingdon.

MELLANBY, K. (1967). "Pesticides and Pollution." Collins, London.

MELLANBY, K. (1972). "The Biology of Pollution." Arnold, London.

PERRING, F. (ed.) (1970). "The Flora of a Changing Britain." Classey, Middlesex.

SAUNDERS, P. J. W. and WOOD, C. M. (1973). Sulphur dioxide in the environment: its production, dispersal and fate. *In* "Air Pollution and Lichens" (B. W. Ferry, M. S. Baddeley and D. L. Hawksworth, eds), pp. 6–37. University of London Athlone Press, London.

2 | Changes in our Native Vascular Plant Flora

F. H. PERRING

Biological Records Centre, Monks Wood Experimental Station,
The Nature Conservancy Abbots Ripton, Huntingdon

Abstract: It is shown that some changes in distribution have taken place apparently independently of the influence of man: fluctuations in the distribution of orchids and other species with light, easily dispersed seeds are demonstrated. The sea-shore is a relatively undisturbed habitat and evidence is produced which indicates natural advances and retreats in the limits of distribution of more localized species. Several new species of marshy habitats have been recorded for the first time in the British Isles along the western fringe and the suggestion is made that migrating birds have been responsible for long distance dispersal from Greenland and America.

Man has had a destructive effect on our flora: he has been responsible for the extinction of twelve species and the population of many of our rare species is only about one-third of what it once was. Losses have mainly been due to the destruction of the habitat by ploughing and drainage, and a sequence of changes in agricultural practice has affected the balance of our arable weed flora. However, the area covered by many native species has increased through man's activity: cliff plants have spread to walls and several local species have become widespread following the construction of the railways.

INTRODUCTION

Reliable recording of vascular plants in this coutry began towards the end of the sixteenth century but the first published account of the flora of an area was "Iter in Agrum Cantianum" by Thomas Johnson, published in 1629, which described a botanical journey through Kent made earlier the same year. The first book with a claim to be a County Flora is "Catalogus Plantarum circa Cantabrigiam nascentium" published by John Ray in 1660. It was the result of 9 years' work and gives localities of plants, which in numerous cases can still be found in the same place today. Thus began a series of County Floras which still progresses and now covers every county of England except Lincolnshire,

Systematics Association Special Volume No. 6, "The Changing Flora and Fauna of Britain", edited by D. L. Hawksworth, 1974, pp. 7–25. Academic Press, London and New York.

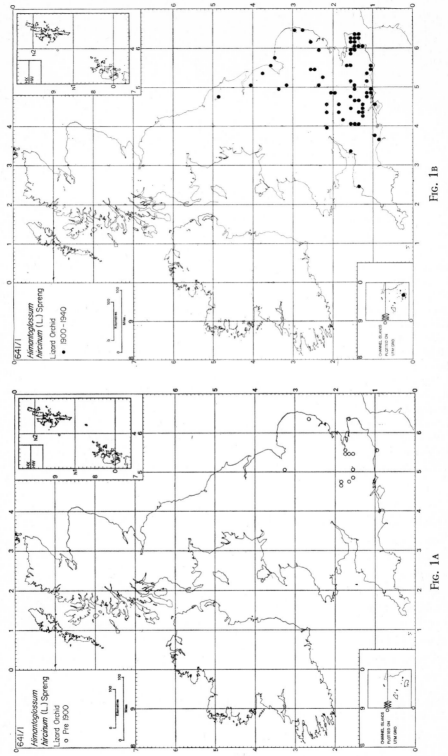

FIG. 1(A–B). Changes in distribution of *Himantoglossum hircinum*. A, Pre-1900; B, 1900–40.

FIG. 1A

FIG. 1B

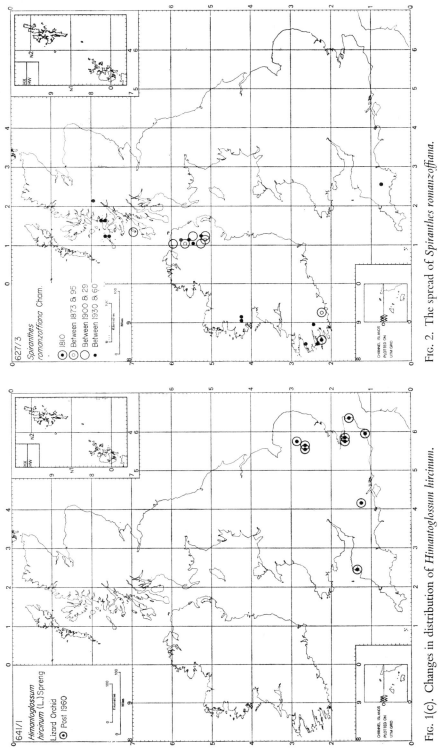

FIG. 2. The spread of *Spiranthes romanzoffiana*.

FIG. 1(c). Changes in distribution of *Himantoglossum hircinum*. c, Post-1960.

641/1

Himantoglossum hircinum (L.) Spreng

Lizard Orchid

⊙ Post 1960

627/3

Spiranthes romanzoffiana Cham.

⊙ 1810
◎ Between 1873 & 95
◯ Between 1900 & 29
● Between 1930 & 60

and many in the rest of the British Isles. Several counties have had more than one and in Cambridgeshire there have been five—1660, 1785, 1860, 1939 and 1964. In such an area recording has been thorough and continuous. In contrast several Irish counties were not looked at for the first time until the end of the nineteenth century when R. L. Praeger visited them in preparation for "Irish Topographical Botany" (1901), and have never had a resident botanist.

Between 1954 and 1960 the Botanical Society of the British Isles (BSBI) co-ordinated the largest single exercise in recording ever undertaken by field botanists in this country, and collected lists from every 10 km square which were the basis of the "Atlas of the British Flora" published in 1962 (Perring and Walters, 1962). This provides a base line against which we can compare all records from the past. In some areas we can look at changes over 300 years but in others it is less than 60 years.

Since 1966 my colleague at Monks Wood, Miss M. N. Hamilton, has been co-ordinating the resurveying, through BSBI members, of the 300 rarest species in the British Flora and I am grateful to her and to them for the data which form the basis of much of this paper. The nomenclature used follows Dandy (1958, 1969).

CHANGES NOT INFLUENCED BY MAN

Undoubtedly the majority of changes which have taken place have been the direct or indirect consequence of the activities of man, but there are a few for which such an explanation is unnecessary and unlikely, changes which might have taken place in the absence of man from the British Isles.

The best documented account is the changes in the distribution of the Lizard Orchid (*Himantoglossum hircinum*) by Good (1936). A native of west and central Europe it is particularly abundant in France, southern Germany and Switzerland. Before 1900 it was established in Britian in one locality only, near Dartford in Kent. It had also appeared, for short periods, in twenty other localities almost all on the North Downs. However, between 1900 and 1933 a further 129 localities were discovered. The majority failed to persist for more than 1 or 2 years and often the record was a single specimen. Good could not discover any single feature of the environment which could account for this sudden increase in the frequency of the species except a change in climate. Since 1940 there has been a marked decline, in the 1950s there were about 12 localities and today only 9 are known (see Fig. 1 A–C).

Another orchid which has spread steadily is the Irish Lady's-tresses, *Spiranthes romanzoffiana* (Fig. 2). First noted in west Cork in 1810, it was found again in that county in 1873. It was noticed in Northern Ireland in 1892 on the

shores of Lough Neagh, Co. Armagh. By the 1950s it had spread to the mainland of Scotland and in 1957 was found in south Devon. It has now been
recorded from at least 25 localities.

Less spectacular, but of a similar nature, was the appearance of the Dense-
flowered Orchid (*Neotinea intacta*) in the Isle of Man in 1966: this species was

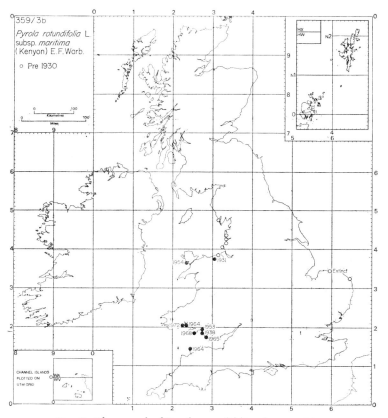

FIG. 3. The spread of *Pyrola rotundifolia* subsp. *maritima*.

previously only known from the west of Ireland and this new locality suggests
long distance dispersal of over 150 miles.

There are two subspecies of the Round-leaved Wintergreen, *Pyrola rotundi-*
folia: subsp. *maritima* occurs on dunes. Before 1930 it was known only from
Norfolk and Lancashire but in 1938 it was found in Kenfig Burrows, Glamorgan,
and by 1972 had spread to seven localities in dunes round the Welsh coast
and one in north Devon (Fig. 3).

These orchids and the wintergreen are all species with very light seeds which could be carried in the wind; the habitats they have occupied in their new localities are hardly affected by man and the natural long distance dispersal is the most likely explanation. For *Spiranthes romanzoffiana* this could be the means by which it first reached this country. It is a North American species

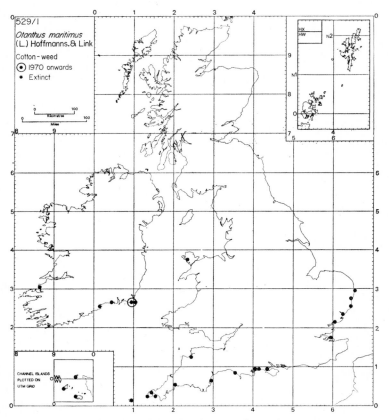

FIG. 4. Past and present distribution of *Otanthus maritimus*.

not known from elsewhere in Europe and it seems reasonable to suppose it may be a recent, natural introduction by seed from across the Atlantic.

Over the same period there have been some remarkable declines in light-seeded species, the most severe being in Greater Broomrape (*Orobanche rapum-genistae*). This species was known from 319 of the 10 km squares before 1930. In the period 1930–50 there were records from only 45, and we only have records from 22 squares post-1950. Although loss of habitat of this heathland species may account for some of this decline, the majority cannot be explained in this way and it seems possible that during the recording period we have witnessed the end phase, but not the build-up, of a species which may have gone through a similar cycle to *Himantoglossum*.

Several species of orchid have also declined. The Military Orchid (*Orchis militaris*) occurred in about eighteen 10 km squares in the chalk of the Chilterns

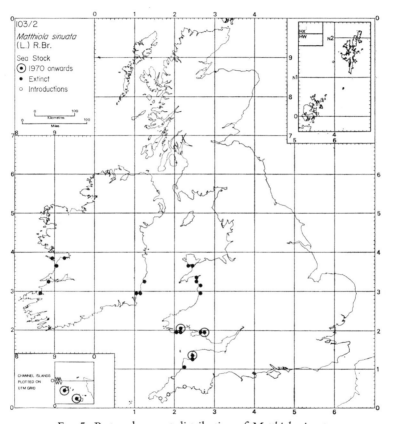

FIG. 5. Past and present distribution of *Matthiola sinuata*.

in the nineteenth century but then disappeared and was indeed thought to be extinct until in 1947 a colony was discovered in Buckinghamshire at least 25 years after its last previous record. Since then it has appeared in two further localities, the first involving a jump of almost 100 miles, or even further if one assumes the source of seed was not the solitary British population, but the Continent of Europe where it is much more widespread. Here we are witnessing not the rise and fall of *Himantoglossum*, but a fall followed by a rise.

Another group amongst which change is constant and which may be independent of man includes the sea-shore species. Unfortunately most of the changes seen in the last 300 years have been losses. One species, the Sea Purslane (*Halimione pedunculata*) once occurred in fourteen 10 km squares in salt marshes on the east coast of England, but has not been seen since 1934. There are records of Cottonweed (*Otanthus maritimus*) from 26 squares in the British Isles (Fig. 4) but it is now known from only one, extensive, colony in S.E. Ireland. There have been no records from Great Britain since 1936. Sea Knotgrass (*Polygonum maritimum*), recorded from 10 squares in England and from the Channel Isles, was thought to be extinct in England until recently, but still occurs in small quantity on one Cornish beach, as well as persisting on one Channel Island. There are records of the Purple Spurge (*Euphorbia peplis*) from 28 squares, but today it is known from only one Channel Island. Similarly the Sand-Spurrey (*Spergularia bocconi*) whilst still plentiful in Guernsey and Jersey, has not been seen in England since 1959. The Sea Stock (*Matthiola sinuata*) has been reported from 25 squares, but only 5 localities are known today (Fig. 5).

Whilst it cannot be denied that human pressure on the pleasure beaches of the south and west, where many of these species occur, may have made some contribution to their disappearance, this has not been the main factor: many of the last records date from the eighteenth and nineteenth centuries. Apart from *Halimione* they are representatives of the Mediterranean element and are at the edge of their range. It seems likely that the populations have always been subject to change: destroyed in one place by a storm of unusual violence, but appearing elsewhere as a result of sea-borne dispersal of their specially adapted seeds. This fluctuation in populations is particularly well demonstrated by the Sea Pea (*Lathyrus japonicus*) which has appeared or reappeared in 4 localities on the east coast of England and Scotland during the last 10 years whilst disappearing from others. If all records made in the course of time are compared with present-day distribution it would appear that there has been a decline in this species whereas it is probable that the population and overall distribution have been more or less constant, but shifting.

Lathyrus japonicus is a member of the Oceanic northern element in the flora: another species with a similar distribution, the Oyster Plant (*Mertensia maritima*), really does seem to have declined, in a northerly direction (Fig. 6), and reports received indicate that populations are smaller even within the present distributional range.

A final group of species where changes of distribution may have been independent of man includes several which have been recorded for the first time in the British Isles relatively recently, and mainly along the western fringe. The first record of the Canadian St John's Wort, *Hypericum canadense*, dates from 1906 when it was collected in Co. Galway, but the specimen was wrongly identified and the species was rediscovered in the field and recognized

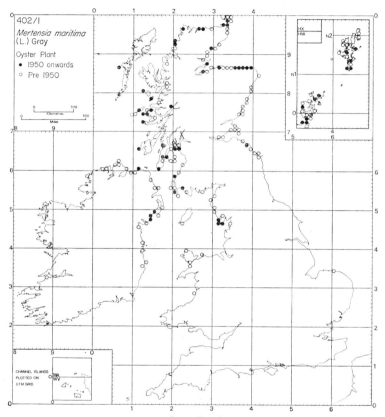

FIG. 6. Retreat northwards of *Mertensia maritima*.

for the first time by Webb in 1954 (Webb, 1958). Several more localities have since been found in the same general area, but in 1968 a new locality was found in west Cork over 200 km further south. The only other European locality is in Holland where it was first collected in 1909 (Jonker, 1959). *H. canadense* grows in wet acid conditions at all of its European sites and as an undoubted native in similar habitats in North America. It is not a species of ruderal or man-modified habitats normally associated with man's activities and I have argued elsewhere

(Perring, 1963, 1967) that the most satisfactory interpretation is that it is a relic which survived the Pleistocene in Ireland. But the force of this argument is diminished when one questions whether it could have survived in Holland during the last glaciation and appreciates that its recent appearance in a well botanized corner of west Cork indicates that long distance dispersal may be happening now.

The argument has been further undermined by the amazing discovery in July 1971, by Mary J. P. Scannell (Scannell, 1973) of *Juncus planifolius* in the Carna area of west Galway where it was found on a track on the south side of Lough Truscan. The nearest known localities for this species are all in the Southern Hemisphere: Australia, Tasmania, New Zealand, Hawaii and in the southern part of Chile from 20° to 45° South. The nearest locality to Galway is 8000 miles away. Scannell believes this to be a man-made introduction, but it is difficult to imagine the human vector which travels from the Southern Hemisphere and deposits viable seed in a suitable habitat beside a lough in a remote corner of western Ireland. This raises again the possibility that migrating birds play a part in plant distribution. The evidence has been reviewed by Gillham (1970). She points out that many rush seeds are mucilaginous and when moistened could stick to birds' feet in the absence of mud, and are so light they would not affect flight in any way. It is also known that propagules of water plants may remain in a mallard's stomach for at least 70 h. Migrating American birds do occasionally go off course and end up on the wrong side of the Atlantic. As Gillham remarks at the end of her paper: "Over the course of millions of years, plant disseminules may need to arrive only once to become part of the flora. With odds against successful passage at a million to one, it would still be more surprising that they had not achieved this than that they had."

Two species of Marsh Yellow-cress, *Rorippa*, are now recognized in the British Isles following the work of Jonsell (1968): the diploid *R. islandica* and the tetraploid *R. palustris*. The former is a very rare species known from only about seven localities all but one from coastal districts in the north and west (Fig. 7) where it grows on bare ground, subject to winter flooding. Randall (1974) has recently shown that these localities are linked by being the winter feeding grounds of geese, and suggests that these geese have been responsible for introducing this species from Iceland where it is abundant in the Mývatn district, a region famous for its rich bird-life, especially waterfowl (Jonsell, 1968). Jonsell points out that this is an autogamous species so that the arrival of a single seed in favourable circumstances could be sufficient to initiate a population.

third, Spreading Hedge-parsley (*Torilis arvensis*) which has finely divided leaves which are often a protection against herbicides, appears to be a victim of another recent development, stubble-burning and early ploughing. Reports suggest it is now very difficult to find. There is a group of weeds including the Fluellens (*Kickxia* spp.) and Night-flowering Catchfly (*Silene noctiflora*) which, like the

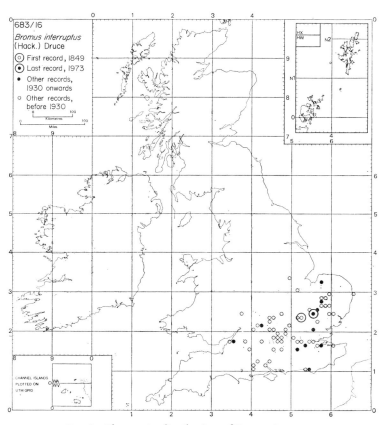

FIG. 9. Changes in distribution of *Bromus interruptus*.

Torilis, flowered in the stubble in September and October; one fears they may also show declines when the next survey of the British flora is made.

Another species entirely associated with agriculture has both increased and decreased during the last 125 years. The Interrupted Brome (*Bromus interruptus*) was first collected in Britain in 1849 in a sainfoin field in Cambridge-shire. It was at first assumed that it had been introduced with imported seed (Druce, 1897), but no specimens have been found except more recently in

Holland where it is regarded as an introduction (Hubbard, 1954) and it is now
apparent that *B. interruptus* was a neo-endemic. Its spread in Britain (see Fig. 9)
was fairly rapid and reached its maximum extent in the first two decades of
this century when it was recorded from most counties south and east of a line
from the Humber to the Severn. It was probably distributed in sainfoin seed

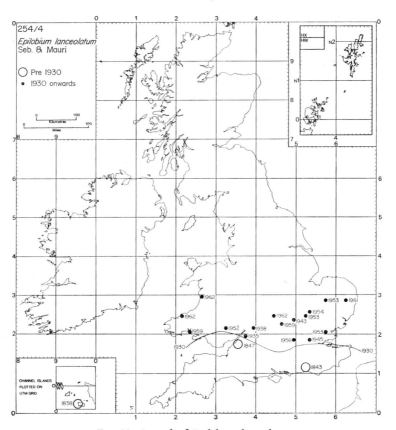

FIG. 10. Spread of *Epilobium lanceolatum*.

along with two other characteristic weeds associated with that crop, Meadow
Brome (*Bromus commutatus*) and Fodder Burnet (*Poterium polygamum*). Since
1930 the decline has been dramatic and only six records have been traced since
1936. Today only one is known at the edge of a field a few miles from the first
locality. Though the reason for this decline may never be known it is probable
that this too was a victim of cleaner seed. But it is a remarkable species whose
origin, spread and decline may have all been dependent on man.

Ploughing, draining and arable cultivation have affected the flora continuously since the Bronze Age, and more recent artifacts have also contributed to changes in distribution. The construction of stone or brick walls and quarries has created artificial rock faces of all aspects throughout the country and, during the period of recording, many species of ferns, with light wind-blown spores, have been spreading on to these habitats. Rustyback (*Ceterach officinarum*) was originally confined to limestone rocks in the west of the British Isles, but is today found almost throughout, and recent records suggest it is still spreading into the last remaining vacant areas in England. Other wall ferns, the spleenworts *Asplenium adiantum-nigrum*, *A. ruta-muraria* and *A. trichomanes*, have had a similar history and several other species including Maidenhair Fern (*Adiantum capillus-veneris*), Bladder-fern (*Cystopteris fragilis*) and Limestone Fern (*Gymnocarpium robertianum*) are frequently recorded on walls and bridges many miles from their centres of distribution.

A flowering plant which has also spread, partly on walls and partly in quarries is the Spear-leaved Willowherb (*Epilobium lanceolatum*) (see Fig. 10). It was first recorded in this country in 1843, but is regarded as an overlooked native rather than an alien. Until 1930 it was known only from an area south of the River Thames and the estuary of the River Severn. However, in the last 40 years it has been recorded from many other localities further north. Willow herbs all have seeds with parachutes attached which are easily distributed in the wind. Other species in the genus, including Rosebay Willowherb (*E. angustifolium*), have also spread. Until the middle of the nineteenth century this was a remarkably rare species but since then there has been a population explosion the cause of which is not understood, and it has come to occupy a wide range of man-made niches including recently felled woodland and railway embankments. The building of a network of canals and railways undoubtedly had a significant effect on the distribution of our flora, providing not only new habitats but a dispersal mechanism at the same time. They have been dealt with in detail by Lousley (1969) but to mention a few native species affected in this way, Floating Water-plantain (*Luronium natans*) was once confined to small acid-water lakes and tarns in the Welsh mountains, but it got into the Shropshire Union Canal and is now widespread in canals in the Cheshire Plain. The distribution of the genus *Potamogeton* has been altered by their dispersal through the canal system. This has brought together species which had previously not occurred in the same habitat, so that hybrids arose which have subsequently spread and now occur in the absence of one or both parents.

Railway building was remarkable because it simultaneously created a continuous strip of open habitat in the ballast and imported in the material used, the seeds to grow in it. Then as soon as trains began to run they created air currents to suck or blow propagules up and down the line. Amongst native species the most notable spread was made by the Small Toadflax (*Chaenorhinum minus*), an annual previously an arable weed of calcareous soils, and in Ireland by a local species of open limestone grassland, Fine-leaved Sandwort (*Minuartia hybrida*).

Undoubtedly the major effect of railway building was to provide a route for the spread of aliens such as the Oxford Ragwort (*Senecio squalidus*). This was first recorded at Oxford in 1794 and persisted there and in Cork for many years without further spread until the arrival of the railway. It has subsequently been dispersed throughout the railway system of both countries, and from the railway stations into the towns. But this is venturing into the realm of aliens which would be the subject of several lengthy papers.

REFERENCES

DANDY, J. E. (1958). "List of British Vascular Plants." British Museum (Natural History), London.

DANDY, J. E. (1969). Nomenclatural changes in the "List of British Vascular Plants". *Watsonia* **7**, 157–178.

DELVOSALLE, L., DEMARET, F., LAMBINON, J. and LAWALRÉE, A. (1969). Plantes rares, disparues ou menacées de disparition en Belgique: L'appauvrissement de la flore indigène. *Ministère de l'Agriculture, Administration des Eaux et Forêts, Service des Réserves Naturelles domaniales et de la Conservation de la Nature, Travaux no.* **4**.

DRUCE, G. C. (1897). "The Flora of Berkshire." Oxford.

FRYER, J. D. and CHANCELLOR, R. J. (1970). Herbicides and our changing arable weeds. *In* "The Flora of a Changing Britain" (F. Perring, ed.), pp. 105–118. Classey, Middlesex.

GILLHAM, M. E. (1970). Seed dispersal by birds. *In* "The Flora of a Changing Britain" (F. Perring, ed.), pp. 90–98. Classey, Middlesex.

GOOD, R. (1936). On the distribution of the Lizard Orchid (*Himantoglossum hircinum* Koch). *New Phytol.* **35**, 142–170.

HESLOP-HARRISON, J. (1953). The North American and Lusitanian elements in the Flora of the British Isles. *In* "The Changing Flora of Britain" (J. E. Lousley, ed.), pp. 105–123. Classey, Middlesex.

HUBBARD, C. E. (1954). "Grasses." Penguin Books, Harmondsworth.

JONKER, F. P. (1959). *Hypericum canadense* in Europe. *Acta bot. neerl.* **8**, 185–186.

JONSELL, B. (1968). Studies in the north-west European species of *Rorippa* s. str. *Symb. Bot. upsal.* **19** (2), 1–222.

LOUSLEY, J. E. (1970). The influence of transport on a changing flora. *In* "The Flora of a Changing Britain" (F. Perring, ed.), pp. 73–83. Classey, Middlesex.

OCKENDON, D. J. (1968). Biological Flora: *Linum perenne* ssp. *anglicum* (Miller) Ockendon. *J. Ecol.* **56**, 871–882.

PERRING, F. H. (1963). The Irish Problem. *Proc. Bournemouth Nat. Sci. Soc.* **52**, 1–13.

PERRING, F. H. (1967). The Irish Problem. *In* "Pflanzensoziologie und Palynologie" (R. Tüxen, ed.), pp. 257–268. Internationalen Vereinigung für Vegetationskunde, Den Haag.

PERRING, F. H. (1970). The last seventy years. *In* "The Flora of a Changing Britain" (F. Perring, ed.), pp. 128–135. Classey, Middlesex.

PERRING, F. H. and WALTERS, S. M. (1962). "Atlas of the British Flora." Nelson, London and Edinburgh.

PRAEGER, R. L. (1901). "Irish Topographical Botany." Dublin.

RANDALL, R. E. (1974). *Rorippa islandica* (Oeder) Barbás *sensu stricto* in the British Isles *Watsonia* **10**, in press.

SCANNELL, M. J. P. (1973). *Juncus planifolius* R.Br. in Ireland. *Ir. Nat. J.* **17**, 308–309.

WEBB, D. A. (1958). *Hypericum canadense* L. in western Ireland. *Watsonia* **4**, 140–144.

WELLS, T. C. E. (1968). Land-use changes affecting *Pulsatilla vulgaris* in England. *Biological Conservation* **1**, 37–43.

3 | Changes in the Bryophyte Flora of Britain

F. ROSE

Department of Geography, King's College, University of London

and

E. C. WALLACE

2 Strathearn Road, Sutton, Surrey

Abstract: Changes in the bryophyte flora of Britain may be either apparent (due to increased survey work or changing taxonomies) or real (due to increases, decreases or introductions of species). While some native species have undoubtedly increased over the last century, it is among certainly introduced species that the most spectacular increases have occurred.

Most real changes which have taken place have been decreases correlated with deforestation, drainage, air and water pollution, and changes in agricultural practices. Some mire species (e.g. *Paludella squarrosa*) have totally disappeared and some formerly widespread epiphytes (e.g. *Antitrichia curtipendula*) have now become very rare and more restricted in their distribution.

Many declines appear to be due largely to widespread air pollution, but felling of old trees and changes in forestry practice have also played a major part. It is clear that many bryophytes will continue to become rarer in the future and that some further species will become extinct in Britain.

INTRODUCTION

Changes in the bryophyte flora of a long botanized country such as Britain fall into two categories: the apparent changes based on the state of recording at particular times, and the real changes.

The apparent changes concern increases in the number of species recorded, due either to (a) more thorough field surveys of remoter areas, or (b) increasingly critical concepts of taxa and the application of modern taxonomic techniques

Systematics Association Special Volume No. 6, "The Changing Flora and Fauna of Britain", edited by D. L. Hawksworth, 1974, pp. 27–46. Academic Press, London and New York.

F. Rose and E. C. Wallace

which may lead to the recognition of new taxa within established ones. Previously recorded taxa may appear to have become commoner as a result of more thorough field studies in previously underworked areas, and, conversely, the lack of modern field work in areas formerly well worked may lead to apparent decreases in the abundance of species.

Real changes, in contrast, concern actual increases or decreases in the quantity or distribution of taxa with the passage of time.

HISTORICAL OUTLINE

Our knowledge of the changes that have occurred in the British bryophyte flora is entirely dependent on the state of our knowledge of this flora (both taxonomically and biogeographically) in the past and present.

The first localized records of bryophytes in the British Isles were apparently those of Johnson (1629, 1632), who recognized six plants which were almost certainly mosses on Hampstead Heath and in eastern Kent. The identity of Johnson's plants is uncertain for he based his names on those of de L'Obel (1576) and his specimens are no longer extant. The systematic recording of British bryophytes really began with Dillenius (1741) who accepted a considerable number of species, incorporated earlier listings (e.g. Johnson, 1629, 1632; Ray, 1660, 1690) and presented synonyms. Because of nomenclatural difficulties it is difficult to assess the precise number of British bryophytes known to Dillenius. Turner and Dillwyn (1805) made an attempt to record the rarer bryophytes of England and Wales on a county by county basis but although their work provides invaluable information on the localities of rare species (many of which persist in the sites they listed) it omits species they regarded as common.

The two most important early nineteenth-century works on British bryophytes are undoubtedly Hooker and Taylor's (1818) "Muscologica Britannica" and Hooker's (1816) "British Jungermanniae" which between them listed 259 species of mosses and 82 species of hepatics. The growth of the Victorian railway network, the use of the microscope, and the increase in scientific knowledge to which these led meant that in the first edition of the "Census Catalogue of British Mosses" of 1907 and the "Census Catalogue of British Hepatics" of 1905 it was possible to record 619 species of mosses and 249 species of hepatics for the British Isles.

In the third edition of the "Census Catalogue of British Mosses" (Warburg, 1963) and the fourth edition of the "Census Catalogue of British Hepatics" (Paton, 1965a) 660 species of mosses and 284 species of hepatics were recorded, respectively. Up to April 1973, 24 species of mosses were added to the British list and three deleted, and five species of hepatics were added (see Table I). The totals now (April 1973) stand, therefore, at 681 species of mosses and 289 hepatics.

TABLE I. Additions and deletions in the British bryophyte flora since the Census Catalogues of Warburg (1963) and Paton (1965)

MOSSES: 24 additions (*ca.* 66 in the last 30 years)	MOSSES: 3 deletions
Amblystegium saxatile	*Fissidens minutulus*[b]
Aongstroemia longipes	*Orthotrichum shawii*[c]
Bryum bornholmense	*Weissia crispata*[d]
B. riparium	
B. ruderale	
B. tenuisetum	HEPATICS: 5 additions
B. violaceum	(18 in the last 30 years)
Dicranella staphylina	
Dicranum leioneuron	*Herberta borealis*
Eriopus apiculatus[a]	*Lophozia perssonii*
Fissidens celticus	*Riccia cavernosa*[e]
Grimmia agassizii	*Solenostome levieri*
G. borealis	*S. oblongifolium*
Hygrohypnum polare	
Hypnum vaucheri	
Leucobryum juniperoideum	
Mielichhoferia mielichhoferi	
Orthotrichum gymnostomum	
Pohlia crudoides	
P. lutescens	
P. pulchella	
Seligeria oelandica	
Tortula freibergii	
T. vectensis	
Weissia perssonii	

[a] An almost certainly introduced antipodean species excluded from the totals; [b] a synonym of *F. viridulus*; [c] a synonym of *O. striatum*; [d] a mixture of *Weissia* species; [e] most records of *R. crystallina* in the British Isles are this species.

From these figures it is clear that the majority of the bryophytes now known in Britain were first recorded during the nineteenth century, but that in the present century (particularly during the last 25 years) a considerable number of additions have been made, largely due to a renewal of intensive interest in

bryology. What is not evident from these figures is the enormous increase
in our knowledge of the detailed distribution (and ecology) of British bryophytes in recent years as a result of intensive field work. The numbers of new
vice-county records listed each year in the "Transactions of the British Bryological Society" (and its successor, the "Journal of Bryology"), however,

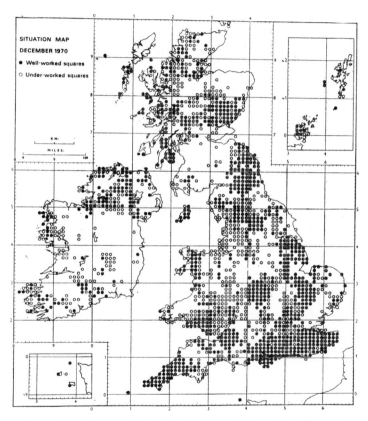

FIG. 1. British Bryological Society's Distribution Maps Scheme situation map December
1970 (after Smith, 1971).

testify to this increase in knowledge as does the accumulation of records in
the British Bryological Society's Mapping Scheme on a 10 km square basis.
Smith (1971; Fig. 1) provides a report of the progress of the first ten years of
this scheme when records had been obtained for some 2000 of the 3500 10 km
squares in Britain although 1050 of the 2000 were still considered as underworked. It is of interest to note from the map published by Smith that the

parts of Britain underworked for bryophytes are largely the same as those underworked for lichens, namely (1) Cardiganshire, (2) north central Wales, (3) much of Essex, Suffolk and east Norfolk, (4) north Wiltshire, (5) the north-east Midlands and south Lancashire, (6) most of south-west and central Scotland up to and including Fifeshire, (7) Lewis and Harris, (8) Caithness, (9) the area from Aberdeen to Inverness in north-east Scotland, and (10) much of central and eastern Ireland.

Some of these areas, particularly (3), (4) and (5) appear less attractive in terms of a potentially interesting cryptogamic flora than many other areas, while other areas we have listed (7–10) are more inaccessible. Consequently, while the distribution of many bryophytes of more exacting ecological require-ments is probably sufficiently well known already, at least in outline, that of many commoner species of wider ecological tolerance is less well known, their distributions as presently known reflecting that of investigators rather than of the bryophytes themselves. It is clear that the British Bryological Society and the British Lichen Society will need to concentrate on these areas in the future.

RECENT ADDITIONS

In addition to the species listed in Table I as additional to the most recent Census Catalogues, there have been many other additions to the British bryophyte flora in the last thirty years or so. Notable examples include the remarkable chlorophyll-lacking saprophytic hepatic *Cryptothallus mirabilis*; *Telaranea sejuncta*, an hepatic previously considered to be confined to the tropics and South America, has been recorded from several sites in Ireland and one in Cornwall; *Gongylanthus ericetorum*, a "mediterranean" hepatic found in the Lizard area of Cornwall; *Campylopus introflexus* (formerly confused with *C. polytrichoides*) has spread widely since its first recorded appearance in Sussex in 1941; *Tortula stanfordensis*, previously known only in California; and *Fissidens celticus*, a small moss widespread on unstable loamy soil of stream beds described as new to science (Paton, 1965b). These additions may be divided into three categories:

(1) Apparently new arrivals in Britain, possibly introduced accidentally by man (e.g. *Eriopus apiculatus*, *Telaranea murphyae*, *Tortula stanfordensis*).

(2) Species apparently overlooked until recent years, either (a) new to science (e.g. *Dicranella staphylina*, *Fissidens*, *celticus*), or (b) known already elsewhere (e.g. *Cryptothallus mirabilis*, *Leucobryum juniperoideum*).

(3) Critical segregates formerly included in "aggregate" species (e.g. species of the *Bryum erythrocarpum* group).

A number of species (11 hepatics and 14 mosses) have been incorrectly recorded for the British Isles at various times (Table II). In some cases this has been due to erroneous recording of a taxon that exists abroad and in others because taxa themselves have been shown to be unsatisfactory so that they are now treated as synonyms of other recorded species.

It is probable that some other names may have to be added to this list as a result of further studies.

TABLE II. Species deleted through erroneous recording or now treated as synonyms of other taxa

Anthoceros crispulus	*Orthotrichum shawii*
A. dichotomus	*Pellia borealis*
Barbula rufa[a]	*Plagiochila ambagiosa*
Bazzania triangularis	*Rhacomitrium sudeticum*
Bryum purpurascens	*Riccia ciliata*
Campylium hispidulum	*R. michelii*
Cephalozia affinis	*Sphagnum amblyphyllum*
Cinclidotus riparius	*S. centrale*
Cynodontium gracilescens	*S. fallax*
Fissidens minutulus	*Ulota nicholsonii*
Fontinalis dalecarlica	*Weissia crispata*
Fossombronia loitlesbergeri	
F. mittenii	
Moerckia hibernica	

[a] It is possible that this species once occurred in Britain but the records are uncertain.

EXTINCTIONS

In starting to consider extinctions we are beginning to discuss real changes which have occurred in the British bryophyte flora. Several species have, however, been reported as extinct only to be refound after a long interval. An example of this is *Scapania calcicola*, found in 1900 at Craig-an-Lochain-Lairige, near Killin, Perthshire, by Macvicar; it was not seen again after 1923 until 1962 when it was rediscovered in the same locality (Perry, 1967). Similarly, *Mnium medium*, collected on Ben Lawers in 1899 and 1902, was not recorded in Britain again until 1967 when it was found fertile on Bidean nam Bian, Glencoe (Duckett and Little, 1968).

In many cases it is very difficult to be certain that a species is really extinct, though the probability of extinction is far higher in the case of certain species with highly restricted requirements (e.g. some species of base-rich mires).

Table III lists the bryophytes (20 mosses) that have not been seen in the British Isles for many years but it must be borne in mind that any, or even all, of them might be refound. On the other hand there are some species which are known to be exceedingly rare and approaching extinction, and a few of them may now have become extinct.

TABLE III. Bryophytes not seen in Britain for many years and apparently extinct

Anomobryum juliforme	*Leskea nervosa*
Anomodon attenuatus	*Neckera pennata*
Cynodontium fallax	*Paludella squarrosa*
Fontinalis dolosa	*Paraleucobryum longifolium*
F. hypnoides	*Scapania crassiretis*
Grimmia crinita	*S. parvifolia*
G. elatior	*Sphagnum obtusum*
Gyroweisia reflexa	*Tortella limosella*[a]
Helodium blandowii	*Trematodon ambiguus*[b]
Lescuraea saxicola	*Weissia mittenii*

[a] Not seen since its discovery last century; [b] only found once in 1883 in Perthshire.

Meesia longiseta should perhaps be added to the list of extinctions for although it has never been recorded as living in Britain, subfossil remains have been discovered in late post-glacial peat deposits at Holme Fen, Huntingdonshire (Dickson and Brown, 1966).

DECREASED SPECIES

Some of the species which have declined in Britain in the past 100 years are listed in Table IV. In some cases the decline has become apparent only recently, while in others it has undoubtedly been continuing since early man began to clear the primaeval forests and drain the bogs and fens. In this and the following section it is convenient to consider the apparently extinct species (Table III) together with those that have declined.

There is no doubt that a large proportion of our bryophyte flora is showing decreases in both distribution and abundance. It is possible to compare the vice-counties from which species have been recorded with those in which they still occur but in most cases this tends to provide a misleadingly optimistic picture.

The situation is, however, much worse in small densely populated countries such as Belgium and the Netherlands which do not retain the extensive, thinly populated, and relatively unexploited areas which Britain still possesses in, for example, the Scottish Highlands. Delvosalle *et al.* (1969) consider that 94

TABLE IV. Some bryophytes that have diminished in the last century over much, or all, of Britain

(A) Air pollution probably the main cause
Cryphaea heteromalla[a]
Frullania tamarisci (as an epiphyte)[a]
Neckera pumila[a]
Orthotrichum lyellii[a]
O. obtusifolium[b]
O. pallens
O. pulchellum[a]
O. schimperi
O. speciosum
O. stramineum
O. striatum
O. tenellum[a]
Tortula laevipila[a]
T. papillosa[a]
Ulota bruchii[a]
U. crispa[a]

(B) Air pollution and the felling of ancient trees probably both responsible
Antitrichia curtipendula[a]
Pterogonium gracile[a]

(C) Changes in road use and wall construction
Aloina brevirostris
A. rigida
Pterygoneuron lamellatum
P. ovatum

(D) Drainage of bogs and fens and the disappearance of ponds
Amblyodon dealbatus
Camptothecium nitens
Cinclidium stygium
Dicranum undulatum
Drepanocladus aduncus
Sphagnum imbricatum

(E) Changes in land use
Splachnum ampullaceum
Tetraplodon mnioides

(F) Causes not well understood
Anomodon longifolius
Jungermannia lanceolata
Sematophyllum demissum
Tayloria lingulata
T. longicolla

[a] Decreases confined to areas of more severe air pollution; [b] only one post-1945 British record (from Angus).

mosses and 20 hepatics are now extinct in Belgium and that a further 30 mosses and 4 hepatics are very rare, in general decline, and are more or less directly threatened with extinction, out of a total of 600 bryophytes recorded for Belgium. A further 58 species in Belgium, though not appearing directly endangered, are now very rare or local and so are only likely to survive in the long term with the aid of positive conservation measures for their habitats.

In Britain, in contrast, only 20 mosses appear to be extinct (Table III) while we have listed 33 mosses and 2 hepatics as seriously diminished. This perhaps understates the case, however, as many other species characteristic of lowland "natural" habitats (e.g. woodlands, heaths, calcareous grasslands, rocks, fens, bogs, fresh water, coastal dunes) are often now restricted more or less precariously to the remaining small "oases" of uncultivated terrain, particularly in lowland Britain.

THE CAUSES OF DIMINUTION AND EXTINCTION

In 1800, although largely deforested already, lowland Britain was essentially agricultural with industry still on a very small scale and usually domestically orientated. With the exception of London, the towns covered very small areas and air pollution was negligible except in London and its limited suburbs and in the near vicinity of the new industrial towns such as Birmingham, Leeds and Manchester. Large areas of unenclosed common land remained, and agricultural practices did not involve the use of pesticides, herbicides or artificial fertilizers. The coastline, except at a few points, was still largely a desolate waste.

The changes in the environment that have occurred since that time have been outlined by Mellanby in Chapter 1. The ways in which these changes have affected bryophytes since that time are reviewed in the following paragraphs.

1. Air Pollution

Pollution of the air by sulphur dioxide has undoubtedly been the main factor reducing the frequency and abundance of epiphytic bryophytes, although the continued felling of old trees and the clearance and modification of forests has certainly contributed to this reduction. As is the case with the epiphytic lichens (see pp. 51–58; Hawksworth *et al.*, 1973), all epiphytic bryophytes except the most tolerant (e.g. *Dicranoweisia cirrata*, *Hypnum cupressiforme*) have been eliminated over a wide zone of England from the London area, southern East Anglia and the Lower Thames Valley up through the Midlands to Lancaster in the north-west and Tyneside in the north-east, and also in industrial South Wales and central Scotland.

There are fewer obligatory epiphytic species of bryophytes as compared with the lichens, so that the damage to the bryophyte flora in these regions has perhaps been less drastic numerically as compared with lichens. Figure 2 illustrates the changes that have occurred in the distribution of a particularly sulphur dioxide-sensitive moss, *Antitrichia curtipendula*. This moss has decreased in range more than most epiphytic bryophytes (or even lichens) being of similar sensitivity to lichens of zone 10 of Hawksworth and Rose (1970). *Antitrichia curtipendula* continues to maintain much of its old distribution and abundance in the Scottish Highlands, where it occurs on rocks as well as on trees. This species has even diminished in the woods long famous for it on Dartmoor (Black Tor Copse

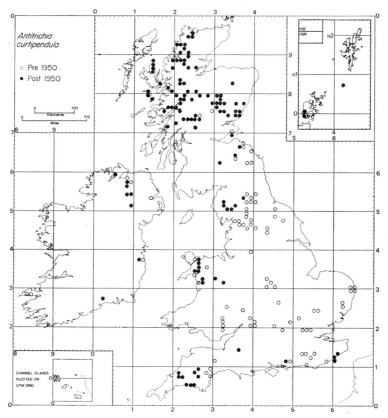

FIG. 2. British distribution of *Antitrichia curtipendula*. A moss once frequent and widespread on old trees in lowland Britain, now greatly declined due to air pollution, forest clearance and the felling of old trees, except in the Scottish Highlands where it is still common on trees and on rocks. There are also unlocalized pre-1950 records from vice-counties 9, 10, 12, 39, 40, 44 and 46.

and Wistmans Wood). It was plentiful in these woods until about 1950 at least but is now very rare in both sites. In south-east England this species is now confined to north-facing chalk grasslands (Heyshott and Didling, Sussex), and under *Prunus spinosa* scrub on shingle beaches (Denge Beach and Hythe Beach, Kent) where it also appears to have declined.

Coker (1967) demonstrated the extreme sensitivity of *Orthotrichum lyellii* to sulphur dioxide in the laboratory and so it is perhaps not surprising to note that the epiphytic species of this genus (excluding *O. diaphanum* and to some extent *O. affine*) have shown comparably devastating declines to *Antitrichia* in areas of lowland Britain where pollution of the air by sulphur dioxide is important, although *O. lyellii* has been able to persist in many sheltered parklands and open woodlands in the extreme south of England, south of London.

The decline in fertility of mosses that persist in areas of moderate air pollution may also be attributable to sulphur dioxide. *Orthotrichum lyellii*, *Neckera complanata*, *N. pumila* and *Rhytidiadelphus loreus*, for example, are all known to have fruited quite well in places like Berkshire and Oxfordshire up to the mid-nineteenth century (Jones, 1953), and Swinscow (1959) showed that specimens collected by W. H. Coleman in the 1840s indicated that *Rhytidiadelphus loreus* and many other species fruited in Hertfordshire at that time. These species, where they occur at all in the southern Midland counties today, never produce sporophytes.

2. Road Use and Wall Construction

The tarring of even minor roads, road widening and the consequent removal of hedgebanks, and the pollution by diesel oil and fumes on major roads has greatly affected the water relations and even the existence of many roadside banks, which were (and still are in many areas, e.g. The Weald) an important habitat for bryophytes in lowland areas. Carbon monoxide from petrol engines appears to have comparatively little effect on bryophytes as is the case with lichens, but the high lead levels that must accumulate on banks by major roads may well be detrimental to bryophytes.

The compacted soil of little-used unmetalled minor roads seems at one time to have been an important bryophyte habitat which has now largely disappeared. The mud-capping of stone walls with material scraped off the roads was once a common practice which has now been discontinued entirely. This change appears to be largely responsible for the enormous decrease in frequency of *Aloina brevirostris*, *A. rigida*, *Pterygoneuron lamellatum* and *P. ovatum*; all species once common on mud-capped walls.

3. Drainage of Bogs and Fens and the Disappearance of Ponds

The drainage of both acid and alkaline mires has resulted in the apparent extinction of at least two British mosses within the last century—*Helodium blandowii* and *Paludella squarrosa*. These species appear to have been relicts of late or early post-glacial times, which like the vascular plant *Scheuchzeria palustris*, were unable to survive disturbance and drainage of their ancient relict habitats, being also (unlike many mire species) apparently unable to colonize newly created habitats with high water tables in such mires (e.g. peat cuttings). *Scorpidium turgescens* (see Birks and Dransfield, 1970) was formerly widespread in lowland eutrophic mires in Britain as is evident from subfossil remains in Cambridgeshire, Middlesex, Northamptonshire, Worcestershire, Berwickshire and Co. Louth, dating from full-glacial or late-glacial times. It is probable that *Scorpidium turgescens* survived until 100–200 years ago, like *Helodium* and *Paludella*, in suitable sites in lowland Britain until drainage and peat cutting destroyed them; this species still occurs, however, in one site in the Scottish Highlands. *Meesia longiseta*, another late-glacial relict species of some continental European mires, however, died out in Britain before botanical recording began (Dickson and Brown, 1966).

Camptothecium nitens and *Cinclidium stygium* are other apparently late relicts of eutrophic mire habitats but these two species have been able to persist even in lowland Britain in a few fens in Norfolk although they are now extremely rare.

Drainage, burning and peat cutting of ombrogenous bogs ("hochmoore") have certainly been the main factors responsible for the decline of *Sphagnum imbricatum* (formerly, as is evident from subfossil remains, one of the major peat-forming species of ombrogenous raised and blanket bogs) and *Dicranum undulatum*. *S. imbricatum* may also have been reduced, along with other species of its genus, by air pollution in the case of the southern Pennine blanket bogs. *D. undulatum*, which still exists in relics of lowland raised bogs even in Cheshire, Cumberland and Shropshire, does not appear to have been lost from many vice-counties, but subfossil remains indicate that it used to be much commoner than it is today, as is the case at Holme Fen (see Dickson and Brown, 1966).

The massive exploitation of Irish raised and blanket bogs for the production of milled peat for use as fuel for power stations has had serious effects on the status of some bog species. It is unfortunate that some of the best examples of these bogs could not have been conserved; some reserves have been made but these are unfortunately not in the best sites from the bryological standpoint.

4. *Pasture Practice on Heaths and Fens*

The decline in pasturage of unimproved wet-heath and fen sites by cattle has undoubtedly led to a decline in lowland areas of Britain of such species as *Splachnum ampullaceum* and *Tetraplodon mnioides* which grow on decaying animal dung in wet peaty situations. *S. ampullaceum* seems now confined in lowland Britain to the New Forest and one site in Suffolk where the old land-use persists. Both these species have, however, become extinct in Belgium and northern France (see Delvosalle *et al.*, 1969).

5. *Forest Clearance and Management*

Forest clearance and management have played a major part in the reduction of our bryophyte flora. It is possible that some species of bryophytes which still persist in some central European forests became extinct at least in lowland Britain before recording began. Reductions in the frequency of even some common species such as *Isothecium myosuroides* and *Mnium hornum* have also occurred from this cause.

The conversion of deciduous woodland to coniferous plantations has drastic effects on the bryophyte flora, although several species characteristic of the pine forests of the Scottish highlands have appeared in the Breckland of East Anglia since pine forests were planted on the formerly open heathland (e.g. *Dicranum polysetum*, *Ptilium crista-castrensis*).

6. *Water Pollution*

Aquatic species, such as those of the genus *Fontinalis*, have declined in many parts of Britain as a result of the pollution of rivers. This factor may have been responsible, at least in part, for the extinction of *F. dolosa* and *F. hypnoides*. *Orthotrichum rivulare* and *O. sprucei*, characteristic of silt-covered tree roots by rivers below the flood-level, may also have been diminished by water pollution.

7. *Agricultural Chemicals*

Modern herbicides and fertilizers, together with practices such as stubble-burning and reploughing soon after harvesting, have led to the reduction of mosses that occur on fallow arable land, just as they have affected many flowering plant weeds. No extinctions appear to be attributable to this cause, however. Several new species (e.g. *Bryum* spp., *Dicranella staphylina*) have been described from this specialized habitat and many others continue to be common in suitable places.

8. *Public Pressure*

The increasing pressure of recreational use on such habitats as chalk grass-lands, heaths, and popular mountainous areas, has undoubtedly affected some bryophytes. In some cases, however, public pressure has had a beneficial effect on the bryophytes for, since the decline in sheep grazing and the myxomatosis epidemics, chalk grasslands on the Downs tended to develop coarse grass or scrub to the detriment of the short open turf species characteristic of such grasslands. The increase of scrub (particularly hawthorn) causes many bryophytes to disappear or become etiolated in the longer grass. Moderate trampling, such as that which occurs on parts of Box Hill, Surrey, tends to encourage such species by maintaining short turf.

Increased incidence of heathland fires, largely due to increased public pressure, has, however, certainly reduced the heathland bryophyte flora of some areas.

9. Other Factors

A number of other species of bryophytes have shown considerable declines or even perhaps extinction (e.g. *Anomodon attenuatus*) from causes which are less clear than those enumerated above. Some of these are on the edge of their European ranges (e.g. *Anomodon attenuatus, Jungermannia lanceolata*) and may always have been rare, so that over-collecting may also have been important in causing the local disappearance of some hepatics from limited, probably relict habitats, in places such as the Sussex sandrocks (e.g. *Jamesoniella autumnalis, Jungermannia lanceolata, Plagiochila spinulosa*) which are still very rich in other local species.

The arctic-alpine bryophyte flora of Ben Lawers has also probably been diminished by over-collecting. Commercial moss-collecting is now also a threat to species in both bogs and chalk grasslands as rare species are likely to be collected indiscriminately together with commoner ones.

The various factors which have probably contributed to a reduction in the quantity of many of our "atlantic" bryophytes are discussed by Ratcliffe (1968). The opening up of old forest canopies, resulting in drying out, and the spread of *Rhododendron ponticum* in some Kerry woodlands, are two of the additional factors Ratcliffe discusses in detail.

INCREASED SPECIES

Increases in bryophyte species are often less obvious than decreases, and more difficult to prove as inconspicuous plants may have been present but simply overlooked in the past for various reasons. In general real increases have been relatively few as compared to the decreases which are known to have occurred; nevertheless there are some spectacular examples. Examples of species which have increased in Britain are given in Table V.

Bryum argenteum, like the lichen *Lecanora dispersa*, seems to have adapted itself well to urbanization. This species has become common in built-up areas throughout Britain, particularly where nutrient-rich dust settles on the ground, and it appears to be almost immune to pollution by sulphur dioxide (as is

L. dispersa). It is, however, difficult to show the increase of a species like *B. argenteum* cartographically.

Dicranum strictum, a species of decaying wood and stumps, was at one time apparently rare and restricted to a few Midland counties and a small area in Speyside. In Fig. 3 the apparently rapid expansion in range of this species since 1943 is indicated. This pattern in both space and time resembles that seen in the epiphytic lichens *Parmelia laciniatula* and *Parmeliopsis ambigua* (see pp. 53–56).

TABLE V. Bryophytes that are increasing their ranges in Britain

Campylopus introflexus	Increase widespread, even in Ireland
Dicranoweisia cirrata	Commoner everywhere except in the western Highlands and parts of south-west England and Wales
Dicranum montanum and *D. strictum*	Expansion in the Midlands, southern and south-eastern England
Nowellia curvifolia	Increasing southwards and eastwards
Orthodontium lineare	Now very widespread except in western Highland Britain
Ptilidium pulcherrimum	Increasing southwards
Tortella inflexa	Increasing in south-east England
Tortula stanfordensis	Originally known in vice-county* 1; now in 17, 19, 21, 34, 35, 36 and 64

These two lichen species and *D. strictum* do not penetrate East Anglia to any great extent and all become much rarer westwards. It is possible that air pollution patterns are important limiting factors in these cases, the species being tolerant of moderate air pollution and possibly being selectively favoured by this in competition with other species.

The epiphytic species *Dicranoweisia cirrata* has also increased since the late nineteenth century. Boswell (Jones, 1953) only discovered it in one site in Oxfordshire in 1879 and was unable to report it from Berkshire. If it had been as frequent in these areas then as it is today Boswell would certainly have indicated this. By 1900 this species was common in the Newbury area but it was not reported in Cambridgeshire until about 1928. Between 1907 and 1963 this species was added to 23 vice-counties in the British Isles and since 1963 has been reported from a further 5. This species was formerly largely confined to south-east England, eastern Scotland and eastern Ireland but is now common in most parts of the British Isles except in the western Highlands of Scotland,

---

* For names and boundaries of vice-counties see pp. 425–428.

parts of south-west England, and the New Forest where it is still rather local. As in the case of *Dicranum strictum* this expansion is similar to that seen in *Parmeliopsis ambigua* and again this may be a species that has increased in areas of moderate sulphur dioxide pollution.

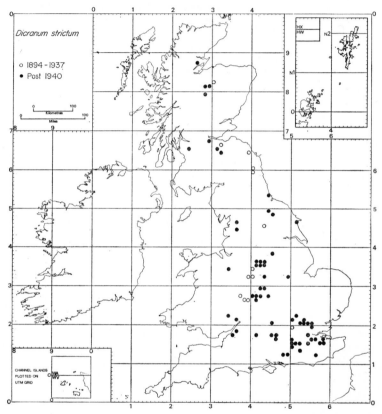

Fig. 3. British distribution of *Dicranum strictum*. A moss found on decaying wood, stumps, and on bark of living trees, formerly confined to a few sites in the west Midlands and Speyside, which is now spreading in northern, central and south-eastern England, particularly in areas of moderate air pollution.

The largest increases, however, have been in mosses introduced into the British Isles. *Orthodontium lineare*, for example, appears to have entered this country on timber from South Africa (where it is native) most probably at Birkenhead docks early this century (Watson, 1922; Margadant and Meijer, 1950). Shortly afterwards it was discovered in the Cheshire and Yorkshire Pennines and has now spread throughout England and much of Wales and has even invaded the Netherlands, Denmark and Germany, apparently from Britain. In many Midland and south-east English counties this is now one of the commonest bryophytes on rotten wood, tree bases, and acid peat soils. It always bears sporophytes and is able to withstand air pollution by sulphur

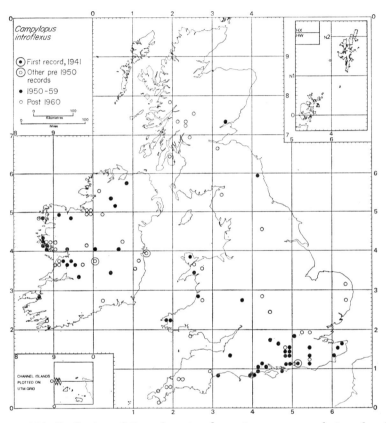

FIG. 4. British distribution of *Campylopus introflexus*. A moss apparently introduced into Britain prior to 1941 when it was first found in West Sussex, now very common and still spreading all over the British Isles on heathland and on acid peaty soil and stumps.

dioxide so well that it is plentiful in some very polluted parts of the Midlands where few other bryophytes are able to survive.

The distribution of another introduced species, *Campylopus introflexus*, is shown in Fig. 4 which is based on the map published by Richards (1964) with some additional records added. The true *C. introflexus* was first discovered on a heath near Pulborough, West Sussex, by J. B. Marshall in 1941 (Richards, 1963). Previously there had been some confusion between this species and the native *C. polytrichoides*, a mediterranean-atlantic moss long known in south-west England, South Wales and south-west Ireland, which had generally been named as "*C. introflexus*". *C. introflexus* is native to certain tropical mountainous regions (e.g. in East Central Africa) and appears to be an introduction in western Europe. Since 1941 this species has spread explosively throughout the British Isles on heathland, dried out blanket bogs, tree stumps, and acid peaty soils generally. *C. polytrichoides*, in contrast, has remained static in its old localities on rocks and dry heaths (mainly near the sea) in the extreme south-western parts of the British Isles.

Tortula stanfordensis is also an alien species in Britain and appears to be in an early stage of expansion at the present time. It will be interesting to observe its distribution in future decades.

Tortella inflexa, present at Folkestone, Kent, in 1904 but not then recognized, was first recognized on chalk stones in woodland in Sussex in 1957, and is now common on chalk and limestone areas throughout south-east England (Wallace, 1972).

CONCLUSIONS

From the preceding sections it will be evident that there have been many changes in the real distribution, and even in the presence, of species of bryophytes in Britain, particularly in the last 100 years. Some of these changes represent gains of new species or expansions in range of previously well established but more local ones, but the majority of real changes have been decreases in abundance and(or) fertility due to man-made changes in the environment.

It is clear that further species of bryophytes will become even rarer in the future, and almost certain that more species will become extinct, at least in lowland Britain. On the other hand, apparent changes will certainly continue to add taxa to the British lists by the discovery of previously unrecognized "critical" species, and probably also of further non-critical species in more remote areas. Some of the species which are thought to have become extinct in Britain may also be refound in relatively unworked parts of highland Britain.

ACKNOWLEDGEMENT

We are very grateful to Dr A. J. E. Smith for his assistance in the compilation of data for the maps presented here.

REFERENCES

BIRKS, H. J. B. and DRANSFIELD, J. (1970). A note on the habitat of *Scorpidum turgescens* (T. Jens.) Loeske in Scotland. *Trans. Br. bryol. Soc.* **6**, 129–132.

COKER, P. D. (1967). The effects of sulphur dioxide on bark epiphytes. *Trans. Br. bryol. Soc.* **5**, 341–347.

DELVOSALLE, L., DEMARET, F., LAMBINON, J. and LAWALRÉE, A. (1969). Plantes rares, disparues, ou menacées de disparition en Belgique: L'appauvrissement de la flore indigène. *Ministère de l'Agriculture, Administration des Eaux et Forêts, Service des Réserves Naturelles domaniales et de la Conservation de la Nature, Travaux no.* **4**.

DICKSON, J. H. and BROWN, P. D. (1966). Late post-glacial *Meesia longiseta* Hedw. in S.E. England. *Trans. Br. bryol. Soc.* **5**, 100–102.

DILLENIUS, J. J. (1741). "Historia muscorum." Sheldonian Theatre, Oxford.

DUCKETT, J. G. and LITTLE, E. R. B. (1968). *Mnium medium* B. & S. in Britain. *Trans. Br. byrol. Soc.* **5**, 452–459.

HAWKSWORTH, D. L. and ROSE, F. (1970). Qualitative scale for estimating sulphur dioxide air pollution in England and Wales using epiphytic lichens. *Nature, Lond.* **227**, 145–138.

HAWKSWORTH, D. L., ROSE, F. and COPPINS, B. J. (1973). Changes in the lichen flora of England and Wales attributable to pollution of the air by sulphur dioxide. *In* "Air pollution and lichens" (B. W. Ferry, M. S. Baddeley and D. L. Hawksworth, eds), pp. 330–367. University of London Athlone Press, London.

HOOKER, W. J. (1816). "British Jungermanniae." London.

HOOKER, W. J. and TAYLOR, T. (1818). "Muscologica Britannica." London.

JOHNSON, T. (1629). "Iter plantarum investigationes ergo suscepti in agrum Cantianum anno 1629 Julii 13." London.

JOHNSON, T. (1632). "Enumeratio plantarum in Ericeto Hampsteadiano locisque vicinis crescentium." London.

JONES, E. W. (1953). A bryophyte flora of Berkshire and Oxfordshire. II. Musci. *Trans. Br. bryol. Soc.* **2**, 220–277.

L'OBEL, M. DE (1576). "Plantarum sen stirpium historia." Antwerp.

MARGADANT, W. D. and MEIJER, W. (1950). Preliminary remarks on *Orthodontium* in Europe. *Trans. Br. bryol. Soc.* **1**, 266–274.

PATON, J. A. (1965a). "Census catalogue of British Hepatics", 4th edition. British Bryological Society, London.

PATON, J. A. (1965b). *Fissidens celticus* sp. nov. *Trans. Br. bryol. Soc.* **4**, 780–784.

PERRY, A. R. (1967). Notes on the genus *Scapania*. II. *S. calcicola* (Arn. & Perss.) Ingham. *Trans. Br. bryol. Soc.* **5**, 237–244.

RATCLIFFE, D. A. (1968). An ecological account of Atlantic Bryophytes in the British Isles. *New Phytol.* **67**, 365–439.

RAY, J. (1660). "Catalogus plantarum circa Cantabrigiam nascentium." Cambridge.

RAY, J. (1690). "Synopis methodica stirpium britannicarum." London.

RICHARDS, P. W. (1963). *Campylopus introflexus* (Hedw.) Brid. and *C. polytrichoides* de Not. in the British Isles. *Trans. Br. bryol. Soc.* **4**, 404–417.

RICHARDS, P. W. (1964). Distribution maps of *Campylopus polytrichoides* and *C. introflexus* in the British Isles. *Trans. Br. bryol. Soc.* **4**, 741–742.

SMITH, A. J. E. (1971). [Progress map of the Bryophyte Mapping Scheme of the British Bryological Society]. *Trans. Br. bryol. Soc.* **6**, 331.

TURNER, D. and DILLWYN, L. W. (1805). "The Botanist's Guide through England and Wales", 2 vols. Phillips and Fardon, London.

WALLACE, E. C. (1972). *Tortella inflexa* (Bruch) Broth. in England. *J. Bryol.* **7**, 153–156.

WARBURG, E. F. (1963). "Census Catalogue of British Mosses", 3rd edition. British Bryological Society, London.

WATSON, W. (1922). A new variety of *Orthodontium gracile* Schwaegr. *J. Bot., Lond.* **60**, 139–141.

4 | Changes in the British Lichen Flora

D. L. HAWKSWORTH

Commonwealth Mycological Institute, Kew, Surrey

B. J. COPPINS and F. ROSE

Department of Geography, King's College, University of London

Abstract: Between the periods in which British lichens have been studied most intensively some drastic changes have taken place in the composition of the flora in some regions of the British Isles. Although only about 40 species (*ca.* 3 % of the total flora) have not been refound this century, many have disappeared from large areas of central and eastern England. The major factor causing this decline is pollution of the air by sulphur dioxide which has affected both the distributions of species and the floras of particular sites. Air pollution appears to have had an overriding effect but in areas where this is not an important factor changes may also arise from bark eutrophication, agricultural sprays, maritime pollution, woodland management, quarrying and mining, changes in man-made substrates, heathland management, overcollecting, public pressure, and possibly water pollution and other organisms. Of these factors woodland management has had the most deleterious effect as many species are restricted to ancient woodlands. Recent climatic fluctuations appear to have had little effect on the British lichen flora. Unless trends in air pollution emissions are modified a continuing depletion of our lichen flora will occur and many more species are likely to become extinct in Britain. Many small invertebrates live on and amongst lichens and so a declining lichen vegetation is a matter of some concern to biologists generally and not only to lichenologists.

INTRODUCTION

Before considering changes which have occurred in the British lichen flora it is first of all necessary to review the evidence available on the state of the British lichen flora at different times. The first localized British records are probably those of Johnson (1632) who recorded a number of species in the Hampstead area of London. The first attempt at a systematic list of British

Systematics Association Special Volume No. 6, "The Changing Flora and Fauna of Britain", edited by D. L. Hawksworth, 1974, pp. 47–78. Academic Press, London and New York.

lichens was, however, that of Dillenius (*in* Ray, 1724) who listed 91 species. Most of the specimens referred to in this work are preserved in the Sherard herbarium in Oxford. In the "Historia muscorum" Dillenius (1741) was able to extend his earlier work and discussed 190 species most of which he considered to occur in Britain; the specimens figured in this work are also preserved in Oxford. Following the death of Dillenius in 1747 little interest was shown in British lichens and Hudson (1762) accepted only 85 species. Lightfoot (1777), however, recorded 103 species from Scotland and interest in the group began to increase so that 39 years later Withering (1801) was able to treat 226.

Although field work was very limited in the eighteenth century, largely because of the difficulties of travel, many species were discovered in sites from which they have since been lost. The rarer species known from different counties in England and Wales were noted by Turner and Dillwyn (1805) but their listing of, for example, 81 species from Sussex but none from Kent illustrates the unequal coverage of Britain for lichens at this time.

Towards the middle of the nineteenth century more data were becoming available as botanists interested in cryptogams began to travel more widely and have more contact with each other. The thirty years 1860–90 saw a level of activity in the study of British lichens not to be approached again until the mid-1960s. The extensive herbarium collections and literature which arose from this period provide a reasonably comprehensive view of the lichen flora and vegetation of many parts of Britain at this time and enable comparisons with the present lichen flora to be made on a sound basis. An excellent review of the studies of British lichens into this period was provided by Smith (1922) and the numbers of species recorded at various times are indicated in Fig. 1.

Following the deaths or retirements of many of the most active licheno-logists in this period interest in British lichens began to decline, with a few notable exceptions. A "Lichen Exchange Club of the British Isles" was estab-lished in 1907 but had a very small membership and ceased to function in 1914. The twenty years 1930–50 was probably one of the least active in British lichenology when those keenly interested in the group could be counted on the fingers of one hand. In 1953, however, a "Lichen Study Group" was established, and in 1958 the "British Lichen Society" was founded. Since the foundation of this Society, and particularly since about 1964, an interest in British lichens compar-able only to that achieved in Sweden in the period 1930–50 has been attained.

In 1964 the Society started a Distribution Maps Scheme using the 10 km square units of the Ordnance Survey's National Grid. A very considerable amount of information has now been obtained through this Scheme (see Seaward, 1973) although the intensity of recording in different parts of Britain varies markedly. Most parts of Ireland and southern Scotland remain very little known, but within the rest of the British Isles only Aberdeen and Moray, Caithness, Cheshire, south Lancashire, Shropshire, Staffordshire, Sutherland and parts of central and southern Wales are particularly underworked at the present time. The average number of species recorded from 10 km squares is 57 but the numbers vary very considerably. While squares with about 60

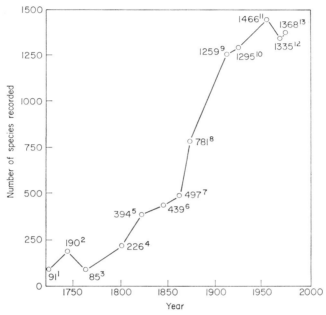

FIG. 1. Number of lichen species recorded as British 1724–1973; [1]Ray (1724), [2]Dillenius (1741), [3]Hudson (1762), [4]Withering (1801), [5]Gray (1821), [6]Hooker (1844), [7]Mudd (1861), [8]Leighton (1871), [9]Horwood (1912), [10]Smith (1918, 1926), [11]Watson (1953), [12]James (1965, 1966), [13]present.

species in the Midland counties, for example, may be well worked, squares with totals of less than about 150 in southern England, western Scotland and Wales are probably incompletely known. These field data, in conjunction with recent herbarium material and published papers, are enabling the present lichen flora to be contrasted with that in the past and real changes which have occurred are now becoming apparent.

At the present time there are probably in excess of 200 000 herbarium specimens of British lichens preserved in national, provincial and private herbaria, and about 1400 published books and papers including localized British records.

ADDITIONS, DELETIONS AND EXTINCTIONS

The increasing numbers of species listed as British between 1724 and 1953 (Fig. 1) were largely a result of increased field work, changing species concepts and a failure to study material critically (i.e. apparent changes). The reduction in the number of species from 1466 in 1953 (Watson, 1953) to 1335 in 1966 (James, 1965, 1966) arose from taxa being examined or re-examined critically for the first time showing that many species had been incorrectly determined or were synonyms of others.

TABLE I. Lichen species now thought to be extinct in the British Isles
(not seen since 1900)

Alectoria nitidula	*Lecania fuscella*
A. trichodes agg.	*Lecanora achariana*
Arctomia delicatula	*L. farinaria*
Arthonia arthonioides	*L. populicola*
A. atlantica	*L. pruinosa*
A. elegans	*Lecidella pulveracea*
Bacidia fuscorubella	*Melanaria lactescens*
B. subturgidula	*M. urceolaria*
Baeomyces carneus[a]	*Nephroma resupinatum*
Caloplaca furfuracea[b]	*Parmelia glabra*[a]
C. haematites	*Pertusaria bryontha*
C. pollinii	*Physcia ciliata*
Cetraria cucullata	*Physconia muscigena*
C. juniperina	*Ptychographa xylographoides*
Cetrelia olivetorum	*Ramalina breviuscula*
Cladonia destricta	*R. capitata*[a]
C. stellaris (= *C. alpestris*)	*R. minuscula*[a]
C. turgida	*Tornabenia atlantica*
Hypogymnia vittata	*Umbilicaria arctica*
Lecanactis amylacea	*U. rigida*[a]

[a] Correctly identified but British status in need of further investigation.
[b] Record in need of further investigation.

The rise since James's check-list to 1368 species in February 1973 reflects a great deal more work than these figures might superficially imply. In this period of seven years 54 species have been added to the British flora of which eight are newly discovered and described, six result from taxonomic splits, and three had lain unrecognized amongst nineteenth-century collections. The remaining 38 species are ones known outside the British Isles but discovered in Britain for the first time. Furthermore, about fifteen species new to science have yet to be formally described, a reflection of the intensive field work carried out in Britain in recent years.

More critical taxonomic work has shown that 22 species in James's check-list are synonyms of other species, while about 10 represent non-lichenized Ascomycotina (including obligately lichenicolous fungi). It is probable that more of the "endemic" crustose species described from poor material last century will also eventually prove to be synonyms of other better known species. James listed 90 of the more important erroneously reported species although six of these are in fact now known to occur still in Britain (*Alectoria sarmentosa* subsp. *sarmentosa*, *Chaenotheca trichialis*, *Cladonia nylanderi*, *Parmelia tinctina*, *Peltigera lepidophora* and *Physcia sciastra*).

Of the species correctly recorded from the British Isles (but omitting many small crustose species which have probably simply been overlooked) only 40 (*ca.* 3% of the total flora) have not been refound this century and appear to be extinct (Table I). It is of interest to note that some of the species which have not been refound were close to or at the northern (e.g. *Caloplaca haematites, Tornabenia atlantica*) or southern (e.g. *Alectoria nitidula, Cetraria cucullata, C. juniperina, Cladonia stellaris, Hypogymnia vittata, Umbilicaria arctica*) European limits of their generally Mediterranean or boreal ranges, respectively, in Britain.

CHANGES IN POPULATIONS AND DISTRIBUTIONS

Many species recorded in the British Isles, although they have not yet become extinct, have undergone drastic changes (most frequently reductions) in their ranges within the British Isles over the last hundred years. In most cases these changes are attributable to man's influence. The following sections review our knowledge of the effects of various factors which appear to have caused local or national changes in our lichen flora and(or) vegetation and assess their relative importance.

1. Air Pollution

The most important air pollutant affecting lichens in Britain is sulphur dioxide which is now considered to be largely responsible for the decline to extinction of many species in large areas of central and eastern England in particular. Sulphur dioxide in solution at pH values and concentrations equivalent to those encountered in nature in polluted areas has been shown experimentally to cause reductions in the net assimilation rates and respiration rates of many species. Fluorides have drastic effects also but these tend to be only of a very local nature (as is the case with most particulate air pollutants) as they are rapidly removed from the air. Smoke appears to have very little

c

direct effect on the occurrence of particular species but may lead to excessive eutrophication of bark (see p. 59). Lichens are able to accumulate heavy metals to exceedingly high concentrations (see James, 1973) but the effect of these on the distribution of species remains obscure. A comprehensive review of the effects of air pollutants on lichens has recently been presented by Ferry *et al.* (1973).

The changes in the lichen flora of England and Wales that appear to be attributable to pollution of the air by sulphur dioxide have been discussed by Hawksworth *et al.* (1973) and so this factor is treated only briefly here in comparison to its relative importance. It should be noted that many of the other factors discussed in the following sections are most important in areas where the mean winter sulphur dioxide levels exceed about 35 $\mu g/m^3$, when the reproductive and(or) regenerative capacities of many species able to withstand the ambient sulphur dioxide levels appear to be depressed.

(a) *Corticolous species.* The corticolous lichens in Britain have been more intensively studied in the last decade than those growing on other substrates. A relationship between the deterioration of the lichen vegetation in and around urban and industrial areas and air pollution was first suggested in Britain in 1859, but only in the last six years has it been possible to relate the lichen vegetation at a particular site to the mean sulphur dioxide levels of the air. As a result of extensive field work, a consideration of former distributions, and the correlation of these data with information from volumetric recording gauges, Hawksworth and Rose (1970) were able to extend the earlier work of Gilbert (1968) and construct a 0–10 zone scale for corticolous species in England and Wales related to the mean winter sulphur dioxide levels. Lichens start to appear on trees at mean winter sulphur dioxide levels of about 150 $\mu g/m^3$ (zone 2) but unaffected communities do not appear until these levels are less than about 30 $\mu g/m^3$ (zone 10). The construction of scales of this type, their application and correlation with sulphur dioxide levels, is discussed by Hawksworth (1973a).

Air pollution has caused reductions in the lichen flora in about two-thirds of England and Wales (see Hawksworth and Rose, 1970, for map). As the corticolous lichen vegetation in parts of Britain which are not subject to levels of sulphur dioxide detrimental to lichens has remained relatively unchanged over the last century, however, air pollution has so far completely eliminated few, if any, species from the British flora. This situation is in marked contrast to that in the Netherlands which had lost 27% of its epiphytic lichens by 1954 (Barkman, 1969). When individual counties are considered, however, the changes in the number of corticolous species known are often well marked. Of

154 corticolous species reliably reported from Leicestershire and Rutland, for example, Sowter and Hawksworth (1970) found that only 58 had been reported since 1960. Similarly, of 50 corticolous species recorded for the London area only nine have been seen in recent years (Laundon, 1970; Coppins, unpublished). In contrast, of the 290 corticolous species reported from Sussex, 249 (86%) still occur in that county.

It is now clear that ancient parklands with many extremely old trees of *Quercus* and *Ulmus* in areas of England with mean winter sulphur dioxide below about 30 μg/m³, characteristically have 80–130 or more corticolous species per km². The lichen floras of such established parklands decline as all major urban and industrial areas are approached (Hawksworth *et al.*, 1973). Some of these sites were intensively studied in the nineteenth century and as the ancient trees still standing today (over *ca.* 250 years) would have been mature and amongst those studied, a comparison of their past and present lichen floras is of a particular interest. In the case of Epping Forest, for example, of 118 epiphytic lichens recorded at various times over the last 170 years, 90 are now extinct; this Forest is known to have had communities of zones 9–10 in 1784–96, 7 in 1865–68, 5(–6*) in 1881–82, 4(–5) in 1909–19, and 3(–4) in 1969–70 (Hawksworth *et al.*, 1973). Some more recent changes have also been reported: for example, Bookham Common, near Leatherhead, Surrey, had 48 corticolous and lignicolous species in 1953–56 but only 36 have been found (many in much reduced quantities) in 1969–73, a decline of 25% (Laundon, 1973a).

A comparison of the past and present distribution of corticolous species in Britain indicates that about 90 have become very rare to extinct in areas of England which now experience mean winter sulphur dioxide levels of over about 65 μg/m³. Maps which show the declines of *Anaptychia ciliaris*, *Evernia prunastri*, *Lobaria pulmonaria*, *Parmelia caperata*, *P. perlata*, *Ramalina* species and *Usnea* species have been published by Hawksworth *et al.* (1973), of *U. articulata* by Hawksworth (1973b) and of *Alectoria fuscescens* by Hawksworth (1972a). The map for *Usnea* species has been updated and included here (Fig. 2) to illustrate the type of change which has occurred.

In contrast to the large number which have declined, only two corticolous species have increased dramatically as a result of increasing air pollution levels, *Lecanora conizaeoides* and *Parmeliopsis ambigua*. *L. conizaeoides*, discussed further by Hawksworth *et al.* (1973) and Laundon (1973), appears to have arisen in the period 1839–71, had spread widely by the end of the nineteenth century, and

---

* Zone numbers in parentheses refer to the parts of Epping Forest furthest from London beyond Epping town.

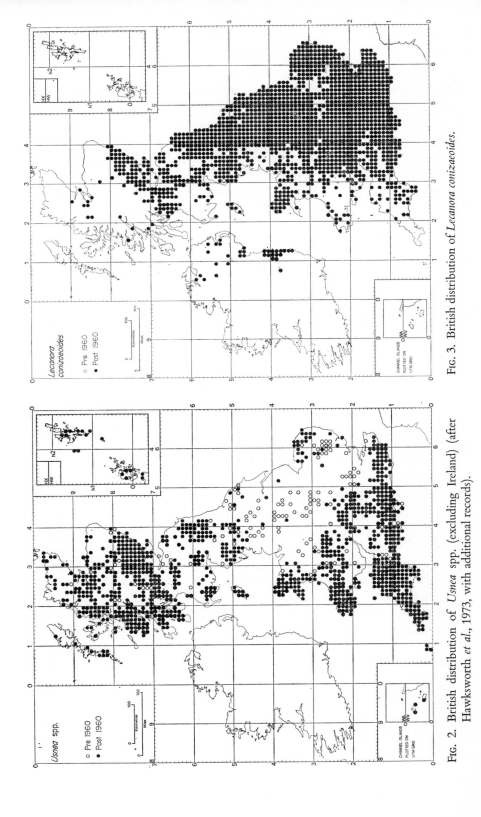

FIG. 3. British distribution of *Leacnora conizaeoides*.

FIG. 2. British distribution of *Usnea* spp. (excluding Ireland) (after Hawksworth *et al.*, 1973, with additional records).

has now become exceedingly common and widespread in areas of Britain with mean winter sulphur dioxide values in the range 55–150 $\mu g/m^3$ (Fig. 3). From its behaviour in relatively unpolluted parts of Britain it is evident that this species has a poor competitive ability but its almost unwettable thallus appears to render it particularly resistant to sulphur dioxide in solution (i.e. the form in which this gas is most toxic to lichens). *Parmeliopsis ambigua* was known up to the end of last century as a species of the bark of coniferous trees and *Betula* in central and eastern Scotland with a very few sites on lignum and *Pinus* (Ightham, Kent) in England and Wales. This species, which has a yellow-green foliose thallus, is unlikely to have been overlooked in lowland Britain last century but is now widespread on trees such as *Acer*, *Fagus*, *Fraxinus*, *Quercus* and *Sorbus* (i.e. species on which it did not formerly occur) in many parts of England and Wales (Fig. 4) where mean winter sulphur dioxide levels are in the range of about 55–65 $\mu g/m^3$. Where this species does occur in less polluted parts of England (e.g. Devonshire) it almost always occurs on decorticate wood and not bark. Sulphur dioxide is known to cause the acidity of tree bark to increase (see, e.g. Grodzińska, 1971) and the spread of this species onto deciduous trees in lowland Britain appears to be due to this effect which lowers the bark pH, making it comparable to that encountered in unpolluted areas only on coniferous trees and *Betula* (see Barkman, 1958).

Some species which have not been eliminated from or which have spread into moderately polluted areas, have different frequencies in these areas as compared to those they show in unaffected regions. This is particularly true of some species of zones 5–6 of Hawksworth and Rose (1970) which are often much commoner at these levels than at zones 8–10; e.g. *Parmelia saxatilis*, *Platismatia glauca* and *Pseudevernia furfuracea* favoured by increased bark acidity, *Buellia punctata*, *Catillaria griffithii*, *Chaenotheca ferruginea* and *Lecidea scalaris* probably favoured by reduced competition.

Reduction in both luxuriance and fertility are seen in species approaching their limits of tolerance to air pollutants. Specimens of *Usnea ceratina* are commonly 25–45 cm long in areas with mean winter sulphur dioxide levels below about 30 $\mu g/m^3$, for example, but usually less than 10 cm long in more polluted areas. Similarly *U. subfloridana* grows to 4–8 cm in unpolluted areas but is usually less than 1 cm long at its closest stations to air pollution sources. Ascocarp production is also inhibited and Hawksworth *et al.* (1973) list 23 species which fail to produce ascocarps in areas of England which now have mean winter sulphur dioxide levels of over about 40 $\mu g/m^3$. This decline in ascocarp production is seen in, for example, *Evernia prunastri* (Fig. 5), which although still widely distributed in Britain (see Hawksworth *et al.*, 1973, for

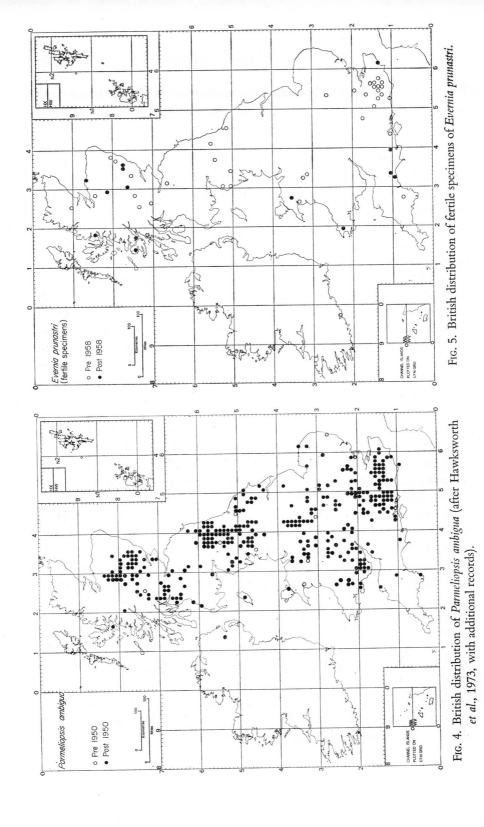

Fig. 5. British distribution of fertile specimens of *Evernia prunastri*.

Fig. 4. British distribution of *Parmeliopsis ambigua* (after Hawksworth *et al.*, 1973, with additional records).

map), now only produces ascocarps in the most unpolluted areas. In this case fertility is related to luxuriance as 93% of the 46 fertile gatherings examined exceeded 4 cm in length. In some instances a decreasing frequency of ascocarp production is related to an increased production of asexual reproductive structures (particularly soredia) but even the frequency of these declines in some species when they are growing close to their limits of tolerance.

In moderately polluted areas, particularly zones 4–6 of Hawksworth and Rose (1970), the richest communities tend to be restricted to the most ancient, and often isolated, trees. This appears to be due to some species losing their ability to colonize younger trees. The reasons for this are not yet clear, although it seems probable that a decreased viability of the propagules together with younger trees having a lowered unfavourable bark pH (which might prevent the germination and establishment of propagules) are responsible.

(b) *Saxicolous species.* Comparable trends to those seen in corticolous lichen communities occur in those on rocks and walls under air pollution stress. On siliceous rocks the effects are comparable to those seen on non-eutrophicated barked trees and many of the foliose species involved at mean sulphur dioxide levels over about 65 μg/m^3 are the same (see Gilbert, 1968, 1970). Relatively little is known of the effects of lower levels of sulphur dioxide than this on siliceous rock communities but observations in Derbyshire and Leicestershire indicate that some *Umbilicaria* species (*U. deusta, U. polyphylla, U. polyrrhiza, U. pustulata* and *U. torrefacta*) cannot tolerate mean winter sulphur dioxide values over about 55–70 μg/m^3 (depending on the species). Late eighteenth and early nineteenth century collections indicate that species of this genus have decreased in luxuriance in Derbyshire since that time, Most *Stereocaulon* species appear to be fairly tolerant of sulphur dioxide air pollution, appearing at about 60–70 μg/m^3. *S. pileatum* is more tolerant than other species of its genus in Britain having extended its range on man-made substrates in very polluted areas (see p. 66), a fact which may sometimes also be associated with increased lead pollution (see James, 1973).

Calcareous rocks exert a neutralizing effect on the ionic forms of sulphur dioxide in solution which appear to be most toxic to lichens. Consequently communities on calcareous rocks are less markedly affected by this pollutant than those on either trees or siliceous rocks. Sulphur dioxide is not known to eliminate many species from calcareous rocks but it may have a profound effect on the establishment of some species. Laundon (1967), for example, showed that *Caloplaca heppiana* only occurred on calcareous tombstones erected prior to 1900 in a London churchyard, i.e. it had assumed a relict status, being unable to colonize more recently erected tombstones.

In some areas with mean winter sulphur dioxide levels over about 150 $\mu g/m^3$ a few species normally characteristic of siliceous rocks colonize calcareous stonework and asbestos cement (e.g. *Bacidia umbrina, Catillaria chalybeia, Lecanora muralis, Lecidella scabra, Micarea lignaria*). The extension in range of some pollution tolerant lichens onto man-made substrates such as asbestos cement is discussed further below (pp. 66–68).

(c) *Terricolous species.* The effects of air pollution on terricolous lichens in Britain are very incompletely known. It is probable that the reduction of many species of *Cladonia* to almost indeterminable crusts of basal squamules in polluted areas is due to this factor. *Lecanora conizaeoides, Lecidea granulosa, L. uliginosa* and *Lepraria incana* are amongst the most tolerant species on acid peats. *Cornicularia aculeata* and some *Peltigera* species (e.g. *P. canina, P. rufescens, P. spuria*) are also able to persist in polluted areas, the latter often occurring in short turf of regularly cut lawns (e.g. tennis courts) in suburban areas.

Lichen communities on calcareous soils tend to be less modified than those on more acidic soils.

2. Bark Eutrophication

The eutrophication of normally acidic bark by both natural and artificial fertilizers tends to promote the growth of algae, particularly *Pleurococcus* species, to the detriment of any lichens that might be present. The effects of agricultural fetilizers on lichen communities have received relatively little attention but James (1973) found that powdered superphosphate encouraged the growth of *Pleurococcus* to such an extent that the lichen thalli present were smothered by it. If eutrophication is severe species-poor communities dominated by, for example, *Buellia canescens, Physcia adscendens, P. orbicularis, Xanthoria candelaria, X. parietina* and *Pleurococcus*, result. If the degree of eutrophication is more moderate and of natural origin, however, the lichen communities may even be encouraged and communities dominated by "nitrophilous" species (e.g. *Anaptychia ciliaris, Candelaria concolor, Physcia aipolia*) develop.

Some species characteristic of rather basic (eutrophicated) bark now appear to be much rarer than they were last century even in areas not affected by air pollution (e.g. *Physcia clementii, P. tribacioides, Teloschistes flavicans*) and it is probable that this may at least be partly due to excessive eutrophication from artificial fertilizers or to roadside trees no longer being splashed with dust enriched with horse dung as they were before the advent of the motor car. Some of the richest eutrophicated bark communities today tend to be confined to particularly ancient *Fraxinus* and *Ulmus* trees, suggesting that they are now often of relict status. The reduction in the numbers of trees with moderately eutrophicated barks may at least in part have contributed to the extinction of species such as *Caloplaca haematites* and *Tornabenia atlantica* in Britain (Table I).

Base enrichment by rock salt used on roads in the winter months, however, appears to eliminate almost all lichens within the splash zone of passing traffic. Soot deposits may also encourage the growth of *Pleurococcus* species by reducing the acidity of the bark and adding nutrients to it and may be one reason why this alga is so common on trees in areas affected by severe smoke pollution.

3. Agricultural Sprays

In addition to fertilizers which cause eutrophication of bark as discussed in the preceding section, some other agricultural fungicidal, herbicidal and pesticidal sprays appear to affect the lichen vegetation. Orchards formerly had very rich lichen communities composed of species characteristic of eutrophicated bark. In south-west England apple orchards used to contain abundant material of *Teloschistes flavicans* (Fig. 6). Blight (1876), for example, noted that this species ". . . grows on thorns and apple trees in the district [Lands End]; many specimens with fructifications . . .". Today this species is very rare even in south-west England and now most frequently occurs near the sea on rocks or heather stems. *T. chrysophthalmus* (last seen on the English mainland in 1966 on a hedge near Start Point, Devonshire, which has since been cut down; Fig. 6) and *Pseudocyphellaria aurata* formerly occurred in orchards in south-west England but both are now confined to the Channel Islands and the Isles of Scilly in Britain.

The deterioration of the lichen flora of apple orchards seems to be largely attributable to the use of agricultural sprays. Fungicidal sprays which are copper based have been used to control lichens abroad (Hawksworth, 1972b) but while there is no evidence that sprays have been applied with the specific intention of removing lichens from orchard trees in Britain, their use in the control of pathogenic fungi and insects (lichens are harmless to their host trees in Britain) may have led to the incidental destruction of the lichen communities including rare species such as those mentioned above. In well managed orchards today in areas not significantly affected by air pollution lichens are often rare or completely absent due to spraying. Orchards in Devonshire which have been derelict for many years often support luxuriant lichen communities but lack *Teloschistes* species which seem to have been eliminated to such an extent that their ability to spread into such orchards has been greatly reduced.

In some parts of East Suffolk and Norfolk (e.g. around Eye, Debenham, Halesworth and Diss), where air pollution is still fairly low, many corticolous lichens have largely disappeared. It seems probable that this is at least partly due to agricultural sprays drifting on the wind and affecting much larger areas than those for which they were intended.

4. Water Pollution

Siliceous rocks in fast flowing streams or shallow margins of lakes where the water is agitated characteristically support a lichen vegetation composed of species restricted to submerged and(or) inundated rocks.

In Derbyshire the two richest sites for aquatic lichens are the River Derwent at Slippery Stones (with e.g. *Lecanora lacustris, Lecidea hydrophila, Rhizocarpon lavatum*) and Burbage Brook in the upper parts of Padley Wood (with e.g. *Dermatocarpon fluviatile*). Samples of water collected from these localities on 27 December 1972 had nitrate-nitrogen levels of 0·45 and 0·36 ppm (measured with a nitrogen electrode), respectively. The River Derwent flows through a reservoir and agricultural land to Grindleford Bridge where it is joined by Burbage Brook, where no lichens characteristic of aquatic habitats were noted (one frequent on damp rocks occurs on boulders by and in it, however) and this gave a nitrate-nitrogen reading of 0·91 ppm the following day. Ammonium-nitrogen (tested for with Nessler's reagent) was not present in measurable amounts indicating that there was no significant source of nitrogen from urban sewage or farm effluents in these three samples.

These very preliminary studies might suggest a relationship between nitrate-nitrogen and the occurrence of some aquatic lichens but the situation is evidently very complex. Hawksworth (1972c) found that the large freshwater lake of Slapton Ley in South Devon lacked aquatic lichens on apparently suitable rocks and pebbles in its margins whereas several aquatic species of *Verrucaria* occurred commonly in some of the streams flowing into it (e.g. that in Slapton Wood). The Ley has periodic algal blooms indicative of eutrophication and also tends to have a somewhat turbid muddy nature. Troake and Walling (1973) found that in the period January–December 1972 the stream with the richest aquatic lichen flora had nitrate-nitrogen levels in the range 3·97–(5·98)–8·1 ppm, whilst Slapton Ley itself had levels in the range 0·00–(1·82)–4·81 ppm.

It is consequently evident that nitrate-nitrogen is not the limiting factor in Slapton Ley but that the turbid nature of the water is perhaps more important. As the species at Slapton differ from those in the Derbyshire sites studied it may be that different species are affected by different levels of nitrate-nitrogen but it is clear that further comprehensive studies must be carried out before any generalizations as to the effects of water pollution on aquatic lichens can be made.

5. Maritime Pollution

The effects of oil spillages on marine (sublittoral) and maritime (littoral) lichens can be spectacular, many species disappearing entirely from severely affected rocks. Studies by Brown (1972) suggest that at least in the case of *Lichina pygmaea* it is the solvent emulsifiers used to break up and disperse oil slicks that are responsible for the damage rather than the crude oil itself. Field observations in South Devonshire indicate that when rocks have been denuded as a result of very serious oil spillages the lichen vegetation does not start to recover for three to four years.

It is possible that untreated sewage and some industrial wastes discharged into the sea might affect marine lichens but these have not yet been investigated.

6. Woodland Management

We have already mentioned that ancient parklands with many very old trees commonly have 80–130 or more corticolous species per km^2 in areas which are relatively unaffected by air pollution. Numbers of this order would presumably have occurred in deciduous woodland throughout the British Isles before they began to be affected by man. The outstanding example of a relatively intact ancient woodland flora in England is the New Forest, Hampshire, which has some 255 corticolous and lignicolous lichen species (Rose and James, 1974). Although the older unenclosed woodlands of the New Forest were managed for many centuries up to about 200 years ago timber extraction seems to have been in the form of selective rather than clear felling so that a continuity of trees of mixed ages has been maintained.

From a consideration of the data from the New Forest and other woodlands whose histories are well known it has become clear that some species are now restricted to sites which have had a continuity of forest for many centuries (Rose, 1974); examples of some of these lowland "old forest" (or "native wood-land") indicator species are listed in Table II. As in the case of species used as indicators of air pollution levels by Hawksworth and Rose (1970), some species are indicators of old forest in some parts of Britain but not in others (e.g. *Enterographa crassa*, *Normandina pulchella* and *Rinodina roboris* in East Anglia but not in southern and western England and Wales). Upland woods in central and eastern Scotland have not yet been studied from both the historical and lichenological standpoints but there are indications that *Alectoria capillaris*, *Cavernularia hultenii* and *Cetraria pinastri* may be "old forest" indicators in the more boreal areas of central Scotland, whilst *Menegazzia terebrata*, *Parmelia endochlora* and *P. sinuosa* may be so in the west. It should be noted that some species which are "old forest" indicators are not necessarily encountered on the most ancient trees in the site.

Rose (1974) has proposed an "Index of Ecological Continuity" which provides a numerical assessment of the degree of modification woodlands have undergone based on the numbers of "old forest" species present.

TABLE II. Examples of lowland "old forest" indicator lichen species

Anaptychia leucomeleana[s]	*Mycoporellum sparsellum*[s]
A. obscurata[sw]	*Nephroma* spp.
Arthonia cinereopruinosa[w]	*Ochrolechia inversa*[wsw]
A. didyma	*Pachyphiale cornea*
A. stellaris[ws]	*Pannaria* spp.
Arthothelium ilicinum	*Parmelia arnoldii*[w]
Biatorella ochrophora[ws]	*P. crinita*[ws]
Bombyliospora pachycarpa[s]	*P. dissecta*[ssw]
Caloplaca herbidella	*P. horrescens*[ssw]
Catillaria atropurpurea	*P. taylorensis*[w]
C. pulverea[w]	*Parmeliella* spp.
C. sphaeroides	*Peltigera collina* [ssw]
Catinaria grossa	*Pertusaria velata*[s]
C. laureri[s]	*Porina coralloidea*[ws]
Graphina ruiziana[ssw]	*P. hibernica*[ssw]
Haematomma elatinum	*Pseudocyphellaria* spp.
Lecidea cinnabarina	*Rinodina isidioides*[ssw]
Leptogium burgessii[w]	*Schismatomma niveum*[s]
L. cyanescens[w]	*Sticta* spp.
L. saturninum[w]	*Thelopsis rubella*
Lobaria spp.	*Thelotrema lepadinum*
Lopadium pezizoideum	

[s] south of Severn–Thames line and southern Ireland; [sw] Cornwall, Devon and south-west Ireland; [w] western Britain, southern England west from Hants., and Ireland; [ssw] south-west and North Wales, southern England and southern Ireland; [ws] = [w] and [s].

The decline in many of the larger "old forest" species in lowland Britain has been spectacular. As any one of these species is unlikely to have increased its number of localities for several centuries an estimate of its decline can be derived from an analysis of its 10 km square grid records by the method of Perring (1970). This type of analysis is only likely to be realistic for the larger well-known species and should clearly not be applied to more recently recognized species (e.g. *Schismatomma niveum*) which were not formerly reported. *Lobaria pulmonaria* (mapped by Hawksworth *et al.*, 1973) has been known from

177 grid squares in England and Wales but has disappeared from 87 of these representing a decline of 51%.

Woodland management is not the only factor to be considered here, however, as many of the extinctions may be primarily due to pollution of the air by sulphur dioxide. Some indication of the effect of habitat destruction and modification can be obtained from considering records made in parts of southern England which are still relatively unaffected by air pollution. Such an anlysis has been carried out in Table III in which the decline in the number of records of

TABLE III. Number of 10 km square records of some "old forest" species south of the Ordnance Survey lateral grid line "15" (see text for further explanation)

Species	Total number of records	Post-1960 records	% decline
Pannaria rubiginosa	13	1	92
Parmeliella plumbea	15	2	87
Lobaria amplissima	17	7	60
Catinaria grossa	11	4	64
Lobaria scrobiculata	32	12	63
Pannaria pityrea	24	9	62
Peltigera collina	26	14	46
Lobaria pulmonaria	74	41	45
Nephroma laevigatum	25	15	40
Pertusaria velata	10	7	30
Sticta limbata	37	30	19
Thelotrema lepadinum	75	66	12
Pachyphiale cornea	72	70	3

Lobaria pulmonaria is seen to be 45% and the most marked decline (92%) to have occurred in *Pannaria rubiginosa* (Fig. 7). From Table III *Pachyphiale cornea* and *Thelotrema lepadinum* appear to have scarcely declined at all, reflecting their ability to survive and recolonize younger trees within old, yet managed woodland (e.g. "coppice-with-standards"), as long as free standing trees remain (e.g. along rides).

The effect of changing management practice is seen in the case of St Leonard's Forest, in a relatively air pollution-free part of Sussex. During the last century, and particularly in the last thirty years, much of the Forest has been felled, cut up, enclosed and cultivated, replanted with coniferous trees, or drained. Although some mature trees still survive the most ancient ones have been lost. Thirty-six "old forest" indicator species were found here by Borrer, some of

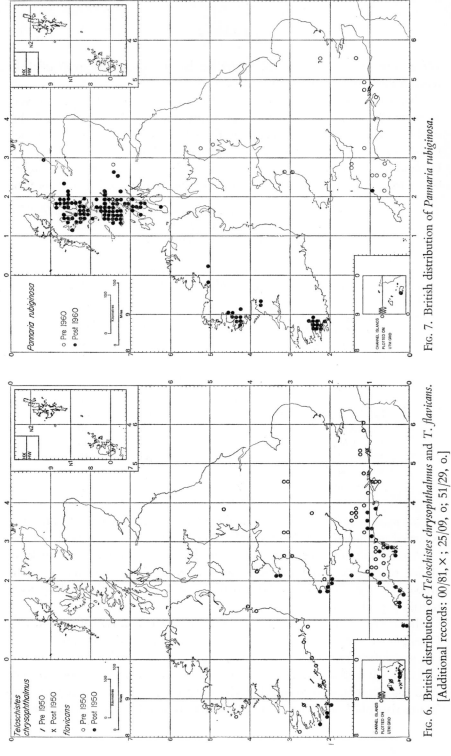

Fig. 7. British distribution of *Pannaria rubiginosa*.

Fig. 6. British distribution of *Teloschistes chrysophthalmus* and *T. flavicans*.
[Additional records: 00/81, ×; 25/09, o; 51/29, o.]

whose records are cited in Turner and Dillwyn (1805). The present lichen flora includes 107 epiphytic lichens, but the 36 "old forest" indicator species found by Borrer have disappeared (Rose, 1974). Their loss appears to be attributable to the extensive management and disturbance of the Forest.

Where clear felling has taken place all the "old forest" indicator species are lost. Some woodlands in polluted areas are, however, known to have lost this element because of air pollution effects even before they were felled (e.g. Buddon Wood, Leicestershire; see Sowter, 1950). In about 1850–60 Bloxam discovered *Lobaria laetevirens*, *L. pulmonaria*, *Parmeliella plumbea*, *Sticta limbata* and *S. sylvatica*, for example, in Appuldurcombe Park, Isle of Wight. Trees suitable for this assemblage of species no longer occur in this Park and the otherwise very rich lichen flora suggests that felling has been responsible for their loss. Clear felling and replanting with coniferous trees has drastic effects on the lichen flora as the species characteristic of coniferous trees in central and eastern Scotland seem to be unable to establish themselves in plantations in southern England so that exceedingly poor lichen floras result.

7. Heathland Management

The drainage of agricultural land and peat mosses has probably had little effect on the lichen vegetation in most parts of Britain as field observations on peat of varying degrees of dryness in Scotland and Shetland suggest that although drainage tends to reduce the luxuriance of *Cladonia* species it does not usually eliminate them. *Icmadophila ericetorum*, however, is characteristic of moist peat and its loss from, for example, Charnwood Forest, Leicestershire (Hawksworth, 1971) may be partly attributable to drainage. The same may be true for *Cladonia strepsilis* and *Pycnothelia papillaria* in some parts of Britain.

The practice of burning grouse moors in the southern Pennines leads to an impoverished lichen flora consisting of only a few often poorly developed species of *Cladonia* and dominated by crustose lichens such as *Lecidea granulosa* and *L. uliginosa* (Hawksworth, 1969) and this effect may also be seen in other areas where moor burning is carried out. From a study of some heathlands in Yorkshire, Coppins and Shimwell (1971) found that the development of *Cladonia* communities including species such as *C. impexa* required at least twenty years; grouse moors are burnt much more frequently than this (usually every 10–15 years).

Relatively dry heathlands are also subject to encroachment by *Betula*, *Molinia caerulea*, *Pinus*, *Pteridium* and *Ulex*, all of which tend to crowd out any terricolous lichen communities that might have been present.

8. Man-made Substrates

The changing building and fencing materials and practices used by man have led to both some reductions and some extensions of range of British lichens. Last century fencing was mainly of untreated wooden posts and rails but these have gradually been replaced by more durable creosoted wood, iron or wire fences. This has had little effect on the overall distributions of most species recorded on decorticate wood but a few which were restricted to this habitat and old wooden barns (e.g. *Lecanora farinaria*, *Lecidella pulveracea*) have now become extinct in Britain (Table I), and others (e.g. *Cyphelium notarisii*, *C. tigillare*) are now much rarer than they were last century. The loss of *Lecanora farinaria* is particularly unfortunate as this species appears to have been endemic to the British Isles.

In contrast, wooden fencing is now being used more widely in Caithness, Orkney, Shetland and Sutherland than it appears to have been in the nineteenth century and this has led to increased frequencies of some species in these areas (e.g. *Alectoria fuscescens*, *Pseudevernia furfuracea*).

The loss of mud-capped walls (see Chapter 3) has probably led to the decline of *Lecania nylanderi* in East Anglia and other inland sites.

In south-east England old roofing thatch was a rich source of interesting *Cladonia* species. While most of these still occur it is probable that the frequency of some of them (e.g. *Cladonia bacillaris*, *C. gonecha*) has declined considerably. *Lobaria scrobiculata* formerly occurred on thatch in Devon and Suffolk.

Perhaps the most dramatic changes in the distribution and frequency of species on man-made substrates are those seen in some saxicolous species. Asbestos cement has a characteristic lichen flora which includes a number of species which were formerly rare or absent in lowland Britain. Crombie (1894) indicates that *Lecanora muralis* occurs on " . . . rocks, boulders, and walls, sometimes on flints, tiled roofs . . . in maritime and upland districts". In maritime and upland areas this species characteristically occurs on bird-perching stones but it has now become exceedingly common on man-made substrates, particularly asbestos cement, slates and stonework throughout central and eastern Britain (Figs 8 and 9) as it is also very tolerant of air pollution. *Candelariella medians* has spread in a comparable way (Hawksworth, 1973c). Similarly, *Lecidea lucida* is now widely distributed on old brickwork in lowland Britain, and *Lecanora dispersa* is ubiquitous on concrete and even occurs throughout central London. Hawksworth (1969) noted that *L. dispersa* could achieve an 85% cover on concrete in Leicester in five years. *Stereocaulon pileatum*, formerly known only from siliceous rocks in upland parts of Britain, has spread onto sandstone and brickwork in lowland Britain (cf. p. 57).

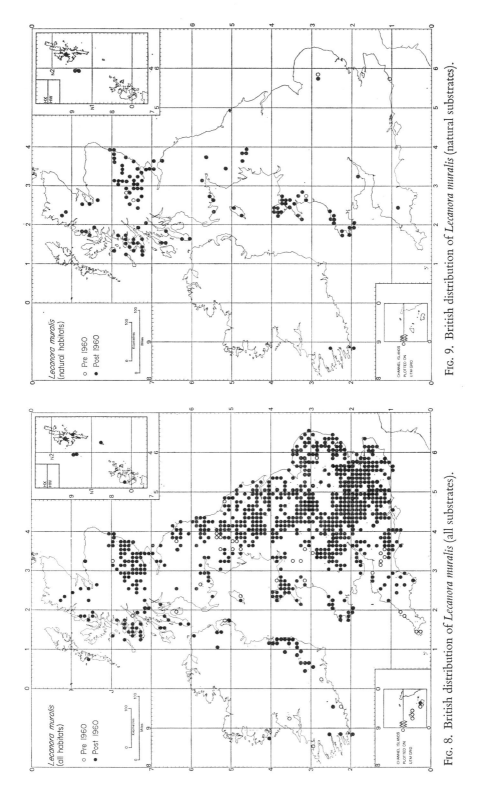

Fig. 8. British distribution of *Lecanora muralis* (all substrates).

Fig. 9. British distribution of *Lecanora muralis* (natural substrates).

The introduction into an area of rock types different from those that naturally occur in it has also undoubtedly led to extensions in the range of some species. This effect is particularly marked in churchyards with tombstones of different rock types. Many of these churchyards have very rich lichen floras (e.g. Trotton church, Sussex, with 85 saxicolous species) which must represent range extensions and increased frequencies of many of the species on them into these areas (e.g. *Caloplaca teicholyta*, Fig. 10). A list of some species occurring in artificial saxicolous habitats in south-east England where natural saxicolous substrates are very rare or absent is given in Table IV.

9. Quarrying and Mining

Quarries, in addition to completely destroying habitats, sometimes produce dust which can have local effects on the lichen flora. Slight occasional deposits of limestone dust tend to have small eutrophicating effects but large amounts forming a whitish layer over bark appear to exert a smothering effect which eliminates almost all corticolous lichens. The effects of limestone dust do, however, require further investigation.

Spoil heaps from lead and fluorspar mines in limestone areas can provide well-drained habitats on which *Cladonia* and *Peltigera* species are able to grow more luxuriantly than in adjacent limestone grassland (cf. Hawksworth, 1969), while those from lead mines in siliceous rock areas often support well developed stands of *Stereocaulon* species (cf. p. 57).

Shingle extraction for making concrete at Denge Beach, Kent, and consequent disturbance of the *Prunus* scrub, may well have contributed at least in part to the loss of, for example, *Lobaria pulmonaria* from this site.

10. Public Pressure

The advent of the motor car has led to many more people visiting parts of Britain than ever before. Terricolous lichens are subject to damage by trampling and Rhodes (1931) noted that *Cladonia uncialis* may be " . . . trampled underfoot to a heap of fragments, which are apparently too small to survive". The effect of trampling may have contributed to the decline in the frequency of *Cladonia arbuscula–C. impexa–C. uncialis–Cornicularia aculeata–C. muricata* communities in Charnwood Forest, Leicestershire, as these are now largely confined to areas from which the public are excluded (Hawksworth, 1971). Trampling on the chalk grassland of Beachy Head, Sussex, may have led to the loss of *Lecanora circinatula* from this site where it was last seen in 1882.

Many very rare British "arctic-alpine" lichens occur on the upper slopes of Cairn Gorm and the Ben Macdhui plateau. As a result of the construction of

TABLE IV. Some species occuring in south-eastern England in artificial habitats in areas where their natural saxicolous substrates are very rare or absent

Acarospora fuscata
A. heppii
Arthonia lapidicola
Arthopyrenia salweyi
Bacidia sabuletorum
B. umbrina
Buellia aethalea
B. verruculosa
Caloplaca aurantia
C. aurantiaca[a]
C. cirrochroa
C. ferruginea[a]
C. heppiana
C. lactea
C. saxicola
C. teicholyta
Candelariella aurella
C. medians
Catillaria chalybeia
C. lenticularis
Collema auriculatum
C. crispum
C. tenax
Diploschistes scruposus
Haematomma ochroleucum s. str.
Lecania erysibe
Lecanora atra[a]
L. calcarea
L. campestris
L. contorta
L. crenulata
L. dispersa[a]
L. gangaleoides
L. intricata var. soralifera
L. muralis
L. polytropa
L. subcircinata
Lecidea crustulata
L. fuscoatra var. grisella
L. lucida
L. percontigua
L. sulphurea
L. tumida

Lecidella scabra
L. stigmatea
Lempholemma chalazanellum
L. chalazanodes
Leproplaca chrysodeta
L. xantholyta
Leptogium plicatile
L. schraderi
Ochrolechia parella[a]
Opegrapha chevallieri
O. mougeotii
Pachyascus byssaceus
Parmelia isidiotyla
P. mougeotii
Physcia caesia
P. dubia
P. nigricans
Placynthium nigrum
Polyblastia albida
Protoblastenia monticola
P. rupestris
Rhizocarpon distinctum
R. geographicum
R. obscuratum
Rinodina subexigua
R. teichophila
Sarcogyne regularis
S. simplex
Solenopsora candicans
Staurothele caesia
Thelidium decipiens
T. incavatum
Toninia aromatica
Trapelia coarctata
Verrucaria glaucina
V. hochstetteri
V. muralis
V. nigrescens
V. viridula

Xanthoria aureola

[a] Also corticolous.

ski-lifts this area has become very popular with tourists and damage by skis and newly forming paths are obvious (cf. Watson, 1967). While no lichen is yet certainly known to have been lost from this area because of this factor several appear to have become rarer over the last two decades and a few species which once occurred there have not been refound at all in recent years.

Rock climbing results in almost all lichens eventually being scoured off the rocks. Popular climbing pitches in areas such as the Langdale Pikes in the English Lake District, where the rocks normally support lush lichen communities, are readily recognizable by the almost complete absence of lichens. Some parts of the abundant communities of *Umbilicaria crustulosa* on Raven Crags, Langdale, are being scraped away by climbers' boots. This is particularly unfortunate as this species, first reported from this site by Martindale (1889), is unknown elsewhere in the British Isles. *Umbilicaria* species appear to be particularly susceptible to damage by rock climbing and this factor may have contributed to the decline of members of this genus in Derbyshire and Leicestershire (Hawksworth, 1969; Sowter, 1950; see also p. 57). Similarly, the loss of *Lobaria* species and *Nephroma laevigatum* from the Sussex sand rocks can also be partly attributed to climbing.

Increased public pressure results in a greater frequency of the incidence of fires. The frequency of fires in the Glen More and Rothiemurchus forests in the Scottish Highlands is of particular concern to lichenologists as a number of species unknown elsewhere in Britain occur in this area.

11. Other Organisms

The rich lichen communities which develop in short turf on highly calcareous soils such as those of calcareous sand dunes (e.g. Braunton Burrows and Penhale Sands; see Hawksworth, 1973c), chalk grassland, and grassland "type A" in the East Anglian Breckland, are dependent on the maintenance of short turf by rabbit and sheep grazing. Some of these communities now appear to be less widespread than they formerly were and also less luxuriant. This may be attributed to the trend away from sheep farming, the depletion of the rabbit populations by myxomatosis in the 1950s, and the increased grazing of cattle (which damage the surface and do not crop turf as short as rabbits and sheep). Some species restricted to this habitat may be threatened with extinction in Britain (e.g. *Buellia epigaea*, *Fulgensia fulgens*, *Verrucaria psammophila*).

Slugs and snails graze lichen thalli (Peake and James, 1967) and psocids have been reported as causing extensive damage to lichen stands (Laundon, 1971). The effects of invertebrates on the lichen flora are very difficult to assess but are probably mainly of a local nature. Grazing must, however, have always operated in natural ecosystems but the possible rôle of this factor in the elimination of species from sites where they are down to critical levels as relict populations must be borne in mind.

Cropping of lichens by sheep on very poor pastures in Shetland, North Wales and parts of Scotland damages populations but does not seem to eliminate them as the bases of thalli are normally left to regenerate.

Changes in the vascular plant flora other than those arising from changes in woodland management appear to have relatively little effect on the lichen vegetation except for the incursion of *Pteridium aquilinum* and *Ulex* species into heathland and *Crataegus* scrub into chalk grasslands. The effects on terricolous lichens are similar to those on bryophytes (see Chapter 3) species being almost entirely eliminated.

In common with vascular plants lichens support a considerable number of fungi. Many of these are saprophytes or lichen parasymbionts (obligately lichenicolous fungi) which do not appear to cause a great deal of damage to the host lichen. A few fungi are, however, able to damage particular lichen species (e.g. *Athelia* cf. *epiphylla*, *Homostegia piggotii*, *Phyllosticta cytospora*) but these effects are often extremely local (e.g. confined to particular trees in a site). As in the case of invertebrates (p. 70) fungi are likely to be more important as agents in relict, rather than regenerating, lichen communities.

Fungi which are pathogenic to trees can exert an indirect effect on lichens by causing the death of their host tree. *Ceratocystis ulmi*, for example, by causing the death of *Ulmus* trees (see Chapter 6), may cause a decline in species restricted to that tree (e.g. *Caloplaca luteoalba*).

12. Over-collecting

The collecting of lichens for use in dyeing and medicine in the period up to the middle of the nineteenth century appears to have had little effect on the overall distribution of many lichens which were collected. *Lobaria pulmonaria*, for example, we know from Wise (1894), was formerly collected and sold as a remedy for tuberculosis ("consumption") in the New Forest, but this species is still present in many sites there today.

In employing lichens in dyeing the amount of lichen material required is usually about twice the weight of the article to be dyed. An extensive series of dyeings consequently uses an enormous quantity of lichen material and for this reason should only be carried out in areas where the species required are extremely abundant and not in danger of local extinction.

The most serious result of over-collecting, as is the case with many other groups of plants and animals, is, however, attributable to specialists themselves.

It was customary for many lichenologists in the mid- and late-nineteenth century to collect large quantities of rare species to distribute amongst their colleagues. As a result of this practice there are now larger amounts of some species in herbaria than remain in the sites from which they were collected (e.g. *Alectoria ochroleuca, Anaptychia leucomelaena*). There are for example, at least ten collections and an exsiccatum of *Anaptychia leucomelaena* from cliffs near Babbacombe, Torquay, a site from which this species appears to have disappeared. A few species may even have been eliminated from the British flora as a result of over-collecting by lichenologists (e.g. *Tornabenia atlantica*).

In addition to affecting rarities the collection of relict specimens (in for example moderately polluted areas) can lead to their local extinction. Lichenologists are now fortunately able to name most British lichens in the field and so the need to collect specimens as vouchers or for confirmation is fortunately minimal. The main threat in this respect now is possibly more from newcomers to lichenology starting to study lichens without proper supervision, who tend to collect anything that looks different for naming later. This approach to collecting in areas containing very rare species or relict populations is likely to be detrimental to the lichen flora and vegetation in them.

DISCUSSION

From the preceding sections one might presume that much of the British lichen flora has been extensively modified over the last century in particular. While this is true for many parts of central and eastern England, however, this is not so in areas which are relatively unaffected by pollution of the air by sulphur dioxide where the lichen flora and vegetation do not, on the whole, appear to have changed appreciably (Hawksworth *et al.*, 1973).

The short-term climatic changes which have occurred in Britain over the last century (Lamb, 1970) seem to have had very little discernable effect on our lichen flora. The majority of species with predominantly northern and predominantly southern distributions in Britain have tended to maintain the same overall geographical limits but have become scarce to absent in parts of the country affected by man or his effluents. It is interesting to note that some species which have become extinct in Britain appeared to be on the edges of their geographical limits here (p. 51) but their loss is more likely to be attributable to man-made factors than to climatic variations. It is conceivable that a few extremely southern species which are now very much rarer than they were, such as *Gyalectina carneolutea* and *Parmelia carporrhizans* (both mapped by Hawksworth, 1972c) and *Teloschistes chrysophthalmus* (Fig. 6), may have been influenced to a small extent by the climatic deterioration of recent decades, but in most instances their decline appears to have started much earlier in the century when the climate was somewhat warmer than it is today.

One point which is worthy of consideration in discussing lichens as opposed to other groups of plants and animals is the long life span of individual plants. Few data are available on the actual growth rates of most species in Britain but a well developed mature specimen of almost any species is likely to be 20–40 or more years old, i.e. it has probably withstood several of the short-term cyclic climatic changes Britain has experienced this century. It should be borne in mind also that some species able to survive in a slightly unfavourable climatic period may not be able to regenerate normally under such conditions (i.e. they would persist as relict populations), but would be able to spread again at least locally when appropriate climatic fluctuations occur.

There can now be little doubt that air pollution by sulphur dioxide has been the main factor causing changes in the lichen flora of many parts of Britain. If emissions of sulphur dioxide are not restrained a continuing depletion in our lichen flora may be expected. The British Isles are now the last stronghold of many species in western Europe and a few species known only in western Europe (e.g. *Graphina ruiziana, Gyalectina carneoluteo, Phaeographis lyellii*) may be under threat of total extinction if emissions in south-west Britain in particular are allowed to rise.

Changes in forest management have also led to the decline of many "old forest" lichen species which appear to have only very limited powers of dispersal. The maintenance of such ancient woodlands as persist in Britain (see Rose, 1974) is consequently essential for their survival as it is for many other groups of plants and animals. As some lichens act as reliable indicators of ancient and native woodland they can provide some indication of sites likely to be important for other groups of organisms now confined to this habitat. Woodland clearance, drainage of areas such as the fens, and the metalling of roads has led to a reduction in the water table in many parts of lowland Britain but the effect this has had on the lichen flora and vegetation is in need of investigation. In many cases it is probable that several different factors may have contributed to the loss of a species from particular sites. *Pannaria nebulosa* (Fig. 11), for example, was formerly widespread on well-drained sandy soils on banks, sea cliffs and mud-capped walls, and its decline may be due to a combination of trampling, the repairing of walls, over collecting and air pollution. Further work on the environmental requirements of many of our rarer species needs to be carried out in order to ascertain those which control their persistence.

When changes in the corticolous species on trunks of mature deciduous trees are considered, these may be interpreted in reference to the communities

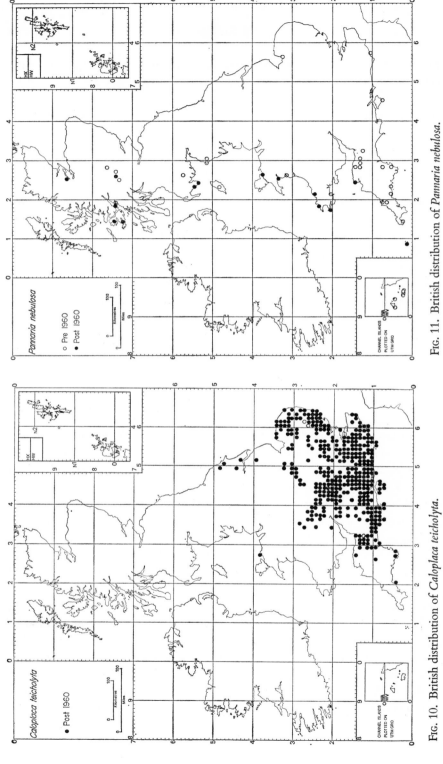

FIG. 11. British distribution of *Pannaria nebulosa*.

FIG. 10. British distribution of *Caloplaca teicholyta*.

Pannaria nebulosa

○ Pre 1960
● Post 1960

Caloplaca teicholyta

● Post 1960

developed (i.e. to phytosociological taxa) as indicated in Fig. 12. Woodland clearance and disturbance, eutrophication and air pollution all contribute to a monotonous, species poor, lichen flora.

A declining lichen flora such as that seen in central and eastern England is not only a matter of concern to lichenologists, as some other organisms live

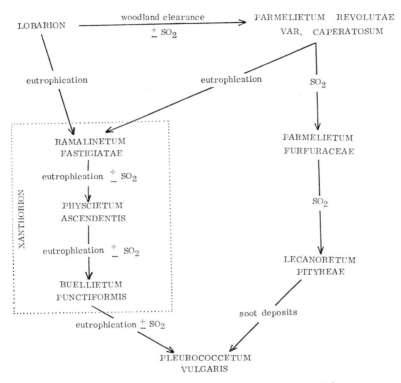

Fig. 12. Relationship between some man-made factors and the lichen communities developed on trunks of mature trees in lowland England.

in, amongst or feed on lichens. These include some Mollusca (Peake and James, 1967), Rotifera order Bdelloida (Pyatt, 1968), Psocoptera (Broadhead, 1958; Laundon, 1971), Acari (Gerson, 1973; Seaward, 1974), etc. The elimination of autotrophs (lichens and algae) by air pollution on *Fraxinus* trunks was found by Gilbert (1971) to lead to a reduction in the diversity of the associated fauna. The extent to which this has had effect on the distribution of invertebrates is uncertain and it is evident that there should be further co-operation between entomologists and lichenologists to ascertain this.

ACKNOWLEDGEMENTS

We are very grateful to Mr P. W. James and Mr J. R. Laundon for their helpful comments on our manuscript; to Dr M. R. D. Seaward for making available data from the British Lichen Society's Distribution Maps Scheme; to Miss D. W. Scott of the Biological Records Centre (The Nature Conservancy) for preparing Fig. 3 and Fig. 8; to Mr J. R. Petch for the Derbyshire chemical analyses reported on p. 60; and to the Athlone Press of the University of London for permission to use Fig. 2 and Fig. 4. One of us (B. J. C.) wishes to thank the Natural Environment Research Council for financial assistance received through a Research Studentship.

REFERENCES

BARKMAN, J. J. (1958). "Phytosociology and Ecology of Cryptogamic Epiphytes." Van Gorcum, Assen.

BARKMAN, J. J. (1969). The influence of air pollution on bryophytes and lichens. *In* "Air Pollution", *Proceedings of the first European Congress on the influence of air pollution on plants and animals, Wageningen 1968*, pp. 197–209. Centre for Agricultural Publishing and Documentation, Waginengen.

BLIGHT, J. T. (1876). "A week at the Land's End." Lake and Lake, Truro.

BROADHEAD, E. (1958). The psocid fauna of larch trees in northern England. An ecological study of mixed species populations exploiting a common resource. *J. anim. Ecol.* **27**, 217–263.

BROWN, D. H. (1972). The effect of Kuwait crude oil and a solvent emulsifier on the metabolism of the marine lichen *Lichina pygmaea. Mar. Biol.* **12**, 309–315.

COPPINS, B. J. and SHIMWELL, D. W. (1971). Cryptogam complement and biomass in dry *Calluna* heath of different ages. *Oikos* **22**, 204–209.

CROMBIE, J. M. (1894). "A Monograph of Lichens found in Britain", **1**, British Museum (Natural History), London.

DILLENIUS, J. J. (1741). "Historia muscorum." Sheldonian Theatre, Oxford.

FERRY, B. W., BADDELEY, M. S. and HAWKSWORTH, D. L. (eds) (1973). "Air Pollution and Lichens." University of London Athlone Press, London.

GERSON, U. (1973). Lichen–Arthropod associations. *Lichenologist* **5**, 434–443.

GILBERT, O. L. (1968). Biological estimation of air pollution. *In* "Plant Pathologist's Pocketbook", pp. 206–207. Commonwealth Mycological Institute, Kew.

GILBERT, O. L. (1970). A biological scale for the estimation of sulphur dioxide pollution. *New Phytol.* **69**, 629–634.

GILBERT, O. L. (1971). Some indirect effects of air pollution on bark-living invertebrates. *J. appl. Ecol.* **8**, 77–84.

GRAY, S. F. (1821). "A Natural Arrangement of British Plants", **1**. Baldwin, Cradock and Joy, London.

GRODZIŃSKA, K. (1971). Acidification of tree bark as a measure of air pollution in southern Poland. *Bull. Acad. Polon. Sci., ser. sci. biol.* II **19**, 189–195.

HAWKSWORTH, D. L. (1969). The lichen flora of Derbyshire. *Lichenologist* **4**, 105–193.

HAWKSWORTH, D. L. (1971). Field meeting at Leicester. *Lichenologist* **5**, 170–174.

HAWKSWORTH, D. L. (1972a). Regional studies on *Alectoria* (Lichenes) II. The British species. *Lichenologist* **5**, 181–261.

HAWKSWORTH, D. L. (1972b). Control of lichens. *Bull. Br. Lichen Soc.* **2** (31), 11.

HAWKSWORTH, D. L. (1972c). The natural history of Slapton Ley Nature Reserve IV. Lichens. *Fld Studies* **3**, 535–578.

HAWKSWORTH, D. L. (1973a). Mapping studies. *In* "Air Pollution and Lichens" (B. W. Ferry, M. S. Baddeley and D. L. Hawksworth, eds), pp. 38–76. University of London Athlone Press, London.

HAWKSWORTH, D. L. (1973b). Some advances in the study of lichens since the time of E. M. Holmes. *Bot. J. Linn. Soc.* **67**, 3–31.

HAWKSWORTH, D. L. (1973c). The lichen flora and vegetation of Berry Head, South Devonshire. *Trans. Proc. Torquay Nat. Hist. Soc.* **16**, 55–66.

HAWKSWORTH, D. L. and ROSE, F. (1970). Qualitative scale for estimating sulphur dioxide air pollution in England and Wales using epiphytic lichens. *Nature, Lond.* **227**, 145–148.

HAWKSWORTH, D. L., ROSE, F. and COPPINS, B. J. (1973). Changes in the lichen flora of England and Wales attributable to pollution of the air by sulphur dioxide. *In* "Air Pollution and Lichens" (B. W. Ferry, M. S. Baddeley and D. L. Hawksworth, eds), pp. 330–367. University of London Athlone Press, London.

HOOKER, W. J. (1844). "The English Flora", **5**(1). Longman, London.

HORWOOD, A. R. (1912). "A Hand-list of the Lichens of Great Britain, Ireland, and the Channel Islands." Dulau, London.

HUDSON, G. (1762). "Flora Anglica." London.

JAMES, P. W. (1965). A new check-list of British lichens. *Lichenologist* **3**, 95–153.

JAMES, P. W. (1966). A new check-list of British lichens: additions and corrections—1. *Lichenologist* **3**, 242–247.

JAMES, P. W. (1973). The effect of air pollutants other than hydrogen fluoride and sulphur dioxide on lichens. *In* "Air Pollution and Lichens" (B. W. Ferry, M. S. Baddeley and D. L. Hawksworth, eds), pp. 143–175. University of London Athlone Press, London.

JOHNSON, T. (1632). "Enumeratio plantarum in Ericeto Hampsteadiano locisque vicinis crescentium." London.

LAMB, H. H. (1970). Our changing climate. *In* "The Flora of a Changing Britain" (F. H. Perring, ed.), pp. 11–24. Classey, Middlesex.

LAUNDON, J. R. (1967). A study of the lichen flora of London. *Lichenologist* **3**, 277–327.

LAUNDON, J. R. (1970). London's lichens. *Lond. Nat.* **49**, 20–69.

LAUNDON, J. R. (1971). Lichen communities destroyed by psocids. *Lichenologist* **5**, 177.

LAUNDON, J. R. (1973a). Changes in the lichen flora of Bookham Commons with increased air pollution and other factors. *Lond. Nat.* **52**, 82–92.

LAUNDON, J. R. (1973b). Urban lichen studies. *In* "Air Pollution and Lichens" (B. W. Ferry, M. S. Baddeley and D. L. Hawksworth, eds), pp. 109–123. University of London Athlone Press, London.

LEIGHTON, W. A. (1871). "The Lichen Flora of Great Britain, Ireland, and the Channel Islands." Shrewsbury.

LIGHTFOOT, J. (1777). "Flora scottica." 2 vols. London.

MARTINDALE, J. A. (1889). "The Westmorland Note Book and Natural History Record." Kendal Natural History Society, Kendal. [Not seen; cited by J. A. Wheldon and A. Wilson, *J. Bot., Lond.* **47**, 448 (1909).]

MUDD, W. (1861). "A Manual of British Lichens." Darlington.

PEAKE, J. F. and JAMES, P. W. (1967). Lichens and Mollusca. *Lichenologist* **3**, 425–428.

PERRING, F. H. (1970). The last seventy years. *In* "The Flora of a Changing Britain" (F. H. Perring, ed.), pp. 128–135. Classey, Middlesex.

PYATT, F. B. (1968). The occurrence of a rotifer on the surfaces of apothecia of *Xanthoria parietina*. *Lichenologist* **4**, 74–75.

RAY, J. (1724). "Synopsis methodicorum stirpium Britannicorum", 3rd edition. London.

RHODES, P. G. M. (1931). The lichen-flora of Hartlebury Common. *Proc. Birmingham nat. Hist. Soc.* **16**, 39–43.

ROSE, F. (1974). The epiphytes of oak. *In* "The Oak Symposium" (F. H. Perring, ed.). David and Charles, Newton Abbot, in press.

ROSE, F. and JAMES, P. W. (1974). Regional studies on the lichen flora of Britain. I: The corticolous and lignicolous lichen flora of the New Forest. *Lichenologist* **6**, in press.

SEAWARD, M. R. D. (1973). Distribution maps of lichens in Britain. *Lichenologist* **5**, 464–466.

SEAWARD, M. R. D. (1974). A note on Oribatid mites and lichens. *Lichenologist* **6**, in press.

SMITH, A. L. (1922). History of lichens in the British Isles. *Trans. S.-E. Un. scient. Socs.* **1922**, 19–35.

SMITH, A. L. (1918, 1926). "A Monograph of the British Lichens", 2nd edition, 2 vols. British Museum (Natural History), London.

SOWTER, F. A. (1950). "The cryptogamic flora of Leicestershire and Rutland—Lichenes." Leicester Literary and Philosophical Society, Leicester.

SOWTER, F. A. and HAWKSWORTH, D. L. (1970). Leicestershire and Rutland cryptogamic notes, I. *Trans. Leicester lit. phil. Soc.* **34**, 89–100.

TROAKE, R. P. and WALLING, D. E. (1973). The natural history of Slapton Ley Nature Reserve VII. The hydrology of the Slapton Wood Stream. *Fld Studies* **3**, 719–740.

TURNER, D. and DILLWYN, L. W. (1805). "The Botanist's Guide through England and Wales," 2 vols. Phillips and Fardon, London.

WATSON, A. (1967). Public pressures on soils, plants and animals near ski lifts in the Cairngorms. *In* "The Biotic Effects of Public Pressures on the Environment (E. Duffey, ed.), pp. 38–45. Monks Wood Symposium No. 3, The Nature Conservancy, Huntingdon.

WATSON, W. (1953). "Census Catalogue of British Lichens." British Mycological Society, London.

WISE, J. R. (1894). "The New Forest: its history and its scenery", 5th edition. London.

WITHERING, W. (1801). "A Systematic Arrangement of British Plants", **4**. Cadell, London.

5 | Changes in the British Macromycete Flora

D. A. REID

Royal Botanic Gardens, Kew, Surrey

Abstract: Changes in the mycological flora involving the larger fungi can be effected in one of two ways. Thus changes may involve the basic number of species comprising the total flora resulting from addition of new species by accidental introduction, natural extensions of range, etc. or, alternatively, by the loss of species due to extinction. Changes may also involve the relative abundance of different species within the community due to alteration in habitat.

Fluctuations involving the basic number of species are difficult to assess for the following reasons: (1) There has been much taxonomic work in Britain during the last 50 years resulting in the description of numerous additional species and recognition of species previously unknown from this country. Most of these were no doubt native species hitherto overlooked, and not the result of changes in the composition of the flora itself. (2) Fungi are very sporadic in occurrence and in their fructification and it is difficult to be sure whether a rare species has become extinct or whether its apparent absence is merely due to lack of the requisite climatic factors necessary to induce fruiting. (3) It is equally difficult to know whether the large number of species reported annually as new British records really represent permanent additions to the flora, or as is much more likely, merely casual occurrences of little significance in relation to the flora as a whole.

Alterations in the relative frequency of individual species may be influenced by changes in agricultural practice, silviculture or forest management.

An evaluation of changes in the British macromycete flora is complicated by a number of factors specific to mycology. The fruitbodies may be small and easily overlooked, or there may be a very short fruiting season with individual fructifications lasting only a few days or even hours. Further, species can seldom be relied upon to reappear in the same locality in two successive seasons, except for those which are lignicolous. There is also the phenomenon of "seasonal flushes", when an uncommon species suddenly occurs in great abundance

Systematics Association Special Volume No. 6, "The Changing Flora and Fauna of Britain", edited by D. L. Hawksworth, 1974, pp. 79–85. Academic Press, London and New York.

either locally or nationally and then disappears in following years, or reverts to its former status.

Hence there is a considerable element of chance whether a rare fungus will be detected when it fruits and whether the fact will be reported. Professional mycologists are few in number and amateurs are seldom in a position to identify rarities since access to a microscope and specialized literature is essential. This, together with the unpredictability of fruiting of rare species from season to season, means that it is merest chance that the right person is at the right place at the right time of year and sometimes even at the right time of day. For these reasons it is perfectly possible for a species to be considered extinct in Britain when it is in fact widespread but rare.

The "fungus" one observes represents the fruiting stage of an extensive unseen vegetative mycelium. The occurrence of "flushes" of a given species in a certain year over large areas of woodland indicates that the mycelium must be widespread in the soil and that special climatic factors are necessary to induce fructification. It must also be borne in mind that some macromycetes have an imperfect conidial state represented by a very inconspicuous mould stage. Such fungi are able to spread by means of conidia in the absence of the more familiar fructification and so may remain undetected for years.

With these reservations change in the macromycete flora may be evaluated from two viewpoints. There are gains and losses of species and there are changes in the relative frequency of species under the influence of changing environmental conditions, usually as a result of human interference.

During the last twenty years there has been a surge of mycological activity, culminating in the publication in 1960 of the "New Check List of British Agarics and Boleti" (Dennis *et al.*, 1960), in which well over a hundred new species were described, and since then many more have been published, bringing the total near to 200. This illustrates a very important point; whereas in many groups of both animals and plants the exact composition in terms of numbers of species is known with some precision, in the fungi we still have little idea of the total composition of the flora, and new species are being described every year. Add to this the very large annual number of records of species not hitherto known from Britain as opposed to species new to science and one gets some idea of the complexity of the problem in evaluating the status of new additions to and changes in the agaric flora alone—especially when there are currently thought to be upward of 1800 species involved.

Of the 200 or so new species recently described from Britain, most have formed part of the flora for a very long time but have not previously been recognized as distinct taxa, although some are quite conspicuous and easily recognizable. The same applies to many species which have only recently been recorded from Britain. These too have often been overlooked in the past owing to their fruiting at odd times of the year or in specialized habitats. Here may be cited as an example the single record of the vernal Discomycete *Discina leucoxantha* from Perth in 1924.

Many other new records, possibly even the majority, represent chance occurrences of species which are not permanent members of the flora and which are extremely unlikely to become established here. Other records represent species which could become permanent additions. However, it is difficult to know except in retrospect to which category a newly reported species belongs, with the exception of plant pathogens. In the rust fungi there is good evidence of several recent introductions which have become permanent members of the mycological flora. *Puccinia lagenophorae*, parasitic on *Senecio* spp., was first reported in 1961 and is now widespread in Britain (see Chapter 6). Similarly, *P. pelargonii-zonalis*, a pathogen of garden geraniums (*Pelargonium* spp.), was introduced in 1965 and is now frequent in southern England.

There are numerous examples of transient species which are unlikely to become established here and some of these are discussed below. *Omphalotus olearius* is a bright coppery-orange, funnel-shaped agaric with decurrent gills. The record from Slinfold, Sussex in 1967 can only be regarded as a casual occurrence since this fungus has a predominantly tropical or subtropical distribution, although it does extend into central and southern Europe and further north along the Atlantic coast of France.

Pycnoporus cinnabarinus, a brilliant blood-red bracket fungus, has a similar range, except that it extends into Scandinavia and becomes rare in western Europe. It has been reported from Epping Forest and also from Murthly, Perthshire, in 1913, but has never become established.

Lyophyllum favrei, a fairly large agaric with a purple cap and yellow gills, known only from Switzerland and first described in 1949, was reported from Druid's Grove, Box Hill, Surrey, in 1955 and 1956. Assuming this very rare species to have a central European distribution it is unlikely to become established in southern England.

A further example is the occurrence of *Tremiscus helvelloides* in Herefordshire in 1972. This fungus is one of the Tremellales or jelly fungi, and has a soft, gelatinous texture, and forms pinkish, spathulate or flared-tubular fruitbodies up to 8 cm high. The only previous localized record was from Sandsend, Yorkshire, 1914. The appearance of this species in these two localities is particularly surprising for in Europe it is typical of montane coniferous forest, although there is a record from the Baltic coast of Germany and it also occurs in Latvia and Estonia. With this type of distribution one cannot but regard the British records as casual occurrences.*

It is perhaps pertinent here to draw attention to the fact that fungi of transient occurrence are often found in atypical situations as exemplified by the records of *Tremiscus*. Again, *Chaetoporus pearsonii* is not uncommon in coniferous forests of central and eastern Europe on fallen wood. Yet the British records comprise a collection from Somerset and one from Sussex. The latter specimen was found on a fallen rotten pine trunk in a hedgerow where the tree had formerly stood as an isolated specimen in the middle of arable farmland.

Ramsbottom (1935), discussing changes in the mycological flora from 1885 to 1935, mentioned several very conspicuous Gasteromycetes. After an interval of nearly 40 years we are in a better position to evaluate their status in the flora. These fungi are so remarkable in shape that anyone finding them would be almost certain to report the fact. Most of these were evidently transient introductions but there have been other more recently introduced species which have become established and spread locally.

Lysurus australiensis was known to Ramsbottom from four British collections, resulting from three separate introductions. Evidently the fungus did not become established and an additional gathering made in a field of barley stubble at West Hagley, Worcestershire, in 1970, was probably an instance of yet another introduction.

Anthurus archeri has a flower-like receptacle which opens in the form of a variable number of bright red arms bearing the spore mass on their inner surface. The bright colour and unpleasant smell attract insects which disperse the spores. This fungus was first reported from a garden in Penzance, 1945; the next record was from an orchard in Duddleswell, near Fairwarp, Sussex, in 1953, where it persisted until 1958, and may still be present. However, the interesting thing is that it appears to have become established and has spread to nearby areas. There was a collection in 1965 from the same general region as the first gathering and yet another in 1967 from a site two miles distant. Thus there is good reason to believe that *A. archeri* may have become a permanent addition to the flora.

Ileodictyon cibarium presents an analogous situation. The receptacle has the bizarre appearance of a white hollow meshwork sphere about 8 cm in diameter. This fungus is found in Australasia and South America; the first report from outside this region was of a fairly long-established locality at Hampton,

---

* See "Notes added in proof" (1) on p. 85.

Middlesex, in 1963. Here again, local spread has occurred, for there was a further report from Thames Ditton, Surrey, in 1971.

Guepiniopsis alpina belongs in a totally different group of fungi, the Dacrymycetales. It forms small, gelatinous, yellow or orange, stalked cupulate fruitbodies up to 4 mm in diameter, on coniferous wood, and was hitherto known only from North and South America and Japan. The recent discovery of herbarium material collected in 1971 from the Isle of Mull raises the interesting problem as to whether this represents a recent introduction new to Britain and Europe or whether it is a species of long standing which has hitherto been overlooked on account of its small size. Certainly it could well become established in Scotland.

So far mention has been made only of species additional to the flora. Transient species represented by single gatherings or by collections separated by very long intervals of time have already been briefly discussed. Such specimens can hardly be said to become extinct. The only genuine example of an established species becoming extinct is *Myriostoma coliforme*, which used to occur locally around King's Lynn, Norfolk, until 1880, but has not been seen for almost 100 years. It is an earth star with many pores instead of a single apical mouth.

There are numerous examples of species which have not been reported for many years, and judged from the literature they appear to have become extinct. However, the early records of some of these species are suspect and they may never have formed part of the flora. In other instances the species may be vernal and overlooked on this account, others may fruit sporadically and can sometimes exist undetected in the vegetative state for years. Hence it is unwise to conclude that extinction has occurred without very careful consideration.

Laxitextum (*Stereum*) *bicolor* was said by Carleton Rea to be rare in 1922, and was not represented by herbarium material subsequent to 1937. There are collections at Kew from near Birmingham, 1919; Haye Park, Herefordshire, 1937; and from St Leonard's Forest, Sussex, 1931. From this evidence the fungus might justifiably have been considered to have become extinct, yet it was rediscovered in several localities in the Birmingham area in 1969. In the light of these new finds it seems probable that *L. bicolor* had been overlooked in these localities during the intervening years.

Many similar examples could be cited. *Battarea phalloides* is a stalked puffball up to 30 cm high with a small head 2–3 cm in diameter. It was reported sporadically from southern England until 1931, but was not seen again until 1944 when a specimen was found at Virginia Water, Surrey. Since then there have been increasing numbers of records: Box Hill, Surrey, 1953, 1954, 1956, 1962; Leigh Woods, Bristol, 1968; West Wickham, Kent, 1956; and near

D

Strode, Kent, 1969. In most instances only single specimens were found but in the West Wickham collection there were at least seven fruitbodies and in the Strode collection there were nine. Judged from lack of records between 1931 and 1944 one might have assumed that *B. phalloides* had become extinct in Britain. This sporadic fruiting of such an unusual fungus again highlights how unwise it is to assume extinction merely from lack of records.

Boletopsis leucomelas is a centrally stipitate terrestrial polypore, widespread in the conifer forests of Europe, and formerly reported from Morayshire and from the Rothiemurchus Forest, Inverness-shire, in 1876, 1901, 1906 and 1938, since when no further records were received until it was found in Rothiemurchus in 1963; again a case of a relatively large fungus being overlooked for 25 years!

Amongst the Discomycetes there are even more striking instances. For example, the brown, pezizoid *Discina perlata* was known only from a single collection at Dyce, Aberdeenshire, in 1888, until it was rediscovered in 1972 in the Black Woods of Rannoch, Perthshire.* However, the fungus was almost certainly present in Scotland during the intervening years and overlooked on account of its undistinguished appearance and vernal fruiting.

There were two collections of *Microstoma mirabilis* from near Ballater, Aberdeenshire, in April 1890, but the species has not been seen since. Again the fungus is probably still to be found in the area but mycologists do not collect in that part of the country in spring.

There is a range of species of varying degrees of rarity which are restricted to the native pine forests of the Scottish Highlands. These fungi are all to some extent endangered by the steadily declining area of the habitat. They are in a sense relict species in Britain and form outliers to their main centres of distribution in Europe, and are probably more sensitive to changes in the environment in these marginal localities.

From fungi at the edge of their range in Scotland there are a series of species which are at the limit of their range in southern England. It is these species which might be expected to react to long term, but slight changes in climate by fluctuation in range. Nevertheless 50 years is too short a period in which to detect such changes.

Clathrus ruber, a hollow, red meshwork sphere, up to 20 cm in diameter, which is a permanent component of the flora of the Scilly Isles, also occurs annually scattered through southern England, but must be considered at its northern limit in this country.

---

* See "Notes added in proof" (2) on p. 85.

From the foregoing discussion it can be seen that there are very considerable difficulties involved in analysing the status of various species in the flora and in deciding whether certain species have really become extinct. Fluctuations in relative frequency of individual species are even more difficult to tabulate. Undoubtedly changes in forest management and in planting programmes have had a profound effect on the mycological flora. In the past 50 years there has been much more definite management of our woodlands in that diseased trees are felled and fallen timber cleared promptly. This has resulted in the reduction in numbers of many wood-rotting,. saprophytic fungi belonging to the "Polyporaceae" and "Thelephoraceae".

Changes in forest practice and the planting of conifers on an ever larger scale has led to an increase of the well known root pathogen *Heterobasidion annosum*. However, the trees are felled at an early age and this seems to be the explanation as to why such fungi as *Sarcodon imbricatum*, at one time widespread in the older established pine plantations of the south, are now virtually confined to the pine forests of the Highlands of Scotland.

The decline in the importance of the horse in agriculture has no doubt contributed to the scarcity of the dung-inhabiting Pyrenomycete *Poronia punctata*, which is now found only in the New Forest as a great rarity.

Changes in farming relating to pasture management also seem to have had an effect, resulting in fewer species of *Hygrophorus* and other grassland fungi such as the Blewit—*Lepista saeva.*

The uses of fungicides and the chemical treatment of seeds and tubers have also been responsible for the reduction in frequency of many plant pathogens, but these are outside the scope of this paper and are discussed by Booth in Chapter 6.

REFERENCES

DENNIS, R. W. G., ORTON, P. D. and HORA, F. B. (1960). New check list of British Agarics and Boleti. *Trans. Br. mycol. Soc.* **43** (Suppl.), 1–225.
RAMSBOTTOM, J. (1935). Changes in the British fauna and flora during the past fifty years. (6) Fungi. *Proc. Linn. Soc. Lond.* **148**, 49–52.

Notes added in proof

1. Since this paper was written further collections have been made from the original locality in 1973 and additionally from: Binton Hill, Warwickshire; St Gwynnos Forest, Glamorgan; from a number of quarries east of Pontsticill, mostly in Brecon (but also over the border in Glamorgan). It would therefore seem that this species may after all be a recent introduction which is spreading rapidly in the west and which may become a permanent addition to our flora.

2. In 1973 it was again found at Rannoch and in this same year there was also a collection from the Radnor Forest in Wales.

6 | The Changing Flora of Microfungi with Emphasis on the Plant Pathogenic Species

C. BOOTH

Commonwealth Mycological Institute, Kew, Surrey

Abstract: Many of the very large number of microfungi have the capacity to cause profound changes in our flora. Nevertheless, collections and authenticated records of microfungi in general are totally inadequate to support any concept of change. The only exceptions are where economic aspects have intruded, as is the case of those microfungi that cause plant diseases. Unfortunately, in these species man's activities have often had a much greater influence than natural causes.

Where collections have been made systematically in a natural environment some intriguing problems both in relationship to distribution and cycles of occurrence suggest themselves but the explanation or solutions to these problems lies in the future and initially in greater support for taxonomic investigations.

PROBLEMS OF IDENTIFICATION AND RECORDING

This paper does not cover microfungi in general because our knowledge of the occurrence and ability to identify microfungi is totally inadequate to enable one to talk about real changes in their distribution or frequency of occurrence. Pirozynski (1968) stated that any attempt to write an essay on the distribution of microfungi was to attempt an impossible task. How much greater, therefore, are the difficulties in talking about changes in the flora. The taxonomy of most micro-organisms and certainly of microfungi is in no sense complete, even for British species, and yet because taxonomy in general is no longer thought worthy of university support, little taxonomic research is carried out. The major reason for this is that mycology in this country, and in many others, is regarded as a minor section of botany and of the botany department. After

Systematics Association Special Volume No. 6, "The Changing Flora and Fauna of Britain", edited by D. L. Hawksworth, 1974, pp. 87–95. Academic Press, London and New York.

approximately 300 years of active collecting and taxonomic investigation of flowering plants by many hundreds of botanists, the universities decided, consciously or subconsciously, that the subject was exhausted. This, in part justifiable, attitude, although many botanists will dispute it, spread throughout the department and influenced greatly the current approach to all taxonomic research, even that related to microfungi where our knowledge is still totally inadequate.

The reasons for this imbalance between higher plants and fungi are not hard to find. In the first place, morphology of flowering plants is largely macroscopic and developmental or anatomical studies are extremely simple compared to similar studies with microfungi. Thus we are in a position today where considerable physiological and biochemical work is being carried out on microfungi and yet no continuation of this work can be pursued by future workers because the names used for the organisms are so often meaningless or impossible to authenticate. The only exception is where workers have obtained their material from, or deposited it with, a national collection.

Grove (1935) speaking of the Coelomycetes, a section of the Deuteromycotina (Fungi Imperfecti), said that at any time of the year an ardent collector could in any one locality collect a number of forms hitherto unrecorded in this country. The situation has not changed. Last autumn R. W. G. Dennis, one of our most avid and ardent collectors of microfungi, seized a handful of *Juniperus* leaves in the last moment before returning to his hotel after a day's collecting. On examination he found he had three new British records and a new species. These from a host genus with only one British species. B. C. Sutton, looking at a common but somewhat unusual substrate, the capsules of sweet chestnut (*Castanea sativa*), found four new species in his first collection of the winter 1972–73. It should be added that those two collectors are experienced mycologists who know their genera and do not apply names lightly.

THE DISTRIBUTION OF MICROFUNGI

With regard to the distribution of microfungi, they can be grouped geographically and there are of course temperate and tropical species, but care should be exercised in making such distinctions. For example, in the microfungi which have adapted to a specialized habitat such as soil fungi or fresh water or marine species, the same species with few exceptions can be collected in both tropical and temperate regions. In spite of this generalization, not all microfungi are generally distributed. In Britain there are some very interesting observations on the distribution of microfungi which are difficult to explain.

Hypoxylon bulliardi on beech is extremely common on the north slopes of Ranmore Common, Surrey, and has been for 100 years, and yet further along the North Downs both east and west, where beech is still the climax vegetation on the chalk, this species cannot be found.

Nectria leptosphaeria, a parasite on the common nettle, occurs, or has been collected, in only four localities in Britain and yet other fungi of nettle such as *Leptosphaeria acuta*, are apparently present wherever its host occurs. Other species such as *Nectria ralfsii*, a Mediterranean species, may have its distribution controlled by temperature as it occurs only in the south-western part of Britain and in the Channel Islands. Some plant pathogens which are also limited by temperature will be discussed later.

It is not only a question of adequate collecting that will determine the distribution of microfungi. Many are constantly associated with their hosts but are only apparent when they form their fructification either on the death of their host or as a result of some other factor.

(1) In our woods one can frequently search in vain for evidence of *Ceratocystis* species, but if one follows a woodman, and if damp weather persists, after a few weeks virtually every chip of wood left after either felling or trimming will be covered by perithecia of the appropriate *Ceratocystis* species.

(2) Similarly, after the flash fires that occur from time to time over the commons in the south of England the young birch that are killed will be covered after a few weeks with perithecia of *Valsa ceratophora*. There is no evidence here of infection spreading across a burnt plot but more likely a spontaneous production of perithecia by a systemic fungus in an adversely affected host.

(3) *Rhizina undulata* apparently requires fire to stimulate spore germination and will spread rapidly from the site of a woodman's fire in coniferous plantations to kill the surrounding trees. There are several other Discomycetes which are only found following fire but they are not usually pathogenic to the surrounding trees.

(4) *Cladosporium resinae* in the modern world is known for its blockage of aircraft fuel lines, as it grows in kerosene. To my knowledge it has never been isolated in nature from soil by any of the orthodox isolation methods and yet when Parbery (1967, 1969a, b) treated soil with cresote or buried creosote-treated matchsticks, he frequently isolated the fungus from soil in Australia and also in several European countries that he visited including Britain. This suggests the fungus has a world-wide distribution and yet previously it was only known in association with coal-tar products, creosoted poles or pine resins. How many other selective species remain to be found?

Microfungi, and for that matter many macrofungi also, can have a profound effect on the occurrence and distribution of higher plants and when this occurs the pattern of the distribution of the fungus is apparent for all to see. However, it is only when these changes have economic importance as in the case of parasites of crop plants or plants of economic importance that we have adequate documentation.

At the present time, we may well be in the middle of one of the greatest changes that has occurred in our flora since the forests were felled in the Middle Ages. This is due to the presence of Dutch elm disease caused by *Ceratocystis ulmi* which may well destroy all our native elms just as the American chestnut (*Castanea dentata*) was destroyed by another microfungus *Endothia parasitica* between 1910 and 1935. The destruction of a host tree will also have an effect on the epiphytes of other groups which are largely restricted to that host (cf. Chapter 4).

CHANGES IN DISEASE PATTERNS

There are 3 major ways in which a change in the disease pattern may arise: (1) importation from overseas; (2) genetical changes either in the fungus or host; and (3) adverse conditions or climatic changes.

1. Importation from Overseas

Our quarantine regulations are designed to prevent the importation of diseased plant material into this country but due to the importation of vast quantities of fruit, vegetables and cereal grains from almost every country in the world there is no evidence that quarantine regulations have prevented any disease that could survive in this country arriving here.

Lettuce rust (*Puccinia opizii*) is an example of what apparently happens to many of the parasites of plants which arrive in this country. This rust first appeared in British records when it was intercepted in a consignment of lettuce from Holland in 1934. Ellis (1951) made the next important record when he found it to be common on lettuce in Norfolk and in the following year he found it in Suffolk. Following these well authenticated records it now appears to have died out.

Many introductions have of course survived and the following are brief records of three examples.

(a) *Cumminsiella mirabilissima* (Peck) Nannf., the rust of *Mahonia aquifolium*, is a species endemic in the western U.S.A. and Canada. It was first recorded in Britain near Edinburgh in 1922 and in 1923 it was found to be widespread in the Tees valley. This was probably the source of its introduction into Britain and Europe. It is known that various plants were imported into this area from Oregon in 1915, and although a *Mahonia* species was not listed amongst them, the rust spores could have been introduced with them on leaf fragments. The subsequent spread of the rust throughout Britain was quite rapid and in 1925 it was found in Denmark and in 1926 in Holland.

(b) The introduction from America of one of our most serious diseases of gooseberries is a similar story: *Sphaerotheca mors-uvae*, the American gooseberry mildew, was first found in Northern Ireland in 1900 and although concern was expressed about the danger of its spreading they were unable to prevent its arrival in England where it was reported to be present in both Worcestershire and Kent in 1906.

(c) Dutch elm disease (*Ceratocystis ulmi*) is believed to have spread from eastern to western Europe early this century. In eastern Europe resistant but not immune varieties of elms occur. As a serious pathogen it was first recorded in France in 1918 and in Britain at Totteridge, Hertfordshire in July 1927. With this interest was aroused and by a survey in 1928 shown to be widespread in southern England. According to Peace (1962) the typical ring coloration in dead elms suggested it had been present in Britain since 1920. In the following ten years or so the disease was not so much typified by its spread (in fact it was considered to be generally distributed in the southern half of England below a line from Chester to Hull) but by sudden "flare-ups" in scattered localities. The disease then subsided. In fact, at a symposium on the changes in the British Fauna and Flora held at the Linnean Society, Ramsbottom (1935) stated that Dutch elm disease had not caused the havoc that was first feared. This is no longer the position and certainly the present flare-up is the most severe. In a recent survey, Gibbs and Howell (1971) estimated that of the elms in all the counties in southern England excluding Cornwall, the most serious infection was in Herefordshire and Gloucestershire with 42% and 38%, respectively, of elms infected and the least in Leicestershire and Berkshire with 1·6% and 2·3% respectively. But we are no longer talking about the spread of a species as by 1930 *Ceratocystis ulmi* was already considered to be generally distributed in southern England. What we are talking about and what is really important is the distribution of a more virulent pathogenic strain of *C. ulmi*, an infraspecific taxon that has arisen by genetical change either in this country or by introduction of a particular genetical strain from abroad.

2. Genetical Changes in the Fungus or Host

Gibbs and Brasier (1973) state that isolates of *C. ulmi* from Britain fall into two groups on the basis of their cultural characteristics. Pathogenicity experiments show one group to be aggressive and the other to be non-aggressive. They also state that North American isolates, where the disease is very serious, resemble the aggressive isolates whereas isolates from northern Europe are non-aggressive. Gibbs and Brasier are extremely lucky if their pathogenic strains are linked with a simple diagnostic cultural character.

Fusarium oxysporum is generally distributed in British soils, apart from those which are extremely acid. This species has a large number of pathogenic strains which in this country can destroy many ornamentals, and also leguminous and solanaceous crops. The fungus also occurs as five or six cultural forms which are sufficiently distinct to have been referred to as separate species in the past and yet there is no recognizable connection between the genes for pathogenicity and those controlling growth, pigmentation or morphology. Thus, it is a sad fact to a taxonomist and others, that all we can do is to name an isolate *F. oxysporum* and yet the name tells the grower or farmer nothing about the physiological capabilities of his isolate or about its possible adverse effect on his crop.

The rusts (Uredinales) and powdery mildews (Erysiphaceae) are the causal organisms of many important plant diseases. Both are for all practical purposes obligate parasites and attack only the living tissues of the host. Obviously, therefore, their distribution is determined by that of their host, but the story is not quite so simple. *Puccinia graminis* var. *tritici* Erikss. & Henn. is able to attack a number of varieties of cereals if these varieties are not resistant to it. One of the major aims of any cereals breeding programme has to be the incorporation of genes resistant to rust or mildew attack. These new varieties will retain this resistance until some mutation in the fungus gives it the ability to break the host resistance and this gives us another physiological race. At the present time over 200 physiological races of *Puccinia graminis* var. *tritici* have been distinguished by their uredinial reaction on the different varieties of wheat (Stakman *et al.*, 1962). In these cases it is not the distribution or change in distribution of *Puccinia graminis* var. *tritici* that is important but the occurrence and distribution of the physiological race.

What does one mean therefore by changes in the distribution of microfungi? Even when a species is present, and its presence or absence is not always immediately apparent, what one really needs to know is its biological potential, or in simple terms what strain one has. This after all is the significant characteristic in its capacity to affect or even destroy its host.

Adaptation from wild to cultivated hosts. Flowering plants have, of course, evolved in competition with parasitic micro-organisms and a balance has been achieved that is seldom seriously upset unless some other factor intervenes. Factors which may upset this balance include climatic changes, the introduction of a new host into an area where it encounters a new range of pathogens, or the introduction of a microfungus into an area where certain naturally occurring hosts carry no resistance to its attack; this appears to be the case with Dutch elm disease.

The most frequent breakdown in this natural balance occurs with breeding of cultivated varieties, usually for increasing yield, which so often cuts out naturally occurring genes for resistance. The epidemic of southern corn leaf blight in the U.S.A. in 1970 was an example of this when the introduction of male sterile strains in the breeding of hybrid varieties cut out the genes for resistance to *Drechslera* (*Helminthosporium*) *maydis*. In Britain we have a number of examples of cultivated varieties which are much more seriously attacked by pathogenic microfungi than are their naturally ocurring relatives. *Gnomonia fructicola* on strawberry is seldom found on the wild species and even when it occurs it appears to cause little damage. *Peronospora farinosa*, the downy mildew of sugar beet, can be a serious disease. It also occurs on the wild *Beta vulgaris* ssp. *maritima* but with a much less serious effect.

3. Adverse Conditions or Climatic Changes affecting Distribution

Adverse conditions may debilitate the host reducing its natural resistance and allowing entry of pathogenic fungi. On the other hand, climatic conditions (in this country usually temperature rather than humidity), are the limiting factors to the distribution of some pathogenic fungi. There are, of course, other considerations such as acid or alkaline soils or salt in salt marshes and on the shore line but the occurrence of any specific microfungi in these habitats is largely governed by the host (Apinis and Chesters, 1964).

With regard to disease spread under weather conditions adverse to the host we have the classical example of the potato blight (*Phytophthora infestans*) of 1845 which resulted in the Irish potato famine. At this time, several weeks of fine warm weather in July were followed by several weeks of cold rainy weather in August and virtually every potato patch in Ireland and England became infected, with the result that the crop was largely destroyed (Large, 1940). What this illustrates to us is the danger of talking about change in the flora of microfungi. In some years one cannot find *Phytophthora infestans* and in other years it is apparent to all.

Cold spells in spring may damage trees buds and shoots and allow the entry of *Botrytis cinerea* which is consequently extremely common and pathogenic in the following summer, causing dieback of the twigs. *Pycnostysanus azaleae* is another example of a species abundant after a cold spring, although not necessarily cold enough for frost damage.

Our summer temperatures may limit some fungi to the warmer southern or south-western part of the country. *Leveillula taurica* is very common in southern Europe and throughout the tropics on a wide variety of hosts, possibly as a series of host-specific strains. In Britain it has only been found on *Helianthus* chiefly in the Exeter area but also in small pockets between Reading and Luton. One also watches with interest the spread of *Fusarium moniliforme* (*Gibberella fujikuroi*) in this country with the increase in maize growing; this species has previously been isolated only from ornamental plants in hot-houses but not in the field.

Fusarium nivale (*Micronectriella nivalis*) is an example of a low temperature pathogen. The effect of this species is very serious on cereals in Scandinavia, especially in Finland. Its effect is still economically important in Scotland but it becomes progressively less so as one moves towards southern England.

Air pollution. Air pollution also has some interesting, although compared to lichens (see Chapter 4) very minor, effects. The Clean Air Acts and the consequent reduction in smoke levels have resulted in a great upsurge of the leaf parasite disease complex in the London suburbs. Rusts are now much more common on ornamentals than they were previously. Black spot of roses (*Diplocarpon rosae*) is now very common whereas 15 years ago one had to travel into Surrey to find it. Not only ornamentals but also weeds are affected; both *Albugo tragopogonis* and *Puccinia lagenophorae* are now extremely common on the Oxford Ragwort (*Senecio squalidus*; see Chapter 5) and *Melampsora euphorbiae* on the Petty Spurge (*Euphorbia peplus*).

Some fungi, such as *Rhytisma acerinum* (tar spot of sycamore), appear to thrive under a certain amount of pollution. This species is of much rarer occurrence in the southwest of England than in East Anglia where the atmosphere is more polluted but it is also virtually absent from towns with high pollution.

Kock (1955) found that oak mildew (*Microsphaera alphitoides*) was also absent from areas of high sulphur dioxide pollution.

The effects of air pollutants on fungi are reviewed in more detail by Nash (1973).

CONCLUSIONS

Thus changes in microfungi are characterized by sudden outbursts and often rapid spread of the pathogen, or even a saprophyte if a host has been severely affected by adverse climatic or other conditions. One can only say with severe reservations that a microfungus is absent. It is much safer to say it has not been recorded or that it has no economic significance.

REFERENCES

APINIS, A. E. and CHESTERS, C. G. C. (1964). Ascomycetes of some salt marshes and sand dunes. *Trans. Br. mycol. Soc.* **47**, 419–435.

ELLIS, E. A. (1951). Flora and fauna of Norfolk. *Trans. Norfolk Norw. Nat. Soc.* **17**, 137–138.

GIBBS, J. N. and BRASIER, G. M. (1973). Correlation between cultural characters and pathogenicity in *Ceratorysis ulmi* from Britain, Europe and American. *Nature, Lond.* **241**, 381–383.

GIBBS, J. N. and HOWELL, R. S. (1971). *Forestry Commission Forest Record* **82**, 1–34.

GROVE, W. B. (1935). "British Stem- and Leaf-Fungi", **1**. Cambridge University Press.

KOCK, G. (1955). Eichenmehltau und Rauch gasschaden. *Z. PflKrankh. PflPath. PflSchutz.* **45**, 44–45.

LARGE, E. C. (1940). "The Advance of the Fungi." Jonathan Cape, London.

MOORE, W. C. (1959). "British Parasitic Fungi." Cambridge University Press.

NASH, T. H. (1973). The effect of air pollution on other plants, particularly vascular plants. *In* "Air Pollution and Lichens" (B. W. Ferry, M. S. Baddeley and D. L. Hawksworth, eds), pp. 192–223. University of London Athlone Press, London.

PARBERY, D. G. (1967). Isolation of the Kerosene Fungus, *Cladosporium resinae*, from Australian soil. *Trans. Br. mycol. Soc.* **50**, 682–685.

PARBERY, D. G. (1969a). Isolation of the ascal state of *Amorphotheca resinae* direct from soil. *Trans. Br. mycol. Soc.* **53**, 482–484.

PARBERY, D. G. (1969b). *Amorphotheca resinae* gen. nov., sp. nov. The perfect state of *Cladosporium resinae*. *Aust. J. Bot.* **17**, 331–357.

PEACE, T. R. (1962). "Pathology of Trees and Shrubs." Clarendon Press, Oxford.

PIROZYNSKI, K. A. (1968). Geographical distribution of fungi. *In* "The Fungi" (G. C. Ainsworth and A. S. Sussman, eds), vol. 3, pp. 487–504. Academic Press, New York and London.

RAMSBOTTOM, J. (1935). Changes in the British fauna and flora during the past fifty years. (6) Fungi. *Proc. Linn. Soc. Lond.* **148**, 49–52.

STAKMAN, E. C., STEWART, D. M. and LOEGERING, W. R. (1962). Identification of physiologic races of *Puccinia graminis* var. *tritici. Sci. Ser. Minn. Exp. Stn,* no. 4691.

7 | Changes in the Seaweed Flora of the British Isles

W. E. JONES

*Marine Science Laboratories, University College of North Wales,
Menai Bridge, Anglesey*

Abstract: Although the extent of the British seaweed flora was well known by the beginning of the twentieth century, a lack of knowledge of local distribution and of quantitative data makes a proper assessment of the changes in the flora impossible. However, it is clear that there have been few losses on a national scale, although pollution and changes in shore substrata have caused local depletion. At the same time a number of species new to the British Isles have appeared. A number of native British species have extended their ranges northwards but few of the new records are of old established European species migrating northwards across the English Channel. *Laminaria ochroleuca* is one of these but others, like *Colpomenia peregrina*, are recent introductions into Europe and have reached Britain as part of a general spread. Others, such as *Bonnemaisonia hamifera* have arrived (in this case from Japan) first in Britain and are spreading from the point of introduction. Recently a number of new species, including two of *Grateloupia*, have been recorded in the region of Portsmouth and the Isle of Wight. The latest arrival is *Sargassum muticum*, which had already spread from Japan to north-west America, presumably brought in with marine materials imported into Europe or transported in some other way by ships. This increase in the frequency of new records is a matter of some concern, particularly where the new species are large plants potentially able to dominate their environments.

INTRODUCTION

It is more difficult to obtain reliable information on changes in the seaweed flora of the British Isles than it is for most other groups of plants, with the possible exception of freshwater algae (Chapter 8) and fungi (Chapters 5 and 6). There are two reasons: firstly, the marine algae are inaccessible: until very

Systematics Association Special Volume No. 6, "The Changing Flora and Fauna of Britain", edited by D. L. Hawksworth, 1974, pp. 97–113. Academic Press, London and New York.

recently, only those plants which grow between tidemarks could be observed directly and then only in reasonably calm weather. Secondly, there has not been, for many years, a body of amateur phycologists comparable with the eager bands of enthusiasts who make a hobby of other groups of plants. This latter point is hard to understand; the algae were a popular field of study for amateur botanists in the last century and Harvey (1846–51) in his "Phycologia Britannica" frequently stresses his indebtedness to his amateur correspondents. Perhaps the early death of Batters and the lack of the keyed flora which he might have produced left a gap which was not filled until 1931 with the publication of Newton's "Handbook of the British Seaweeds" and the "Manx Algae" of Knight and Parke, but by then the fashion had changed and interest was centred on other plants.

Thus the outlines of the flora were well defined by the 1850s as the result of the work of Harvey, his predecessors and correspondents. As their records were partly based on drift material ("rejectamenta" was the contemporary word), the sublittoral plants were not ignored but quantitative data and exact locations with ecological observations were not the fashion. However, the species list was well covered and when later workers, particularly Cotton, Holmes and Batters, had added their exact and detailed observations, the flora was well recorded by the early years of the twentieth century. There were gaps in the coverage, since some early collectors did not move far from their home bases. It may be that the recent report of a new *Cryptonemia* species from Cork (Guiry, Irvine and Farnham, in preparation) falls into this category, since this spectacular plant seems to be of rather local occurrence.

In the early years of the present century, although a few changes were occurring in the flora, the idea of change was not much discussed, so that little can be seen to have altered if a comparison is made between the works of Newton (1931) and Harvey (1846–51). Knight and Parke (1931), however, were able to record an increased number of species in the flora of the Isle of Man since the publication of a list 18 years earlier (Harvey-Gibson *et al.*, 1913). They considered that there had been some real additions to and losses from the flora.

When, four years later, Cotton (1935) reviewed the changes in the marine algal flora over the previous 100 years, he apologized for a lack of detailed records which meant that few changes could be supported by conclusive evidence. It must be confessed that, despite the activity of marine botanists since Cotton's review, our present knowledge is only marginally more detailed. However, there have been a number of interesting changes since that time and it is also possible to comment in greater detail on some of the changes which he considered were going on.

A word of caution is perhaps appropriate at this point. We are inclined in these days to expect change and perhaps ready to believe in it on insufficient evidence. As an example, a supposed change on the English Channel coast received considerable publicity in the years just after the Second World War. This story centred on Worthing and the casting up in late summer of large quantities of drift weed. This was presented as a new and calamitous development and one for which a biological explanation and remedy were sought. Local fishermen, however, saw nothing unusual in the cast and explained (if they were asked) that it had always happened in every summer. As usual, the fishermen knew more than the newspapers; there had been no startling change, merely the pointing out to an unobservant public of a normal occurrence. In fact there are cases of the persistence of algae in some localities which are part of the background against which change should be viewed.

PERSISTENCE OF ALGAL SPECIES

At first sight, the severity of conditions even on fairly sheltered shores, when compared with terrestrial habitats, makes the continued presence of a species in a particular spot rather surprising. This is, however, the normal rather than the exceptional situation. In fact the extreme conditions select a limited, highly adapted flora whose composition is maintained in a particular site by the very severity of the environment. On many shores the limiting factors operate particularly at the time of spore settlement but if spores are present in the water at a particular season, then in some years the conditions will permit settlement— perhaps with success only in a very limited area. Thus, even though the life of littoral algae is usually short, the site will be occupied in some years if not in all. An example of this is the finding of *Cystoseira baccata* (Gmel.)Silva in 1966 at Porth Dinllean in North Wales outside its usual range but apparently in the same spot in which Rees found it in the 1920s (Rees, 1929). Another remarkable case is recorded by Powell (1966) who found plants of *Codium adhaerens* (Cabr.) C. Ag., at the northern-most limit of the species at Loch nam Uamh, Skye, growing on a boulder in exactly the spot where it had been photographed by Robert Adam in 1927. Plants are not always so fortunate, however; Cotton (1935) recorded the complete disappearance of the same species at Swanage.

LOCAL CHANGES

1. Changes Without Obvious Cause

Apart from additions to or losses from the flora of the British Isles as a whole, local fluctuations in the quantity of species or their disappearance and re-appearance occur, probably more frequently than they are recorded.

Local losses are of course inevitable with the spread of coastal industry, the construction of harbour works and variation in the pattern of sediment deposition resulting from the dredging necessitated by the increasing size of coastal ships. Fortunately, such losses have usually been of limited extent; even the extensive losses on the inner shores of Plymouth Sound do not represent losses to the flora of S. Devon as a whole. It is, unfortunately, not always possible to find convincing reasons for the changes observed. As an example of this from my own observations, *Schizymenia dubvi* J. Ag. was surprisingly common on shores in Pembrokeshire in 1958 (Jones and Williams, 1966). This abundance, also observed in the following year, then declined and the plants became scarce, a condition which has continued to recent years with some minor fluctuations. More remarkable is a change observed at St Anne's Head, also in Pembrokeshire. Here, in the late 1940s, R. Williams noted *Pelvetia canaliculata* (L.) Dene & Thur. forming a distinct band on the steep shoreward-facing slope of the Pig Stone, above a deep pool. Such an occurrence is not unusual on a steep slope (Lewis, 1964), the absence of the fucoid algae which would otherwise complete the zonation pattern below the *Pelvetia* being at least partly due to limpet grazing. The shore did not change appreciably in appearance during the 1950s, as confirmed by occasional observations, and was observed at least annually from 1957 onward. But in 1965 a change was observed; *Fucus spiralis* L. was found forming a band below the *Pelvetia* and in the following summer a band of *Fucus vesiculosus* L. was added below the others. Thus a much wider band of algae had appeared and this continued at least as late as 1973, my most recent observation. There were no obvious changes in the local topography or associated fauna which could account for the change but it is interesting to note that the normal zonation pattern developed without a preliminary period of unorganized colonization of the kind observed in the classical clearance and resettlement experiments on the Isle of Man (Lodge, 1948).

Changes in the abundance of algae are difficult to establish unless a foundation exists in the form of a detailed survey and these have been few. One which covered a large part of the British coast was that undertaken by a number of botanists in 1941–42 as part of the search for an agar substitute, concentrating on the availability of *Gigartina stellata* (Stackh.) Batt, and *Chondrus crispus* Stackh. (Marshall *et al.*, 1949). Taking the observations recorded along a section of the coast near Dunbar, E. Booth (*in litt.*) found in 1961 that there had been a very considerable decrease as shown in Table I. It must be remembered that, because of military restrictions, no harvesting followed the survey and the great reduction in quantity twenty years later must therefore have had another cause.

On Anglesey, *Nemalion helminthoides* (Vell.) Batt. has displayed considerable variation. Phillips collected the species at Trearddur Bay in 1895 (Davey, 1953) and there are other records from about this time. I can find no record of its having been present on these shores between the wars and it was certainly absent in the 1950s, although it occurred at that time on Bardsey Island, 35 miles to the south-west. In 1962, however, *Nemalion* was to be found in the

TABLE I. Changes in the abundance of *Gigartina stellata* on the coast near Dunbar in 20 years

Location	1941–42[a]	1961[b]
N. Berwick	600	Very scarce
Seacliff	6500	Little evident
Dunbar	7540	None observed
Dunbar-Skateraw	5280	None observed
Torness point	4500	Occasional plants only (under *Fucus vesiculosus*)

[a] Estimated wet weight of *Gigartina* (and some *Chondrus*) in lb along each stretch of coast (from Marshall *et al.*, 1949).
[b] E. Booth's (*in litt.*) observations.

same localities as those recorded at the turn of the century (Jones, 1964), and has been present, with some annual fluctuation, every summer since then. Assuming Bardsey to be the source of the spores, the recurrence of *Nemalion* might result from a favourable coincidence of wind and tidal currents with good settling conditions at the late summer fruiting period, but what might have caused its disappearance is less easy to explain.

Price and Tittley (1972) mention that a number of changes, some possibly long- or short-term cyclic, have occurred in the distribution of species on the Kent coast; they note as an example the occurrence, on one occasion only in six years, of a luxuriant population of *Taonia atomaria* (Woodw.) J. Ag. This is a species of variable occurrence on Anglesey also; there are some early records and then none until a single one in 1955. However, this is perhaps explicable in this locality by the more recent realization that the species is not uncommon sublittorally in summer at some places on the Anglesey coast (Smith, 1967) and its occasional appearances in the lower littoral probably represent chance extensions of its normal range.

2. Changes due to Alterations in the Substratum

Some local losses may result from changes in the availability of suitable substrata. The most catastrophic of these was the wholesale loss of *Zostera marina* L., about 1933 (Tutin, 1938); *Zostera* bore a number of characteristic epiphytes which were inevitably affected by the loss of their host. They included:

(a) *Cladosiphon zosterae* (J. Ag.) Kylin, now reappearing and recorded on Colonsay (Norton *et al.*, 1969) and the Channel Islands (Dixon, 1961).

(b) *Myrionema magnusii* (Sauv.) Lois which has been recorded on Guernsey (Dixon, 1961) and in the Isles of Scilly (Russell, 1968).

(c) *Rhodophysema georgii* Batt. does not appear to have been recorded since the 1933 disaster and may not have survived.

(d) *Melobesia lejolisii* Rosan. has been recorded only once since 1933, on St Mary's, Isles of Scilly (Russell, 1968).

Other substratum losses have been of some local significance. In the Menai Straits two factors have operated; firstly the increasing height of some banks of silt and shell gravel has blanketed some rocks and so reduced the area available for littoral species such as *Fucus vesiculosus* L. and *Ascophyllum nodosum* (L.) Le Jol. There is still a luxuriant growth of these species in the Straits as a whole, but the belief, sometimes expressed by those who have known these shores for many years, that "there is less seaweed nowadays" may be partly due to this increasing deposition. Secondly, and more seriously, there is the loss resulting from the activity of anglers. In searching for bait, especially shore crabs, the less public spirited of these sportsmen turn over the larger stones on the shore but do not return them to their original positions. The luxuriant fucoid growths, now partially buried, decay. *Elminius modestus* Darwin, which reached the Menai Straits about 1952 (Crisp, 1958) is now well established there and, having a very long breeding season, is well placed to colonize the exposed surfaces of the inverted stones. Once the barnacles are established, successful recolonization by fucoids is more difficult and the result is that large areas of the intertidal flats, particularly east of Beaumaris, have changed from being weed dominated to being barnacle covered, with a consequent loss of cover for fish. This is a danger which must also exist on other shores.

On a smaller scale, the construction of a new harbour wall at Lerwick greatly reduced the area available to the rare northern *Fucus distichus* subsp. *edentatus* (De la Pyl.) Powell, which survives on a remnant of the old wall (Powell, 1963).

3. Pollution Effects

The changes already mentioned at Plymouth, which can be duplicated in many places, are a result of a combination of substratum change and increased pollution. Such effects need not, however, be the direct result of man's activity;

Russell (1964) reported that, on the Isle of May, the increasing population of herring gulls and lesser black-backed gulls had caused considerable changes since his previous visit two years before. *Codium fragile* (Sur.) Hariot and *Eudesme virescens* (Carm.) J. Ag. had disappeared from the polluted upshore pools but the more resistant *Enteromorpha intestinalis* (L.) Link and *Porphyra umbilicalis* (L.) J. Ag. were unaffected, whilst the nitrophilous *Prasiola stipitata* Suhr was thriving. As the British population of refuse-fed gulls continues to increase, changes of this kind, already evident in a number of places, can also be expected in others.

Elsewhere, however, improvements in sewage disposal methods are producing a reverse effect. One of the most encouraging examples is the change in parts of the Thames estuary. There has been, particularly since the late 1950s, a reduction in pollution by domestic sewage and a smaller, but useful, reduction in industrial waste. Presumably as a result of this, marine algae have begun to recolonize shores from which they have been absent for many years. In the survey of the marine algae of the coast of Kent (Price and Tittley, 1972 and personal communication) luxuriant populations of *Fucus vesiculosus* were recorded as far upstream as Crayfordness (near Erith) and of *F. spiralis* L. a little higher, at Belvedere. This is far beyond the limits of these species in the years of unchecked pollution.

Industrial pollution is a more serious problem than domestic sewage. Although the latter can, if sufficiently concentrated, cause an oxygen deficiency serious enough to kill algae, this is most exceptional in the sea. The more usual effect, as Cotton (1935) observed, is to cause a depletion of the more delicate species and a luxuriant growth of others such as *Ulva lactuca* L. and *Enteromorpha intestinalis*.

The effects of industrial wastes can be much more lethal. In the stream of effluent flowing out to sea from the Associated Octel plant at Amlwch, for instance, no plants of *Laminaria hyperborea* (Gunn.) Fosl. can be found less than five years old and the mature plants can be removed from the rocks with uncharacteristic ease (N. Jephson, *in litt.*). There is a lack of epiphytic growth on the plants and a dearth of other algae. All shore algae have been killed in the region of the outfall and even some distance from it only a few stunted fucoids can be found (R. Hoare and K. Hiscock, *in litt.*). Fortunately these effects are localized but the dangers of unrestricted pollution are obvious. Unfortunately, legislation to limit marine pollution is virtually non-existent and local losses will continue as long as chemical pollution can proceed unchecked.

Oil pollution is probably less dangerous, if only because it can be seen. The detergents to disperse oil on the shore were, until recently, more harmful than the oil itself but even if the shore is cleared of all organisms, recolonization will eventually occur (Nelson-Smith, 1968).

The dangers are, firstly, that the sequence of succession in recolonization may lead to a different community pattern and that full restoration to the original condition may take a very long time (Cowell *et al.*, 1970). Secondly, pollution incidents may be frequent enough, particularly near loading and discharge installations, to produce a state of chronic pollution. Fortunately public opinion in Britain is likely to prevent too much complacency on the part of the oil companies, even though the chemical manufacturers are, as yet, showing little concern.

ALGAL MIGRATIONS

Because the British Isles have a considerable extension north and south, a number of species reach the northern limits of their ranges on the British coast while others, rather fewer in number, have their southern limits there (Hoek and Donze, 1967). As weather patterns and sea temperatures change with the passage of time, these limits are also subject to change. Several species seem to have extended their ranges northwards but there are few records of southern migrations. One of the latter was mentioned by Knight and Parke (1931) who recorded the appearance of the northern species *Stictyosiphon tortilis* (Rupr.) Reinke on the Isle of Man where it was common by 1931 though not recorded in the 1913 list (Harvey-Gibson *et al.*, 1913).

The reason usually advanced for northward migrations is the warming of the British climate in the first half of the twentieth century. There are well known examples of species which are believed to have crossed the English Channel but the picture is made less clear on closer examination, when it seems that the majority of these species are recent introductions into Europe from more distant places. It seems best to consider the less spectacular changes first:

1. Extensions of the Range of Species within the British Isles

A few examples have already been mentioned under other headings. As Dixon (1965) emphasized, records must be treated with great caution, for apparent extensions of the territory of a species may be the result of mis-identification. Knight and Parke (1931) also pointed out that many of the species new to the Isle of Man in their list were not necessarily newly arrived there but were small plants which had probably been overlooked by earlier collectors. Occasionally, however, new records are of conspicuous plants or ones for which unavailing searches have been made previously and in these cases the evidence for a migration is strong. Knight and Parke considered that their new record of *Codium adhgerens* was of this kind, as were those of *Asperococcus turneri* (Sm.) Hook. and the *Trailliella* stage of *Bonnemaisonia hamifera* Hariot.

Caution is also needed when considering the distribution of plants which are partly of sublittoral occurrence. Some extend further north in the sublittoral than in intertidal habitats and the collections made over the last twenty years using diving equipment have produced records which have changed our view of the limits of distribution of some species. Both *Taonia atomaria* and *Dictyopteris membranacea* (Stackh.) Batt. fall into this category but the records of both species from Co. Donegal (Brennan, 1945) suggest a real north-westward extension of their range.

The genus *Cystoseira* is of mainly southern distribution and the British species are of rather variable occurrence away from the south-western coasts. Knight and Parke (1931) mention *C. baccata* (Gmelin) Silva, *C. foeniculacea* (L.) Grev. and *C. tamariscifolia* (Huds.) Papenf. as having disappeared at that time, while *C. tamariscifolia* has not been recorded on Anglesey in recent years although there are earlier records. Elsewhere, however, this species has gained ground: Stewart (1962) recorded luxuriant growths on Barra, a change since Newton (1931) described it as rare in Scotland; Roberts (1970) was cautious about records from the Scottish mainland.

2. Northward Spread of Species across the Channel

As will be seen, most of the algae which fit this description are recent introductions to Europe and are more conveniently considered in that category.

Even one of the best known examples, *Laminaria ochroleuca* De la Pyl. may not have been a true native of the French Atlantic coast (John, 1969) but if not, its entry is at least earlier that the 1890s. *L. ochroleuca* seems to have crossed the channel in the early 1940s, plants five years old being dredged from Plymouth Sound in 1948 (Parke, 1948), the first British record. From Plymouth, the species began to spread along the south coast and John (1969) recorded it at localities from Salcombe to the Helford river and on the Isles of Scilly.

3. Introductions from Distant Places

Here can be included those species that arrived suddenly in Europe and spread to Britain and also that reached Britain direct from a distant source.

If a roughly chronological order is adopted, the first example is *Bonnemaisonia hamifera*. This is one of those red algae in which the form of the sexual phase is entirely different to that of the tetrasporophyte. In this case the latter was long known as *Trailliella intricata* and considered to be a separate species. It was the

Trailliella phase which was first recorded in the northward spread, at Studland in 1870 (Westbrook, 1930). The sexual *Bonnemaisonia* phase was first described from Japanese material in 1871 and was first recorded in Britain in 1890 at Falmouth. Westbrook (1930) gives an account of the spread of the species which had reached Cherbourg and western Ireland by 1911 but had not been recorded elsewhere in Europe (Cotton, 1912). The spread has continued, the *Trailliella* phase reaching the Orkney islands in 1929; by this time it had been found near Bergen. The centre of distribution was in Britain; the evidence suggests an accidental introduction, probably from Japan, and this view is supported (Cotton, 1935) by the presence of female plants only of *Bonnemaisonia*, with reproduction by vegetative means. More recently Floc'h (1969) was able to largely confirm the hypothesis although he recorded male plants in Brittany.

Asparagopsis armata Harvey is another red alga with morphologically different sexual and tetrasporangeal phases. The latter, another insignificant creeping filamentous form, was also given a specific name, *Falkenbergia rufolanosa* (Harv.) Schm., until the connection between the phases was established (see Dixon, 1963). *A. armata* had been described from Australian specimens by Harvey (1858–63). Kerslake (1953) summarized its European history from the first record in Algeria in 1923 and then Biarritz in 1928, together with the tetrasporophyte. The latter was collected in Co. Galway in 1939 (Drew, 1950) and the *Asparagopsis* phase in 1941 (de Valera, 1942). This is still an isolated locality. Since then, *A. armata* has been recorded on Lundy (Harvey and Drew, 1949) at Lough Ine and at Falmouth and the Lizard (Drew, 1950). Horridge (1951) collected it in the Scillies and later it occurred in the Isle of Man (Walker *et al.*, 1954). It had reached Scotland by 1959 (Conway, 1960; McAllister *et al.*, 1967). These scattered records, with other coasts devoid of the species in between, suggest some distributional vector other than natural spread by spores or thallus fragments. The fact that Atlantic liners used to call at Galway *en route* from Cherbourg to Liverpool is suggestive (de Valera, 1942) but there is no positive evidence that shipping is involved in the spread.

Antithamnion spirographoides Schiffer is a species of red alga of somewhat problematic origin. It appears to have reached Europe from the Southern Hemisphere, first appearing in Cherbourg in 1910 and spreading to a number of places in Brittany, Normandy, and the Channel Islands by 1930 (Westbrook, 1930) by which time the species had crossed the English Channel and was established near Plymouth and at Salcombe. It extended northwards to Lundy by 1937 (Parke, 1952), to Argyll by 1966 (McAllister *et al.*, 1967) and was collected in the Isles of Scilly in 1967 (Russell, 1968). Sundene (1964) pointed out the plant's ability to survive unfavourable conditions in culture and suggested that this might make it capable of long distance transport attached to ships.

Colpomenia peregrina Sauv., a membranaceous globular brown alga, is one of the best known immigrants, long quoted as an example of the northward spread of a European species. The facts are a little more complex than this. There are now two species in Europe: *C. sinuosa* (Mert.) Derb. & Sol. which was present at least as far back as the 1840s in Spain and *C. peregrina* which was introduced accidentally and first noticed by oyster fishermen at the northern end of the Bay of Biscay in 1906 (Sauvageau, 1918). Their attention was attracted by the peculiar circumstances of the large hollow thalli, gasfilled and buoyant, growing on oysters and, when grown large enough, floating out to sea together with the shellfish. *C. peregrina* was first recorded in Britain in 1907 in Cornwall and Dorset (Cotton, 1908a, b). It spread along the south coast (Cotton, 1911), probably reached the Isle of Man about 1920 (Cotton, 1935) and was common on some Manx shores by 1931 (Knight and Parke, 1931). The distribution is still somewhat discontinuous; the plant occurs in Anglesey but not in Prembroke-shire (Jones and Williams, 1966) which does not support the otherwise attractive suggestion of Blackler (1939) that, in view of the extensive distribution of *Colpomenia* around Ireland, French fishing boats might be a vector. It is equally possible that, as Dixon (1965) suggested, the buoyant thalli might float for long distances under the influence of wind and currents.

Codium fragile (Sur.) Hariot is another comparative newcomer to Britain, distinguishable from the longer established *C. tomentosum* Stackh. and *C. vermilara* (Olivi) Delle Chiaje by microscopic characters. Silva (1955) defined these differences and showed that two subspecies of *C. fragile* were present, subsp. *atlanticum* (Cotton) Silva and subsp. *tomentosoides* (Goor) Silva. Both appeared to have been introduced from the Pacific; subsp. *atlanticum* "within historical times" and subsp. *tomentosoides* much more recently. *Codium tomentosum* and *C. vermilara* are both species with a southern distribution, the former extending south to Morocco and the latter widely distributed in the Mediter-ranean; both are distributed in the south and west of Britain and neither has changed its geographical range appreciably. *C. fragile*, however, has in recent years migrated northwards and eastwards. Subsp. *atlanticum* seems to have been established in south-west Ireland by about 1808 and was collected on the Isle of Iona in 1826. By 1918 it had appeared in Norway but not until 1949 was it recorded in the British side of the North Sea, when Moss (1957) found it on the inner Farne Islands. This seems to have been part of a rapid spread; the subspecies was found growing luxuriantly in the Shetland Islands (Dixon, 1963). Other records are scattered; Silva (1955) lists western Ireland northwards to western

Scotland, the Orkney Islands, Dorset, the Isle of Man and Northumberland. It has also been recorded in Pembrokeshire (Jones and Williams, 1966), on Bardsey Island and on Anglesey.

Codium fragile subsp. *tomentosoides* was recorded in the River Yealm, Devon, in 1939 (Silva, 1955), about 40 years after the first European record in Holland. Like the other subspecies, it has spread north and west, with records in Argyll, Cork and the Channel Islands (Dixon, 1961).

The pattern in both species seems to be one of spreading from a number of widely separated colonization points rather than a gradual extension of the range. This colonization is one of the most impressive recent changes in the British marine algal flora.

In recent years a number of remarkable introductions have been observed along the south coast of England, particularly in the region of Portsmouth and the Isle of Wight. The first of these escaped identification for a number of years from the late 1940s although it was a large, handsome red alga which was plentiful in the drift weed westwards from Littlehampton. It was variously identified, without critical examination, as a *Calliblepharis*, *Halarachnion*, *Dumontia* or others. The possibility of its being something new was overlooked until W. Farnham at Portsmouth made a closer examination. The plants were identified (Farnham and Irvine, 1968) as unusually large specimens of *Grateloupia*, enormous by comparison with the normal *G. filicina* (Lamour.) C. Ag. which is native to Britain but is rather uncommon and easily overlooked. Farnham (*in litt.*) has found the two forms to be quite distinct, even to their growing side by side for an extended period under observation at Bembridge, Isle of Wight. The luxuriant form is almost certainly of foreign origin and has spread some way along the Channel coast, being well adapted to growth in turbid waters. It corresponds closely to *G. filicina* (Lamour) C. Ag. var. *luxurians* Gepp, originally described in Australia and Japan. An examination of herbarium material has shown that the earliest known record of the plant in Britain was at Bembridge in 1947.

Since this discovery, a second new *Grateloupia* has appeared in the same area (Farnham and Irvine, 1973). This is one of the broad, foliose forms and was first collected by Farnham (*in litt.*) in 1968 on the somewhat unpromising shore (algally speaking) at Southsea. Later it was found in a still more unlikely locality, the back of the shingle ridge at Pagham harbour. This plant appears to be *Grateloupia lanceola* J. Ag. (now included in *G. doryphora* (Mont.) Howe) which had its closest previously known approach to Britain on the Atlantic coast of Spain and southwards to Senegal.

As if it were not sufficiently remarkable that two *Grateloupias* new to Britain should appear in the same area over so short a time, yet another algal immigrant has recently been found nearby. Scagel (1956) described the establishment of a species of *Sargassum*, *S. muticum* (Yendo) Fensholt, in British Columbia, Washington and Oregon. It appeared that the plants had been brought across the Pacific with consignments of a Japanese oyster, *Crassostrea gigas*, introduced for commerical purposes. A re-examination of herbarium specimens has shown that *S. muticum* was present in British Columbia in 1944. Conditions along the coast of Canada and the United States seem to favour the Japanese *Sargassum* and Druehl (1973) stated that it was replacing *Zostera* in some sheltered shallow subtidal areas. He also pointed out that the Japanese oyster was also being introduced to the French Coast and forecast that *Sargassum muticum* would soon be found in the English Channel. This has now occurred, very soon after Druehl's warning. *Sargassum muticum* was found in February 1973 at Bembridge in the form of a number of firmly attached plants, some up to 1 m long (Farnham *et al.*, 1973). Harvey (1846–51) recorded *Sargassum* in the "Phycologia Britannica", but only as drift specimens of the pelagic *S. natans* (L.) Meyer of the tropical south Atlantic and there was no question of the establishment of this species in Britain. The Bembridge records bring the limit of *Sargassum* distribution in Europe northwards by 500 miles.

GENERAL OBSERVATIONS

There have been few losses from the British marine algal flora. The trend has been mainly a northward spread of distribution ranges with only a few southward extensions. A number of new species have been recorded: many of these seem to have arrived on the Channel coast and to have spread from there but some have spread from several widely separated centres. It seems doubtful whether climatic warming has been responsible for more than a fraction of these; the majority seem to have been carried to our shores by shipping, either on the hull or in cargo or store although it is perfectly possible that some have been transported attached to drifting objects (Lucas, 1950) or as whole floating plants or portions. Migrations of this kind seem most likely in the case of plants like *Asparagopsis armata* and *Bonnemaisonia hamifera*, both of which have branches specially modified onto other plants or objects in the water. Buoyant plants like *Colpomenia*, *Cystoseira*, etc., can easily be carried long distances.

The migration into Britain of algal species new to the flora has no doubt been going on since the end of the last glaciation but there is reason to think that the rate of introduction has risen in the twentieth century. There has been a change in the pattern of ship movements since steam replaced sail; ships are faster, passage times are shorter and more direct voyages are made without intermediate ports of call. This greatly reduced the dangers inherent in the transport of a plant from one coast to another on the far side of several climatic zones and might go some way towards explaining why so many new introductions seems to be near major ports. There is no direct evidence for this, nor for the suggestion that some introductions may result from the dumping overboard, from vessels arriving at the home port, of seaweed used as packing round shellfish or other seafood. The area of the Solent and Isle of Wight may be favourable areas because of the considerable length of comparatively hospitable coast close to the port of Southampton, whereas other major ports such as London, the Bristol Channel ports and Liverpool are more estuarine and are much further from suitable rocky shores.

But although the vector concerned is uncertain there is no doubt that alien species have arrived and will probably continue to do so. The introductions which are particularly disturbing are those which occur when marine materials are imported in bulk, as when oysters are brought in to restock beds. Because the quantities are large, the amount of introduced alien material is likely to be considerable and the chances of successful establishment increased. There seems to be little control of this activity and introductions of new algae are never desirable even if the results are unlikely to be as far-reaching as the spread of *Spartina townsendii*, *Crepidula fornicata*, or *Urosalpinx cinerea*, *Elminius modestus*.

Even less welcome would be the introduction of seaweeds for commercial exploitation. The possibility of introducing the gigantic *Macrocystis* was once considered and, according to popular belief, prevented by the Admiralty. As with all conservation, the need is for constant vigilance, particularly as so much of the coast of the British Isles is so far unspoiled. Druehl (1973) urged the need for action by the U.S. and Canadian governments; the needs of Europe and the rest of the world are just as urgent.

ACKNOWLEDGEMENTS

I should like to thank a number of my friends for helpful discussion of the problems outlined in this paper and for their permission to include their unpublished observations: Mr E. Booth, Mr W. F. Farnham, Mr R. L. Fletcher, Mr M. Guiry, Mr K. Hiscock, Mr R. Hoare, Dr and Mrs D. E. G. Irvine, Mr N. Jephson, Mr J. M. Price, Mr T. K. Rees and Mr I. Tittley. My thanks are also due to Mr O. Morton whose search of the literature on this topic while a student of this department has proved most useful.

REFERENCES

BLACKLER, H. (1939). The occurrence of *Colpomenia sinuosa* (Mart.) Derb. et Sol. in Ireland. *Ir. Nat. J.* **7**, 215.

BLACKLER, H. (1961). Some observations on the genus *Colpomenia* (Endl.) Derb. et Sol. *In* "Proceedings of the 4th International Seaweed Symposium" (A. D. de Virville and J. Feldmann, eds), pp. 50–54. Pergamon, London.

BRENNAN, A. T. (1945). Notes on the distribution of certain marine algae on the west coast of Ireland. *Ir. Nat. J.* **8**, 252–254.

CONWAY, E. (1960). Occurrence of *Falkenbergia rufolanosa* on the west coast of Scotland. *Nature, Lond.* **186**, 566–567.

COTTON, A. D. (1908a). The appearance of *Colpomenia sinuosa* in Britain. *Bull. misc. Inf. R. bot. Gdns Kew* **1908**, 73–77.

COTTON, A. D. (1908b). *Colpomenia sinuosa* in Britain. *J. Bot., Lond.* **46**, 82–83.

COTTON, A. D. (1911). On the increase of *Colpomenia sinuosa* in Britain. *Bull. misc. Inf. R. bot. Gdns Kew* **1911**, 153–157.

COTTON, A. D. (1912). Marine Algae [Pt. 15 of Praeger, R. L. (ed.), Clare Island Survey]. *Proc. R. Ir. Acad.* **31** (15), 1–178.

COTTON, A. D. (1935). Changes in the British fauna and flora in the past fifty years. (5) Marine Algae. *Proc. Linn. Soc. Lond.* **148**, 45–49.

COWELL, E. B., BAKER, J. M. and CRAPP, G. B. (1970). "The Biological effects of oil pollution and oil-cleaning materials on littoral communities including salt marshes". Paper E-11, F.A.O. Technical Conference on Marine Pollution, Rome.

CRISP, D. J. (1958). The spread of *Elminius modestus* Darwin in north-west Europe. *J. mar. biol. Ass. U.K.* **37**, 483–520.

DAVEY, A. J. (1953). The seaweeds of Anglesey and Caernarvonshire. *NWest. Nat.*, N.S. **1**, 272–289, 417–434.

De VALERA, M. (1942). A red alga new to Ireland: *Asparagopsis armata* Harv. on the west coast. *Ir. Nat. J.* **8**, 30.

DIXON, P. S. (1961). List of the marine algae collected in the Channel Islands during the joint meeting of the British Phycological Society and the Société Phycologique de France, Sept. 1960. *Br. phycol. Bull.* **2**, 71–80.

DIXON, P. S. (1963a). Marine algae of Shetland collected during the meeting of the British Phycological Society, Aug. 1962. *Br. phycol. Bull.* **2**, 236–243.

DIXON, P. S. (1963b). The Rhodophyta; some aspects of their biology. *Oceanogr. Mar. Biol. Ann. Rev.* **1**, 177–196.

DIXON, P. S. (1965). Changing patterns of distribution of marine algae. *In* "The Biological Significance of Climatic Changes in Britain" (C. G. Johnson and L. P. Smith, eds) [*Symp. Inst. Biol.* **14**], pp. 109–115. Academic Press, London and New York.

DREW, K. M. (1950). Occurrence of *Asparagopsis armata* Harv. on the coast of Cornwall. *Nature, Lond.* **166**, 873–874.

DRUEHL, L. D. (1973). Marine transplantations. *Science, N.Y.* **179**, 12.

FARNHAM, W. F. and IRVINE, L. M. (1968). Occurrence of unusually large plants of *Grateloupia* in the vicinity of Portsmouth. *Nature, Lond.* **219**, 744–746.

FARNHAM, W. F., FLETCHER, R. L. and IRVINE, L. M. (1973). Attached *Sargassum* found in Britain. *Nature, Lond.* **243**, 231–232.

FARNHAM, W. F. and IRVINE, L. M. (1973). The addition of a foliose species of *Grateloupia* to the British marine flora. *Br. phycol. J.* **8**, 208–209.

FLOC'H, J-Y. (1969). On the ecology of *Bonnemaisonia hamifera* in its preferred habitat on the western coast of Brittany. *Br. phycol. J.* **4**, 91–95.

HARVEY, C. C. and DREW, K. M. (1949). Occurrence of *Falkenbergia* on the English coast. *Nature, Lond.* **164**, 542–543.

HARVEY, H. W. (1846–51). "Phycologia Britannica", 4 vols. London.

HARVEY, H. W. (1858–63). "Phycologia Australica", 6 vols. London.

HARVEY-GIBSON, P. J., KNIGHT, M. and COBURN, M. (1913). Observations on the marine algae of the L.M.B.C. district (Isle of Man area). *Proc. Trans. Lpool biol. Soc.* **27**, 123–142.

HOEK, C. VAN DEN and DONZE, M. (1967). Algal phytogeography of the European Atlantic coasts. *Blumea* **15**, 63–89.

HORRIDGE, G. A. (1951). Occurrence of *Asparagopsis armata* on the Scilly Isles. *Nature, Lond.* **167**, 732–733.

JOHN, D. M. (1969). An ecological study on *Laminaria ochroleuca. J. mar. biol. Ass. U.K.* **49**, 175–187.

JONES, W. E. (1960). List of algae collected on the Northumberland coast, August 1959. *Br. phycol. Bull.* **2**, 20–22.

JONES, W. E. (1964). Occurrence of *Nemalion* on the coast of Anglesey. *Br. phycol. Bull.* **2**, 381.

JONES, W. E. and WILLIAMS R. (1966). The seaweeds of Dale. *Fld Studies* **2**, 303–330.

KERSLAKE, J. (1953). Occurrence of *Asparagopsis armata* Harv. on the coast of Devon. *Nature, Lond.* **172**, 874.

KNIGHT, M. and PARKE, M. W. (1931). "Manx algae" [L.B.B.C. Memoir XXX]. Liverpool University Press, Liverpool.

LEWIS, J. R. (1964). "The Ecology of Rocky Shores". English Universities Press, London.

LODGE, S. M. (1948). Algal growth in the absence of *Patella* on an experimental strip of fore-shore at Port St Mary, I.O.M. *Proc. Trans. Lpool biol. Soc.* **56**, 78–85.

LUCAS, J. A. W. (1950). The algae transported on drifting objects and washed ashore on the Netherlands coast. *Blumea* **6**, 527–543.

MARSHALL, S. M., NEWTON, L. and ORR, A. P. (1949). "A Study of certain British Seaweeds and their Utilisation in the Preparation of Agar." H.M.S.O., London.

MCALLISTER, H. A., NORTON, T. A. and CONWAY, E. (1967). A preliminary list of sub-littoral algae from the west of Scotland. *Br. phycol. Bull.* **3**, 175–184.

MOSS, B. (1957). *Codium* on the coast of Northumberland. *Br. phycol. Bull.* **1**, 40.

NELSON-SMITH, A. (1968). Biological consequences of oil pollution and shore cleansing. *In* "The Biological Effects of Oil Pollution on Littoral Communities" (J. D. Carthy and D. R. Arthur, eds), *Fld Studies* **2**, Suppl. 1–198.

NEWTON, L. (1931). "A Handbook of the British Seaweeds." British Museum (Natural History), London.

NORTON, T. A., MCALLISTER, H. A., CONWAY, E. and IRVINE, L. M. (1969). The marine algae of the Hebridean island of Colonsay. *Br. phycol. J.* **4**, 125–136.

PARKE, M. W. (1948). *Laminaria ochroleuca* De la Pyl. growing on the coast of Britain. *Nature, Lond.* **162**, 295–296.

PARKE, M. W. (1951). Notes on the Plymouth marine flora; algal records for the Plymouth region. *J. mar. biol. Ass. U.K.* **29**, 257–261.

PARKE, M. W. (1952). "The Marine Algae" [*Flora of Devon*, 2(1), 1–77]. The Devonshire Association, Torquay.

POWELL, H. T. (1963). New records of *Fucus distichus* subspecies for the Shetland and Orkney islands. *Br. phycol. Bull.* **2**, 247–254.

POWELL, H. T. (1966). The occurrence of *Codium adhaerens* (Cobr.) C. Ag. in Scotland and a note on *C. amphibium* Moore and Harv. *In* "Some Contemporary Studies in Marine Science" (H. Barnes, ed.), pp. 591–595. Allen and Unwin, London.

PRICE, J. H. and TITTLEY, I. (1972). The marine flora of the county of Kent and its distribution, 1597–1972. *Br. phycol. J.* **7**, 282–283.

REES, T. K. (1929). Marine algae of the coast of Wales. *J. Bot., Lond.* **67**, 231–235, 250–254, 276–282.

ROBERTS, M. (1970). Studies on the marine algae of the British Isles, 8. *Cystoseira tamariscifolia* (Huds.) Papenf. *Br. phycol. J.* **5**, 201–210.

RUSSELL, G. (1964). Additional marine algal records for the Isle of May. *Trans. Proc. bot. Soc. Edinb.* **39**, 467–471.

RUSSELL, G. (1968). List of marine algae from the Isles of Scilly. *Br. phycol. J.* **3**, 579–584.

SAUVAGEAU, C. (1918). Sur la dissemination et la naturalisation de quelques algues marines. *Bull. Inst. oceanogr. Monaco*, **342**.

SCAGEL, R. F. (1956). Introduction of a Japanese alga, *Sargassum muticum* into the north-east Pacific. *Fish Res. Pap. Wash. Dept. Fish* **1**(4), 49–58.

SILVA, P. C. (1955). The dichotomous species of *Codium* in Britain. *J. mar. biol. Ass. U.K.* **34**, 565–577.

SMITH, R. M. (1967). Sub-littoral ecology of marine algae on the North Wales coast. *Helgoländer wiss. Meeresunters.* **15**, 467–479.

STEWART, W. D. P. (1962). Occurrence of *Cystoseira tamariscifolia* (Huds.) Papenf. on the w. coast of Scotland. *Nature, Lond.* **195**, 402–403.

SUNDENE, O. (1964). The conspecificity of *Antithamnion sarniense* and *A. spirographidis* in view of culture experiments. *Nytt Mag. Bot.* **12**, 35–42.

TUTIN, T. G. (1938). The autecology of *Zostera marina* in relation to its wasting disease. *New Phytol.* **37**, 50–71.

WALKER, M. I., BURROWS, E. M. and LODGE, S. M. (1954). Occurrence of *Falkenbergia rufolanosa* in the Isle of Man. *Nature, Lond.* **174**, 315.

WESTBROOK, M. A. (1930). Notes on the distribution of certain marine red algae. *J. Bot., Lond.* **68**, 257–264.

8 | Changes in the British Freshwater Algae

B. A. WHITTON

Department of Botany, University of Durham, Durham

Abstract: A brief account is first given of our present knowledge of the distribution of freshwater algae in the British Isles, together with the criteria by which changes taking place in this flora may be recognized. The paper then summarizes data concerning lake, pond and river sites whose algae have been studied at long time intervals. Only a few algae are convincing aliens. However, there are many records of population increases, both from particular sites, and, in the case of some eutrophic water species, from many different regions. There are also many records of particular losses at particular sites, but general comments on losses are largely guesswork based on a knowledge of habitat losses.

INTRODUCTION

It is probably harder to give an objective account of the changes which have taken place in the freshwater algal flora during the last few centuries than for the majority of the other groups treated in this volume. No doubt bacteria and protozoa would have presented the greatest problems, but it is significant that no one has attempted to deal with them here. The present chapter attempts to summarize our knowledge of freshwater algae in Britain, together with the few firm facts known about changes in the abundance of various species.

PRESENT KNOWLEDGE OF THE BRITISH FLORA

Anyone used to studying flowering plants, bryophytes or lichens, and who decided to expand his interests to include freshwater algae, might well be shocked at the state of the subject. He would find that not only is there no mapping scheme, but also that for most of the groups there is not even any sort of British flora or published check-list. This is not just a British problem, however, for in probably only three countries, Austria, Sweden and Switzerland, is knowledge of field distribution any better than it is here.

Systematics Association Special Volume No. 6, "The Changing Flora and Fauna of Britain", edited by D. L. Hawksworth, 1974, pp. 115–141. Academic Press, London and New York.

No group of freshwater algae had received sufficient taxonomic study two centuries ago for more than a few scattered records to have been made which can be compared with modern ones. Hassall (1845) provided the first general account of British freshwater algae, but for only a very few species such as *Lemanea fluviatilis* and *Hydrurus* would his taxonomic and location notes both be adequate for anyone to check a record now. Towards the end of the nineteenth century, however, systematic studies on desmids, charophytes and diatoms were sufficiently advanced for detailed surveys to have been made which could be compared with modern ones. The accounts of desmid taxonomy and distribution by W. West and his son G. S. West are still the most thorough that have been made (West and West, 1904–12; Carter, 1923) for any algal group in Britain. The "British Charophyta" of Groves and Bullock-Webster (1920, 1924), although primarily taxonomic, includes many observations on distribution and ecology. Detailed and critical resurveys of the sites reported on by these authors would be of great value, but it is unfortunate that neither of these algal groups has received much interest in recent years in this country. In other groups, such as the blue-green algae, their taxonomy is only just now becoming adequate for full floristic lists to be made of a sort which would be useful to future workers. In still other groups, such as some families within the red algae and the chrysophytes, present taxonomic understanding is inadequate for attempting a comprehensive check-list, even if there were sufficient knowledgeable people to carry out field surveys.

The most thorough studies that have been carried out on the algae of one particular region are those of the lakes of the Lake District. For some of these lakes the data available include both detailed plankton counts made over several decades and also counts of diatom frustules deposited in the lake sediment since the late-Glacial period (see below). Habitats other than lakes have not been dealt with so fully in the Lake District region, but a recent survey of all groups of plants on the Isle of Mull organized by the Botany Department of the British Museum (Natural History) has included a wide variety of habitats within this one area. This will provide a particularly useful base-line for future generations to monitor changes taking place there. Some idea of species diversity in a particular region is given by the fact that on this island of about 800 km², some 400 freshwater (including terrestrial) diatom species have been recorded by P. A. Sims (*in litt.*), and another 360 species belonging to other groups by D. J. Hibberd. These latter were mostly desmids, and a more detailed treatment of the blue-green algae and of the flagellates would certainly have substantially increased this latter figure (Hibberd, *in litt.*). For other regions the information is much more diffuse, and Table I is included here to help anyone interested to start finding out the known records for particular groups or regions. It is obviously far from comprehensive, but many further references can be obtained from the articles mentioned. The Freshwater Biological Association at Windermere (including their annual reports) provides the best place to carry on a more detailed hunt for relevant material. J. W. G. Lund (*in litt.*) holds a somewhat dated card index of British records.

Three other types of information concerning changes are available to potential workers, but these are of unequal value for the various taxa:

(1) Many collections of old material still await further study. For diatoms such material is usually almost as useful as material freshly collected, and more useful than lists of names made at the time of collection. Desmids, many of the Chlorococcales and chrysophytes with siliceous structures are also often preserved sufficiently well for detailed lists of species to be given. Preserved samples of blue-green algae are usually adequate for specific identification if from the plankton, but usually not so if from other habitats. Nevertheless important information such as frequency of heterocysts is easy to interpret. Finally, preserved material of many groups of flagellates is almost worthless unless particular care was taken at the time of collection.

(2) The use of fossil algae is almost entirely restricted to those with some sort of siliceous structure, although identifiable remnants of a few other organisms such as some genera of the Chloroccocales (*Coelastrum, Pediastrum, Tetraedron*), and also the blue-green alga, *Gloeotrichia* (J. Waddington, *in litt.*) may persist in sediments. Most studies so far of lake cores have concentrated on long-term changes in diatom composition, but, as pointed out by Round (1971), there is vast scope for detailed studies not only over periods of hundreds of years, but also over much shorter ones, linking the information with the recent changes recorded in living populations.

(3) In a few cases it is possible to obtain living material of algae representative of populations growing many decades ago. Material of terrestrial or soil blue-green algae have several times (in other countries) been taken from herbarium sheets more than half a century old, and shown to be capable of growth. Stockner and Lund (1970) studied viable algae taken from varying depths in deep-water cores from Ullswater, Esthwaite Water and Windermere. Viable algae were found down to 0·35 m in Esthwaite, corresponding to 262 ± 28 years ago, and to 0·25 m in Windermere, corresponding to 195 ± 12 years. Although some viable cells in the sediments were undoubtedly due to a "disturbance factor", in other cases it seems probable that the algae are very resistant to decomposition, and may remain viable for very long periods, this being especially pronounced in the case of *Melosira italica*. For at least a few

species it is therefore theoretically possible that clones of an alga representing forms living a century apart might be isolated and studied to see whether there is any indication of adaptational changes to factors such as eutrophication or use of algicides.

CHANGES IN FLORA AT PARTICULAR SITES

1. Interpretation of data

For a variety of reasons, there are many problems in making detailed comparisons of sites that have been studied at long time intervals. Taxonomic difficulties have already been mentioned, but there are others which may be less obvious to workers in other groups. Most freshwater algae have a life-cycle which is brief in comparison with higher plants, so marked fluctuations in populations sometimes occur within a week or less. Aquatic habitats such as small pools are often transient, and it may not be possible to re-sample from an identical site, even if the general area has remained undisturbed. Streams and fast-flowing rivers tend to be influenced markedly by more factors than terrestrial environments, and it is therefore not surprising that they often show considerable variation from year to year in their dominant species, even at the same season. Further, the possibility cannot be ruled out that some species may show fluctuations on a long-term time-scale, even though the environment remains relatively uniform. In general, however, long-term waxing and waning of species is suspected to be related to obvious environmental factors, such as changes in climate, land use and dissolved nutrients. Some examples of waxing and waning of diatoms as revealed by sediment cores are discussed by Round (1971). Although there is often argument about particular species, it is becoming clear that some changes in diatom species composition are particularly good indicators of eutrophication.

Interpretation of the ecological significance of invasion into sites by fresh species is further complicated by the relatively poor state of knowledge concerning the geographical distribution of freshwater algae, and the ease with which some can be transported around. Many are almost cosmopolitan, at least within broad climatic zones; on the other hand, others appear to show marked discontinuities in their distribution, and, in the case of the desmids (P. A. Tyler, *in litt.*), distinct endemism to particular regions of the world. Without experimental studies, it is difficult to decide whether a species could grow well at a site and has just never had the chance of being inoculated there. Although not relevant to the British Isles, the account by Castenholz (1973) of the world distribution of thermal spring algae provides an exceptionally detailed argument that not every species is growing at every site where it would be capable of doing so, if inoculated. Thermal spring algae may have special problems in dispersal, but nevertheless if natural habitats which have been in existence for thousands of years have not become inoculated with all the species possible, then many habitats modified by man may also take long periods to reach some sort of species equilibrium. As argued by Talling (1951), chance may play an important role in the inoculation of habitats, and convincing floristic evidence of changes such as eutrophication over a few decades can come only from a consideration of a range of species.

While Table I indicates that there are many sites which were studied in some detail prior to 1960, in few instances has anyone tried to make a critical modern comparison. Among these few examples are ones where no floristic changes have occurred, and ones where changes are quite evident.

Firth and Hartley (1970) visited three small Pennine streams for diatoms 32 years after Firth himself had first collected there. They found no apparent change in the species and varieties present. The longest period at any one site for which a freshwater alga has been observed is probably that of a Durham pond where *Chara hispida* has apparently been the dominant organism since at least 1775 (Wheeler and Whitton, 1971). This continuity has occurred in spite of the fact that the pond has probably undergone considerable nutrient enrichment during the period (Hudson *et al.*, 1971).

2. Lakes

Most of the attempts to compare old and new floristic lists are concerned with a few large lakes which have certainly undergone some nutrient eutrophication. Lough Neagh, the largest lake in the British Isles, has been the subject of phytoplankton surveys in May 1900 and July 1901 (West and West, 1902), 1910–11 (Dakin and Latarche, 1913) and 1968–71 (Gibson *et al.*, 1971). Some indication of the changes which appear to have occurred in this lake are given in Table II, which is based simply on the species names given by the various authors. A detailed comparison of these three surveys would involve lengthy discussion on taxonomic and sampling problems, and would certainly lead to a considerable reduction in the number of species involved. Nevertheless, it would seem clear that the desmid flora has become much reduced, and that there have been marked changes in both the blue-green algae and diatom species composition. Gibson *et al.* comment that the two commonest species of blue-green algae at present, *Oscillatoria redekei* and *Aphanizomenon flos-aquae*, have both appeared since 1911. Further evidence indicating changes in the diatom flora has come from a study of the lake sediment. Batterbee and Oldfield (*in litt.*, quoted by Wood and Gibson, 1973) have found that changes

TABLE I. Some examples of sites with extensive or especially useful algal lists made before 1960

Date of paper	Author	Date of survey	Counties mainly involved	Vice-counties (where-known) (see pp. 425–428)	Main taxa listed	Habitats mainly involved	Whether sites still present
1924	Atkins and Harris	1921–24	Devon	3	B, Chl, E, M, Py	reservoir	✓
1908	Bachman	1904–05	various Scottish	+	B, Cd, Chl, M, Py	lakes	✓
1959	Belcher	ca. 1958	Hertfordshire	20	B	river	✓
1920	Bristol	1915	various	+	Chl, M, X	soil	–
1927	Bristol and Roach		various	+	Chl, M, X	soil	–
1954	Brook	1946–47	Northumberland	67	B, Chl, M	filter-bed sand	✓
1955a	Brook	1946–47	Northumberland	67	B, Chl, M, X	filter-bed stones	✓
1957a	Brook	1955	various Scottish	+	B, Chl, M	lakes	✓
1958a	Brook	1958	various Irish	+	Cd	lakes	✓
1958b	Brook	1952–56	Perthshire	89	B, Chl, Chr	lakes	✓
1924	Butcher		Yorkshire	65		river	✓
1932ab	Butcher	1929–30	Cambridge, Durham, Yorks, others	29, 66, 62, +	B, Chl, M	rivers	✓
1940	Butcher	1932–34	Yorkshire	61	B, Chl	river	✓
1946	Butcher	1935	Hampshire	11, 12	B, Chl, M	streams	✓
1937	Butcher, Longwells and Pentelow	1929–33	Durham, Yorkshire	66, 62, 65	B, Chl, M	river	✓
1913	Dakin and Latarche	1900–11	Antrim	39	B, Cd, Chl, M, Py	lake	✓
1915	Delf	1912–14	Middlesex	21	B, Chl, Chr, Co	ponds	✓

1958	Evans	1953–55	Hertfordshire	20	B, Cd, Chl, Chr, E, M, Py, X	ponds	√
1958	Evans	1953–55	Middlesex	21	B, Cd, Chl, Chr, E, M, Py, X	pond	√
1958	Evans	1953–55	Surrey	17	B, Cd, Chl, Chr, Cr, E, M, Py, X	pond	×
1936–38	Fenton		Midlothian	83		fen, soil	√
1971	Firth and Hartley	1938, 1970	Yorkshire	65	B	streams	√
1939–40	Fitzjohn	1939	Somerset	6	B, Cd, Chl, Chr, M, X	ponds, reservoir	√
1950	Flint	1936–39	Middlesex	21	B, Cd, Chl, M	reservoir	√
1902, 1903	Fritsch	1902–03	Surrey	17	B, Chl, Chr, M	river	√
1905	Fritsch	1903–04	Bucks, Cambridge, Nottingham	24, 29, 56	B, Chl, Cd, E, M	rivers	√
1929	Fritsch	1918, 1920	Devon	4	Chl, M, Ph, R	streams	√
1942	Fritsch and John		various	+	B, Chl, M, X	soil	√
1946	Fritsch and Pantin	1943	Cambridge	29	M	stream	√
1909	Fritsch and Rich	1908	Somerset	6	B, Cd, Chl, Co, M	pond	√
1948	Galliford and Williams	1945–47	Cheshire	58	B, Cd, Chl, E, M	brackish pools	√
1937	Godward	1933–35	Westmorland	69	B, Chr, Chl, M	lake	√
1912	Griffiths	1908–10	Worcestershire	37	B, Cd, Chl, M, Py	small lake	√
1916	Griffiths	1910	Worcestershire	37	B, Cd, Chl, M	pools	√
1922	Griffiths	1920	Berkshire	22	Chl, M, Py	ponds	some
1924	Griffiths		Yorkshire			pond	2 out of 3
1925a	Griffiths	1922	Cheshire, Shropshire, Staffs.	58, 40, 39	B, Cd, Chl, M, Py	small lakes	√

TABLE I.—*continued*

Date of paper	Author	Date of survey	Counties mainly involved	Vice-counties (where known) (see pp. 425–428)	Main taxa listed	Habitats mainly involved	Whether sites still present
1926	Griffiths	1923	Anglesey, Caernarvonshire	52, 49	Cd, Chl, M	lakes	√
1927	Griffiths	1924	Norfolk	27, 28	Cd, Chl, M	lakes, slow river	√
1936a	Griffiths	1929–31	Durham	66	Chr, Py	pond	√
1936b	Griffiths	—	Durham	66	Cd, Chl, Chr, E, M, X	various	some
1936b	Griffiths	—	Northumberland	67, 68	Cd, Chl, M	various	some
1939	Griffiths	1939		104	Cd, M, Py	lakes	√
1924	Griffiths and Cooke	1923	Durham	66	B, Cd, Chl, Chr, Py	pond	×
1920	Grove, Bristol and Carter	1906–19	Staffordshire, Warwickshire, Worcestershire	39, 38, 37	B, Cd, Chl, Chr, E, M, Py, R, X	various	some
1959	Hammerton	1955–58	Somerset	6	B, Chl, M	lake	√
1957	Hayward	1949–51	Caernarvonshire	49	B	bog	√
1922	Hodgetts	1918–21	Warwickshire	38	B, Chl, Co, E, M, X		
1931	Howland	1925–29	Hertfordshire	20	B, Chl, Co, M, X	pond	
1937–39	Jane	1937	Hertfordshire	20	Chl, E, Fl	pools, ponds	some
1967	Karim	1958–59	Middlesex, Surrey	21, 17	B, Chl, Chr, E, M	gravel pits	√

1938	Lind	—	Yorkshire	63	B, Chl, Chr, Co, E, M, X	ponds	✓
1948	Lind	—	Cheshire	58		small lakes	✓
1952	Lind	1946	Inverness	104, 110	B, Cd, Chl, M, Py	lakes	✓
1942	Lund	1936–38			B, Cd, Chl, Co, Cr, E, M	ponds	✓
1953	Lund	—	Lancashire, Yorkshire	69, 64	Chr	lake, stream	✓
1956	Lund	various	Westmorland	69	Chl	lakes	✓
1961	Lund	1948–59	Yorkshire	64	B, Cd, Chl, Chr, Co, E, M, X	lake	✓
1942	Malins Smith	1923–27	Yorkshire	63	B, Cd, Chl, Co, Cr, E, M	bog	✓
1923	Pearsall	1912–13	Cheshire	58	B, Cd, Chr, M, Py	lake	✓
1932	Pearsall	1928	Cumberland, Westmorland	70, 69	B, Cd, Chl, M, Py	lakes	✓
1942	Pearsall and Lind	—					✓
1884	Phillips	1881	Shropshire	40	M	small lakes	✓
1950	Reynolds and Taylor	1938–39, 46	Leicestershire	55	B, Chl, Chr, Cr, E, M, Py	reservoir	✓
1938	Rice	1928–32				river	✓
1925	Rich	1913–15	Leicestershire	55	B, Cd, Chl, Co, Cr, X	various	some
1970	Ridley	1928 et seq.	Yorkshire	65	B, M, X	reservoirs	✓
1953	Round	1949–50			B, Cd, Chl, M	lake (benthic)	✓
1955	Round	1953	Warwickshire	38	B, Chl	ponds (benthic)	✓
1956a	Round	1949–50	Yorkshire	64	B, Chl, M	lake littoral	✓
1965b	Round	1952–55	Brecon, Radnor	42, 43	B, Cd, Chr	reservoirs	✓

TABLE I.—continued

Date of paper	Author	Date of survey	Counties mainly involved	Vice-counties (where known) (see pp. 425–428)	Main taxa listed	Habitats mainly involved	Whether sites still present
1957	Round	1952–55	Brecon, Radnor	42, 43	B, Cd, E, M	reservoirs	✓
1958	Round	1956	Devon	4	B	moist sand	✓
1960a	Round	1957	Yorkshire	64	B	streams	✓
1960b	Round	1953	Cumberland	70	B, Chl, Cd, Chr, E, M	springs	✓
1961a	Round	1948–50	Westmorland	69	M	lakes	✓
1959	Round and Brook	1953	various W. Irish		B, Chl, Cd, Chr, M, Py	lakes	✓
1930	Schroeder	1926	Yorkshire	25	B, Cd, Chl, Co, M	river	✓ similar
1944	Scourfield	ca. 1943	Essex	18		pools	
1938	Southern and Gardiner	1921–23			B, Cd, Chl, Chr, M	lake, river	✓
1959a,b	Swale	ca. 1958	Hertfordshire	20	Chl, R	river	✓
1962	Swale	1948–55, 61			Chl, R	canal	✗
1964	Swale and Belcher	1958–59	Hertfordshire	20	Chl	river	✓
1954	Taylor	1954	Leicestershire	55	B, Chl, Chr, M	reservoir	✓
1939	Wailes	1937–38	Westmorland	69	Cd, Chl, Chr, M, X	lake	✓
1890	West	various	Anglesey, Caernarvonshire	52, 49	B, Cd, Chl, M	various	some
1909	West	1906–08	Warwickshire	38	Py	small lake	✓

1912	West	Clare Is.	1910–11	H 27	B, Cd, Chl, Chr, Co, M, R	various	some
1932	West and Fritsch		—	+	all	various	some
1900–01	West and West	Yorkshire	—	61, 62, 63, 64, 65	all	various	some
1902	West and West	Antrim	1900–01	H 39	B, Cd, Chl, M	lake	√
1909	West and West	various	—	+	Cd	various	some
1951	Williams	Cheshire	1941–42	58	Cd, Chl, Chr, Cr, E	river	√
1937	Woodhead	Anglesey, Caernarvonshire		52, 49		lakes	√
1938	Woodhead and Tweed	Hampshire	1937	11	B, Cd, Chl, E, M, Py, R	ponds, streams	√
1947	Woodhead and Tweed	Caernarvonshire	—	49	B, Cd	pools, streams	√
1953	Woodhead and Tweed	Cheshire, Shropshire, Yorkshire	1948	58, 40, 63	B, Chl, E, M	pools, streams	√
1954–55	Woodhead and Tweed	Angelsey, Caernarvonshire	—	52, 49	Cd, Chl, Co, Chr, E, M, Py	various	√

Abbreviations for taxonomic groups: B—Bacillariophyta (diatoms); Cd—Desmidiaceae+Mesotaeniaceae (Conjugatophyta, part) (desmids); Cha—Charophyta (charophytes); Chl—Chlorophyta (not Cd, Cha, Co) (greens, in restricted sense); Chr—Chrysophyta (golden-browns); Co—Conjugatophyta other than desmids; Cr—Cryptophyta (cryptomonads); E—Euglenophyta (euglenoids); Fl—flagellates: various not included within other phyla; M—Myxophyta (blue-green algae); Ph—Phaeophyta (browns); Py—Pyrrophyta (dinoflagellates); R—Rhodophyta (reds); X—Xanthophyta (yellow-greens).

in the diatoms appear to have become marked since the seventeenth century, and extremely rapid during the last 50 years.

An attempt to evaluate long-term biological changes has been made also for Loch Leven, Kinross, although the various lists of phytoplankton made since 1905 (Bachmann, 1908) have not yet been subjected to a detailed comparison. Several of the changes which are known to have taken place this century appear rather similar to those which have occurred in Lough Neagh. There has been a loss of four desmid species (Morgan, 1970). As an indication of recent eutrophication, Morgan concluded that it would seem clear that there has been an increase in the production of large blooms of blue-green algae. The observation of Bailey-Watts *et al.* (1968) is particularly interesting. A small-celled blue-green alga, *Synechococcus* sp., was recorded in this lake in 1967 for the first time anywhere in the world, and by spring 1968 formed 98% by volume of the phytoplankton crop. While a new "species" of blue-green alga does not cause such surprise to algologists as it would do to flowering plant specialists, nevertheless it indicates that it must at least be rare elsewhere as a dominant.

In addition to these direct observations on the plankton of Loch Leven, Haworth (1972) has provided a detailed study of long-term changes in the diatom flora from frustules taken from the top metre of lake sediment. She found that at the bottom of the core, the diatom taxa are already indicative of a typical eutrophic lake with alkaline, nutrient-rich water. Subsequently there is an appearance of *Melosira ambigua* and *Stephanodiscus hantzschii*, suggesting a still richer environment. In the top 0·2 m there is a sudden change, which it was suggested might correlate with a lowering of the lake level in 1830.

Observations have been made over a long period on the algal vegetation of the Shropshire meres, which certainly have been changing also (C. S. Reynolds, *in litt.*). Total salts have increased detectably in the 16–18 years since the survey of Gorham (1957), while combined inorganic nitrogen levels have gone up markedly, many meres now having in winter 2–10 times the 1955 levels. Associated with these nutrient changes, Reynolds has found several signs of increased algal productivity: summer crops are on the average larger, and diatom growth now seems to cause greater silica depletion than it used to even five years ago. Nevertheless, well-established examples of qualitative changes in the algae are few. Reynolds reports that the desmid, *Closterium tortum*, does seem more widespread now than indicated in the earlier papers. Griffiths (1925) compared the species of blue-green algae found as dominants in 1922 with those of 1884 recorded by Phillips (1884), and suggested that considerable changes had taken place in the flora. However, the variations in population size of the blue-green algae found in four meres (including two used in Griffiths's

TABLE II. Comparison of former and present algal floras of Lough Neagh (for details see text)

Taxon	Total species			Species of 1968–70 not recorded in			Species recorded in 1900–01 not so in 1968–70
	1900–01	1910–11	1968–70	1900–01	1910–11	1900–01 or 1910–11	
Bacillariophyta	20	17	14	10	9	8	16
Chlorophyta (*senso strictu*)	24	5	23	18	21	18	19
Chrysophyta	0	3	1	1	1	1	0
Conjugatophyta (desmids only)	19	?	4	1	3	1	16
Cryptophyta	0	0	1	1	1	1	0
Myxophyta	14	8	11	9	7	5	12
Pyrrophyta	?	4	1	1	1	1	?
Xanthophyta	0	1	0	0	0	0	0

comparison) by Reynolds (1971) over the period 1966–68 indicate that meaningful comparisons concerning bloom-forming blue-green algae can be made only when all the records involved are taken over some years. For instance, in Ellesmere, *Anabaena circinalis* was a major component of the 1884 sample, was not recorded in 1922 or 1966, but became frequent during parts of the summers of both 1967 and 1968. Nevertheless, the abundance in this same lake of *Gloeotrichia echinulata* in 1884 and its absence in 1922 and 1966–68 may perhaps be more than a temporary fluctuation, as this is not one of the blue-green algae usually associated with especially eutrophic conditions (see Whitton, 1973). Reynolds (*in litt.*) suggests that while the levels of combined inorganic nitrogen in the meres are still probably the most important limiting factor for many algae, the situation might well change soon if levels continue to rise as rapidly. Should this in fact occur, it seems probable that qualitative changes in the flora will then take place much more rapidly.

Both studies on diatom frustules in cores and long-term counts on phytoplankton samples have indicated that obvious changes have occurred relatively recently in some lakes of the English Lake District. In a core from the North Basin of Windermere, Pennington (1943) found that two planktonic diatoms, *Asterionella formosa* and *Synedra radians*, were almost confined to the surface. If it is assumed that the rate of mineral sedimentation has been constant during the last few centuries, *Asterionella* has been abundant in this lake since about 1720. This is well before the period of large increases in sewage associated with tourists, so the causes of the invasion are not clear. Possibly improved methods of farming led to some change in the lake (Macan, 1970).

According to Pennington (1943) and Round (1961b), an increase in the abundance of *Asterionella* is also evident in cores from Esthwaite Water, but it started earlier in this lake than in Windermere, which supports the hypothesis that it is related to cultivation rather than sewage (Macan, 1970). An indication of more recent and marked eutrophication in Esthwaite is given by Round (1961b) who compared the present-day diatom flora with that of earlier periods including the uppermost part of the core with *Asterionella*. He commented that the existing flora has certain peculiarities, for example the epipelic diatom flora includes a very high proportion of cells of: *Gyrosigma acuminatum*, *Caloneis silicula*, *Neidium dubium*, *Diploneis ovalis*, *Stauroneis phoenicenteron*, *Navicula pupula*, *N. cryptocephala*, *N. hungarica* var. *capitata*, *N. radiosa*, *Pinnularia cardinaliculis*, *P. gibba*, *Amphora ovalis*, *Nitzschia dissipita*, *N. palea*, *Surirella robusta* and *S. linearis*. Of these, only *Navicula rhynchocephala* is abundant in the uppermost sample of core. This indicates that these eutrophic species have increased very recently and suggests acceleration of the transition already shown by the diatoms of the upper zone (VIIb). The recent behaviour of blue-green algae in this lake differs from that in Lough Neagh and Loch Leven, since growths of these algae have developed during the period 1946 to 1969, when the levels of both phosphate and combined inorganic nitrogen rose (Lund, 1972), while some eukaryotic algae such as *Ceratium hirundinella* have increased markedly.

In contrast to the above two lakes, the oligotrophic Blea Tarn has shown no marked recent changes in its diatom flora. In a study of a core from this lake, Haworth (1969) found that there has merely been a steady trend to more acid-preferring species in the flora. She related this change to inwash of material from the drainage basin.

There have been a number of new records during the last three decades, the period during which regular plankton samples have been taken, in particular lakes of the Lake District, especially Blelham Tarn. In all these cases the algae already occurred somewhere else in the region, but one older record concerns a species which might possibly have been new to the region. *Uroglena (Uroglenopsis) americana*, which had previously not been described at all for the British Isles, was found to be abundant in Windermere in 1932 (Pearsall, 1933). Pearsall commented that in view of the rapidity with which this alga disorganizes after collection, it is possible that it may have been present in previous years in smaller numbers. However the alga could hardly have escaped detection if it had been present in earlier years in the numbers in which it was present in 1932. Nowadays this species is widespread in the Lake District (J. W. G. Lund, *in litt.*).

Few lakes from other regions in Britain have received sufficient observation over a long period for any sort of comment to be made on algal changes. It does however seem probable that a study of the Norfolk Broads (at present being carried out by B. Moss) will eventually provide especially valuable data on stability and change in algal species composition. A paper on the Broads by Griffiths (1927) gives a relatively detailed account of the algae at eight different sites in 1924, while various types of modern evidence indicate that marked changes in the vegetation have taken place subsequently, such as a decline in various macrophyte species (Morgan, 1972).

In addition to studies of lakes subject to steady change over relatively long periods, there are several accounts of the effects on algae when lakes have been subjected to much more abrupt changes, such as addition of artificial fertilizer (Brook, 1957b) and mixing of the water in a thermally stratified reservoir (Ridley, 1970). A particularly interesting example is provided by Oak Mere, Cheshire, for here the abrupt change has subsequently been reversed. In 1966–67 this usually rather acidic lake received an addition of calcium-rich borehole water, and marked increases occurred in values of pH and calcium (among others); at the same time a striking change took place in the plankton flora and its seasonal succession (Reynolds and Allen, 1968). As the plankton in this lake had been described over a number of years (Lind, 1944; Lind and Galliford, 1952; Swale, 1965, 1968) it is reasonable to assume that most of the floristic changes observed were directly associated with the changed water chemistry. The most obvious such change was the appearance of the blue-green algae *Anabaena flos-aquae* and *Microcystis aeruginosa*. Now that this lake has returned to an acidic condition, the changes observed in 1966–67 have largely been reversed (C. S. Reynolds, *in litt.*).

3. Ponds

The only pond in the British Isles whose algal flora has been studied at all thoroughly at long time intervals is Abbot's Pond, Somerset. Observations of Fritsch and Rich (1908) and of Fitzjohn (1939–40) have been compared with modern ones by Moss and Karim (1969). The algal flora showed marked differences in all three periods. These changes were interpreted by Moss and Karim as being due probably to physical effects: deepening of the pond between the time of the first and second observations and increased shelter from trees between the second and third observations. One alga may possibly have invaded right in the middle of the recent study period. *Chrysococcus diaphanus* formed substantial populations in autumn and winter 1966 and spring 1967, but had not previously been observed even as an occasional cell in concentrated samples. There was some circumstantial evidence that it was introduced into the pond, possibly on the feet of wild fowl, during the summer of 1966. Moss and Karim classified the algae found during their own study into three groups based on abundance, in order to see whether species now abundant are more or less likely to be species which were already present in Fitzjohn's study. Their results are summarized in Table III. The authors pointed out that the progressive decreases in similarity with decreased present-day abundance could be coincidental, but may reflect wide ecological tolerances of the most abundant species, and progressive need for more specific niches in the less abundant species.

4. Rivers

Until recently the algal flora of rivers has received much less interest than that of lakes, this being especially true for attached algae. Studies on the plankton of the River Thames provide by far the best instance of reliable observations carried out at long time intervals (in 1902–03 by Fritsch 1902, 1903; in 1928–32 by Rice, 1938; in 1966–68 by Lack, 1971; in 1969 by Bowles and Quennell, 1971). Lack discussed the population increases and species changes which have taken place in the plankton of this river, and suggested that they were almost certainly due to eutrophication. Fritsch and Rice both found *Asterionella* to be common, yet Lack's study revealed an almost total absence of this diatom. Over the period of 60 years, however, there had been an increase in importance of small green algae such as *Dictyosphaerium*, chrysophytes such as *Chrysococcus*

TABLE III. Comparison of species found in Abbot's pond by Fitzjohn (1939) with those recorded in 1967 summarized according to abundance (data from Moss and Karim, 1969)

Abundance category in 1967	Total species found in 1962–67	Species similar to Fitzjohn (1939)	%
Max populations > 10^4 cells/ml	3	2	67
Max populations 10^3–10^4 cells/ml	13	6	47
Max populations 10^2–10^3 cells/ml	61	20	30

and various cryptomonads. Houghton (1972) also reported long-term increases in the algal plankton of the River Stour, Essex. On the other hand a survey of the attached diatoms of the River Tees during 1963–65 (Whitton and Dalpra, 1968) found very few changes from the situation reported for 1929–33 by Butcher *et al.* (1937). In spite of nutrient increases having taken place in this river, there were only five obvious changes in abundance out of the 148 species reported, none of them involving dominant species. The only important floristic difference noted between the two surveys was the presence of the green alga *Enteromorpha* in each year of the modern survey, and its apparent absence during the earlier one.

GAINS AND INCREASES

1. Aliens

New British records for freshwater algae are made so frequently that it is generally assumed that the species concerned have been overlooked previously, or at least have simply increased in abundance due to man's activities. In fact almost the only records which are generally accepted to represent invasion by foreign species are ones in an entirely new habitat created by man, waters artificially heated by effluents.

Swale (1962) summarized the occurrence of three presumed aliens, *Pithophora oedogonia*, *Chara braunii* and *Compsopogon* sp., in the Reddish Canal, Manchester, over periods exceeding half a century (see Weiss and Murray, 1909). Eighteen months after the warm water effluent ceased to enter the canal, none of these species could be found. While these aliens have become extinct, it seems probable that investigation of other waters artificially warmed would reveal interesting records. Swale mentioned that *Compsopogon* had also been observed in the heated pool in the Water Lily House at the Royal Botanic Gardens, Kew, and there are several foreign records for this alga being associated with heated effluents in regions where it is otherwise absent.

2. *Increases*

The discovery in 1971 for the first time anywhere in Europe of the diatom *Hydrosera triquetra* in the River Thames (Gleave, 1972) provides a further convincing example of invasion. Although this species comes from tropical and sub-tropical waters in Asia and Australia, it is now well established in the upper reaches of the tidal Thames (R. Ross, P. A. Sims and I. Tittley, *in litt.*).

Besides the waters whose algal history has been fairly well documented, there are numerous records of lakes, ponds and rivers which have shown obvious increases in algal growth within living memory. Inspection of sites causing comment from anglers or boaters most usually reveals large growths of *Cladophora glomerata* if filamentous algae are involved, or blue-green algae capable of forming any sort of nuisance. While these two types have probably shown the greatest increases due to eutrophication, they are also particularly conspicuous visually. Large population increases in many other species are less likely to be noticed by the layman unless the algae attract attention for some other reason. The small flagellate *Prymnesium parvum* is only noticed by anyone other than the specialist when its presence leads to fish mortality. This toxic species was described for the first time (anywhere) from the Isle of Wight by Carter in 1937, while the first fish death in this country associated with its presence was in 1961 at North Fleet, Kent (J. W. G. Lund, *in litt.*). There have subsequently been several other records of fish death due to it, and it is now abundant in some of the Norfolk Broads such as Hickling Broad (Lund, 1971). Possibly the absence for thirty years of reports of fish-kills is due simply to water authorities then not being much interested in what killed their fish, but this alga does seem a prime suspect as one showing recent increases in abundance.

With such an inconspicuous organism as *Prymnesium parvum* it would be extremely difficult to decide whether a new dense population is just the result of a very large increase at a particular site, or whether invasion has occurred from some other location. The situation is easier with *Enteromorpha*, which forms shining green tubes near or on the water surface. As mentioned earlier, this alga may well have invaded the River Tees between 1933 and 1963, and there are several other sites where it seems likely that it has invaded within relatively recent years (Whitton, unpublished). Perhaps the most convincing way to demonstrate that a conspicuous species is invading new sites is to suggest ones where it might well be expected to arrive, and then wait a decade to see if in fact it does do so. The River North Tyne (Holmes *et al.*, 1972) and Loch Leven (J. W. G. Lund, *in litt.*) are both sites where *Enteromorpha* is now apparently absent. Judging by its behaviour elsewhere, it might well be expected to develop populations at both of these.

There are several records of algae showing adaptational changes to new, but very specialized environments. The question must also be asked whether any of the algae that have shown tremendous increases are also undergoing rapid evolutionary changes as a response to the relatively new, but much more widespread, environment of highly eutrophic waters. While there is no firm evidence to support such a hypothesis, it is worth noting that some of the forms involved, such as *Cladophora*, *Enteromorpha* and several of the blue-green algae, are very variable and a taxonomic worry for algologists. The most significant changes are likely to be at the physiological level, but it is also important to look out for possible morphological changes. Morphological oddities are usually neglected, so one striking example may be quoted. Heaney (1973), during a study of the dense *Oscillatoria redekei* population in Lough Neagh, isolated remarkable filaments from a non-clonal laboratory population of this organism. These filaments, which were otherwise very similar to *O. redekei*, had undergone true branching to form an irregularly triradiate structure. Similar filaments have subsequently been seen during direct observation by C. E. Gibson on a Lough Neagh water sample.

LOSSES AND DECREASES

A few examples have already been given of algae that have disappeared from large lakes and rivers subjected to nutrient eutrophication. None of these represents an important loss, as the species involved are all relatively common elsewhere. However, the question must be raised whether other species which were initially much rarer are now becoming, or have already become, extinct in this country. In view of their large size, it is not surprising that the only real evidence on this problem is for the charophytes. Of the 33 species recognized by Allen (1950), he indicated that seven were either known at that time from a single site or had probably already become extinct. In particular, *Nitella capillaris*, *N. hyalina* and *Lamprothamnion papulosum* had probably all vanished from the sites where G. R. Bullock-Webster had originally recorded them.

Many other species of algae are known from only one or a very few records. No doubt this situation is often simply due to the relatively few specialists in a particular group, or to the fact that a particular habitat has largely been neglected. Nevertheless there are cases where we can be fairly sure that the alga really is very rare. For instance G. S. West reported the presence of the distinctive tufa-forming stream desmid, *Oocardium stratum*, from only three localities in the British Isles (Carter, 1923), and apparently no new localities have been found since. In spite of this rarity, the alga is still present at one, at least, of these sites (Gordale *fide* C. Sinclair, *in litt.*).

While we know little about loss of particular species on a national, or even regional, basis, we do know that many types of habitat rich in interesting algal forms have been especially vulnerable to man's activities. Small ponds and streams in lowland areas have often been drastically reduced in number, even where the land has not been taken for building. For instance, a survey by the Darlington and Teesdale Naturalists' Field Club (Wanless, 1964), showed that, out of 250 ponds on 2½ inch/mile Ordnance Survey sheets near Darlington, 103 had disappeared altogether, and only 52 were considered to be in "good condition". Oligotrophic aquatic sites are probably now very rare in southern England, and perhaps no longer occur at all for some types of habitat such as chalk streams. We may therefore suspect that many algal losses have occurred in particular regions. Algae which formerly were restricted entirely to lowland oligotrophic habitats might well have become extinct nationally without any algologist realizing the situation.

ACKNOWLEDGEMENTS

I am most grateful to Mr J. R. Carter, Dr J. H. Evans, Dr D. J. Hibberd, Dr J. W. G. Lund, Dr B. Moss, Dr C. S. Reynolds and Miss P. A. Sims for the help they have given me in bringing together the scattered information on this subject.

REFERENCES

ALLEN, G. O. (1950). "British Stoneworts (Charophyta)." Buncle, Arbroath.
ATKINS, W. R. G. and HARRIS, G. T. (1924). Seasonal changes in the water and heleoplankton of freshwater ponds. *Sci. Proc. R. Dublin Soc.* **18**, 1–21.
BACHMANN, H. (1908). Vergleichende Studien über das Phytoplankton von den Seen Schottlands und der Schweiz. *Archs Hydrol.* **3**, 1–91.
BAILEY-WATTS, A. E., BINDLOSS, M. E. and BELCHER, J. H. (1968). Freshwater primary production by a blue-green alga of bacterial size. *Nature, Lond.* **220**, 1344–1345.

BOWLES, B. and QUENNELL, S. (1971). Some quantitative algal studies on the River Thames. *Wat. Treat. Exam.* **20**, 35–51.

BRISTOL, B. M. (1920). On the algal flora of some desiccated English soils. *Ann. Bot.* **34**, 35–80.

BRISTOL, B. M. and ROACH, W. A. (1927). On the algae of some normal English soils. *J. agric. Sci.* **17**, 563–588.

BROOK, A. J. (1954). The bottom-living, algal flora of slow sand filter-beds of waterworks. *Hydrobiologia* **6**, 333–351.

BROOK, A. J. (1955). The attached algal flora of slow sand filter beds of waterworks. *Hydrobiologia* **7**, 103–117.

BROOK, A. J. (1957a). Notes on freshwater algae from lochs in Perthshire and Sutherland. *Trans. Proc. bot. Soc. Edinb.* **37**, 114–122.

BROOK, A. J. (1957b). Changes in the phytoplankton of some Scottish hill lochs resulting from their artificial enrichment. *Verh. int. Verein. theor. angew. Limnol.* **13**, 298–305.

BROOK, A. J. (1958a). Desmids from the plankton of Irish loughs. *Proc. R. Ir. Acad.* **59B**, 71–91.

BROOK, A. J. (1958b). Notes on algae from the plankton of some Scottish freshwater lochs. *Trans. Proc. bot. Soc. Edinb.* **37**, 174–181.

BUTCHER, R. W. (1924). The plankton of the River Wharfe. *Naturalist, Hull* **1924**, 175–180, 211–214.

BUTCHER, R. W. (1932a). Studies in the ecology of rivers II. The microflora of rivers with special reference to the algae of the river bed. *Ann. Bot.* **46**, 813–861.

BUTCHER, R. W. (1932b). Notes on new and little-known algae from the beds of rivers. *New Phytol.* **31**, 289–309.

BUTCHER, R. W. (1940). Studies on the ecology of rivers. IV. Observations on the growth and distribution of sessile algae in the River Hull, Yorkshire. *J. Ecol.* **28**, 210–223.

BUTCHER, R. W. (1946). Studies in the ecology of rivers. VI. The algal growth in certain highly calcareous streams. *J. Ecol.* **33**, 268–283.

BUTCHER, R. W., LONGWELL, J. and PENTELOW, F. T. K. (1937). "Survey of the River Tees. Part III The Non-Tidal Reaches." [Technical Paper, Water Pollution Research no. 6.] H.M.S.O., London.

CASTENHOLZ, R. W. (1973). Ecology of blue-green algae in hot springs. *In* "The Biology of Blue-green Algae" (N. G. Carr and B. A. Whitton, eds), pp. 379–414. Blackwell, Oxford.

CARTER, N. (1923). "A Monograph of the British Desmidiaceae", Vol. 5. Ray Society, London.

CARTER, N. (1937). New or interesting algae from brackish water. *Arch. Protistenk.* **90**, 1–68.

DAKIN, W. J. and LATARCHE, M. (1913). The plankton of Lough Neagh: a study of the seasonal changes in the plankton by quantitative methods. *Proc. R. Ir. Acad.* **30B**, 20–95.

DELF, E. M. (1915). The algal vegetation of some ponds on Hampstead Heath. *New Phytol.* **14**, 63–80.

EVANS, J. H. (1958). The survival of fresh water algae during dry periods. I. An investigation of five small ponds. *J. Ecol.* **46**, 149–168.

FENTON, E. W. (1936). The periodicity and distribution of algae in Boghall Glen (Midlothian). *Scot. Nat.* **1936**, 143–148.

FENTON, E. W. (1938). Algae studies from Boghall Glen (Midlothian). IV. Soil algae. *Scot. Nat.* **1938**, 165–172.

FIRTH, R. I. and HARTLEY, B. (1971). A Pennine diatom site. *Microscopy* **32**, 108–113.

FITZJOHN, A. E. (1939–40). The algae of four Somersetshire pools. *Proc. Bristol nat. Soc.*, ser. 4, **9**, 62–65.

FLINT, E. A. (1950). An investigation of the distribution in time and space of the algae of a British water reservoir. *Hydrobiologia* **2**, 217–240.

FRITSCH, F. E. (1902). Preliminary report on the phytoplankton of the Thames. *Ann. Bot.* **16**, 1–9.

FRITSCH, F. E. (1903). Further observations on the phytoplankton of the River Thames. *Ann. Bot.* **17**, 163–167.

FRITSCH, F. E. (1905). The plankton of some English rivers. *Ann. Bot.* **19**, 163–167.

FRITSCH, F. E. (1929). The encrusting algae of certain fast-flowing streams. *New Phytol.* **28**, 165–196.

FRITSCH, F. E. (1931). Some aspects of the ecology of freshwater algae. *J. Ecol.* **19**, 23–272.

FRITSCH, F. E. and JOHN, R. P. (1942). An ecological and taxonomic study of the algae of British soils. I. The distribution of the surface-growing algae. *Ann. Bot.*, N.S. **6**, 323–349.

FRITSCH, F. E. and PANTIN, C. F. A. (1946). Calcareous concretions in a Cambridgeshire stream. *Nature, Lond.* **157**, 397.

FRITSCH, F. E. and RICH, F. (1909). A five years observation of the fishpond, Abbot's Leigh, near Bristol. *Proc. Bristol nat. Hist. Soc. ser.* 4, **2**, 27–54.

GALLIFORD, A. L. and WILLIAMS, E. G. (1948). Microscopic organisms of some brackish pools at Leasowe, Wirral, Cheshire. *NWest. Nat.* **23**, 39–62.

GIBSON, C. E., WOOD, R. B., DICKSON, E. L. and JEWSON, D. H. (1971). The succession of phytoplankton in L. Neagh 1968–70. *Mitt. int. Verh. theor. angew. Limnol.* **19**, 146–160.

GLEAVE, H. H. (1972). *Hydresera triquetra*, a diatom new to European waters. *Microscopy* **32**, 208.

GODWARD, M. (1937). An ecological and taxonomic investigation of the littoral flora of Lake Windermere. *J. Ecol.* **25**, 496–568.

GORHAM, E. (1957). The chemical composition of some waters from lowland lakes in Shropshire. *Tellus* **9**, 174–179.

GRIFFITHS, B. M. (1912). The algae of Stanklin Pool, Worcestershire. *Proc. Birmingham nat. Hist. Soc.* **12** (5), 1–13.

GRIFFITHS, B. M. (1916). The August heleoplankton of some North Worcestershire pools. *J. Linn. Soc. (Bot.)* **43**, 423–432.

GRIFFITHS, B. M. (1922). The heleoplankton of three Berkshire pools. *J. Linn. Soc. (Bot.)* **46**, 1–11.

GRIFFITHS, B. M. (1924). The free-floating microflora or phytoplankton of Hornsea Mere, E. Yorkshire. *Naturalist, Hull* **1924**, 245–247.

GRIFFITHS, B. M. (1925). Studies in the phytoplankton of the lowland waters of Great Britain. III: The phytoplankton of Shropshire, Cheshire and Staffordshire. *J. Linn. Soc. (Bot.)* **47**, 75–98.

GRIFFITHS, B. M. (1926). Studies in the phytoplankton of the lowland waters of Great Britain. No. IV. The phytoplankton of the Isle of Anglesey and of Llyn Ogwen, North Wales. *J. Linn. Soc. (Bot.)* **47**, 355–366.

GRIFFITHS, B. M. (1927). Studies in the phytoplankton of the lowland waters of Great Britain. No. V. The phytoplankton of some Norfolk Broads. *J. Linn. Soc. (Bot.)* **47**, 595–612.

GRIFFITHS, B. M. (1936a). The limnology of the Long Pool, Butterby Marsh, Durham: an account of the temperature, oxygen-content, and composition of the water, and of the periodicity and distribution of the phyto- and zooplankton. *J. Linn. Soc. (Bot.)* **50**, 393–416.

GRIFFITHS, B. M. (1936b). A preliminary list of the freshwater algae of Northumberland and Durham. *Vasculum* **22**, 89–95.

GRIFFITHS, B. M. (1939). The free-floating microscopic plant life of the lakes of the Isle of Raasay, Inner Hebrides. *Proc. Univ. Durham Phil. Soc.* **10**, 71–87.

GRIFFITHS, B. M. and COOKE, R. B. (1924). Ryton Willows Pool. *Trans. nat. Hist. Soc. Northumb.*, N.S. **6**, 38–47.

GROVE, A. J., BRISTOL, B. M. and CARTER, N. (1920). The flagellates and algae of the district around Birmingham. *J. Bot., Lond.* **58**, *Suppl.* 3, 1–55.

GROVES, J. and BULLOCK-WEBSTER, G. R. (1920). "The British Charophyta. I. Nitelleae." Ray Society, London.

GROVES, J. and BULLOCK-WEBSTER, G. R. (1924). "The British Charophyta. II. Chareae." Ray Society, London.

HAMMERTON, D. (1959). A biological and chemical study of Chew Valley Lake. *Proc. Soc. Wat. Treat. Exam.* **8**, 87–132.

HASSALL, A. H. (1845). "A History of the British Freshwater Algae", Vol. I. Taylor, Walton and Maberly, London.

HAWORTH, E. Y. (1969). The diatoms of a sediment core from Blea Tarn, Langdale. *J. Ecol.* **57**, 429–439.

HAWORTH, E. Y. (1972). The recent diatom history of Loch Leven, Kinross. *Freshwat. Biol.* **2**, 131–141.

HAYWARD, J. (1957). The periodicity of diatoms in bogs. *J. Ecol.* **45**, 947–954.

HEANEY, S. I. (1973). "Problems Related to the Growth of Blue-green Algal Populations". Ph.D. thesis, New University of Ulster, Coleraine.

HODGETTS, W. J. (1922). A study of some of the factors controlling the periodicity of freshwater algae in nature. *New Phytol.* **20**, 150–164.

HOLMES, N. T. H., LLOYD, E. J. H., POTTS, M. and WHITTON, B. A. (1972). Plants of the River Tyne and future water transfer scheme. *Vasculum* **57**, 56–78.

HOUGHTON, G. U. (1972). Long-term increases in plankton growth in the Essex Stour. *Wat. Treat. Exam.* **21**, 299–308.

HOWLAND, L. J. (1931). A four years investigation of a Hertfordshire pond. *New Phytol.* **30**, 81–125.

HUDSON, J. W., CROMPTON, K. J. and WHITTON, B. A. (1971). Ecology of Hell Kettles 2. The ponds. *Vasculum* **56**(3), 38–45.

JANE, F. W. (1937–39). Some Hertfordshire flagellates. I–III. *Trans. Herts. nat. Hist. Soc.* **20**, 133–140, 340–451; **21**, 115–122.

KARIM, A. G. A. (1967). Algal flora of certain gravel pits in the Thames Valley, U.K. *Hydrobiologia* **30**, 577–599.

LACK, T. J. (1971). Quantitative studies on the phytoplankton of the River Thames and Kennet at Reading. *Freshwat. Biol.* **1**, 213–224.

LIND, E. M. (1938). Periodicity of the algae in Beauchief Ponds, Sheffield. *J. Ecol.* **26**, 257–274.

LIND, E. M. (1948). The phytoplankton of some Cheshire meres. *Mem. Proc. Manchester lit. phil. Soc.* **86**, 83–105.

LIND, E. M. (1952). The phytoplankton of some lochs in South Uist and Rhum. *Trans. Proc. bot. Soc. Edinb.* **36**, 37–47.

LIND, E. M. and GALLIFORD, A. L. (1952). Notes on the plankton of Oak Mere, Cheshire. *Naturalist, Hull* **1952**, 99–102.

LUND, J. W. G. (1942). The marginal algae of certain ponds, with special reference to the bottom deposits. *J. Ecol.* **30**, 245–283.

LUND, J. W. G. (1953). New or rare Chrysophyceae. II. *Hyalobryum polymorphum* n. sp., and *Chrysonebula holmesii* n. gen., n. sp. *New Phytol.* **52**, 114–123.

LUND, J. W. G. (1956). On certain planktonic palmelloid green algae. *J. Linn. Soc. (Bot.)* **55**, 593–613.

LUND, J. W. G. (1961). The algae of the Malham Tarn district. *Fld Studies* **1**, 85–119.

LUND, J. W. G. (1971). Eutrophication. *In* "The Scientific Management of Animal and Plant Communities", pp. 225–240. 11th Symposium of British Ecological Society.

LUND, J. W. G. (1972). Changes in the biomass of blue-green and other algae in an English lake. *In* "First International Symposium on Taxonomy and Biology of Blue-green Algae, Jan. 1970" (T. V. Desikachary, ed.). University of Madras, India.

MACAN, T. T. (1970). "Biological Studies of the English Lakes". Longmans, London.

MALINS SMITH, A. (1942). The algae of Miles Rough Bog, Bradford. *J. Ecol.* **30**, 341–356.

MORGAN, N. C. (1970). Changes in the fauna and flora of a nutrient enriched lake. *Hydrobiologia* **35**, 545–553.

MORGAN, N. C. (1972). Problems of conservation of freshwater ecosystems. *In* "Conservation and Productivity of Natural Waters" (R. W. Edwards and D. J. Garrod, eds), *Symp. zool. Soc. Lond.* **29**, 135–154.

MOSS, B. and KARIM, A. G. A. (1969). Phytoplankton associations in two pools and their relationships with associated benthic flora. *Hydrobiologia* **33**, 587–600.

PEARSALL, W. H. (1923). The phytoplankton of Rostherne Mere. *Mem. Proc. Manchester lit. phil. Soc.* **67**, 47–55.

PEARSALL, W. H. (1932). Phytoplankton of the English Lakes. 2. The composition of the phytoplankton in relation to the dissolved substances. *J. Ecol.* **20**, 241–262.

PEARSALL, W. H. (1933). *Uroglenopsis americana* in Windermere. *Naturalist, Hull* **1933**, 122–123.

PEARSALL, W. H. and LIND, E. M. (1942). The distribution of phytoplankton in some north-west Irish loughs. *Proc. R. Irish Acad.* **48B**, 1–24.

PENNINGTON, W. (1943). Lake sediments: the bottom deposits of the North Basin of Windermere, with special reference to the diatom succession. *New Phytol.* **42**, 1–27.

PHILLIPS, W. (1884). The breaking of the Shropshire Meres. *Trans. Shropshire Arch. nat. Hist. Soc.* **10**, 1–24.

REYNOLDS, C. S. (1971). The ecology of the planktonic blue-green algae in the North Shropshire meres. *Fld Studies* **3**, 409–432.

REYNOLDS, C. S. and ALLEN, S. E. (1968). Changes in the phytoplankton of Oak Mere, following the introduction of base rich water. *Br. phycol. Bull.* **3**, 451–462.

REYNOLDS, N. and TAYLOR, F. J. (1950). Notes on the algae of Swithland Reservoir, Leicestershire. *Naturalist, Hull* **1950**, 49–55.

RICE, C. H. (1938). Studies in the phytoplankton of the River Thames (1928–1932) 1 and 2. *Ann. Bot.,* N.S. **2**, 539–557, 559–581.

RICH, F. (1925). The algae of Leicestershire. *J. Bot., Lond.* **63**, 229–238, 262–273, 322–330.

RIDLEY, J. E. (1970). The biology and management of eutrophic reservoirs. *Wat. Treat. Exam.* **19**, 374–399.

ROUND, F. E. (1953). An investigation of two benthic algal communities in Malham Tarn, Yorkshire. *J. Ecol.* **41**, 174–197.

ROUND, F. E. (1955). Some observations in the benthic algal flora of four small ponds. *Arch. Hydrobiol.* **50**, 111–135.

ROUND, F. E. (1956a). A note on some communities of the littoral zones of lakes. *Arch. Hydrobiol.* **52**, 398–405.

ROUND, F. E. (1956b). The phytoplankton of three central water supply reservoirs in central Wales. *Arch. Hydrobiol.* **52**, 457–469.

ROUND, F. E. (1957). The benthic algal flora of three City of Birmingham waterworks reservoirs in Central Wales. *Arch. Hydrobiol.* **53**, 562–573.

ROUND, F. E. (1958). Observations on the diatom flora of Braunton Burrows, N. Devon. *Hydrobiologia* **11**, 119–127.

ROUND, F. E. (1960a). The diatom flora of streams around Malham Tarn, Yorkshire. *Arch. Protistenk.* **104**, 527–540.

ROUND, F. E. (1960b). The algal flora of a salt spring region at Manesty, Grange in Borrowdale. *Naturalist, Hull* **1960**, 117–121.

ROUND, F. E. (1961a). Studies on bottom-living algae in some lakes of the English Lake District. V. The seasonal cycles of the Cyanophyceae. *J. Ecol.* **49**, 31–38.

ROUND, F. E. (1961b). The diatoms of a core from Esthwaite Water. *New Phytol.* **60**, 43–59.

ROUND, F. E. (1971). The growth and succession of algal populations in freshwaters. *Mitt. Verh. int. Verein. theor. angew. Limnol.* **19**, 70–99.

ROUND, F. E. and BROOK, A. J. (1959). The phytoplankton of some Irish loughs and an assessment of their trophic status. *Proc. R. Ir. Acad.* **60B**, 167–191.

SCOURFIELD, D. J. (1944). The nannoplankton of bomb-crater pools in Epping Forest. *Essex Nat.* **27**, 231–241.

SCHROEDER, W. L. (1930). Biological survey of the River Wharfe. III. Algae present in the Wharfe plankton. *J. Ecol.* **18**, 302–305.

SOUTHERN, R. and GARDINER, A. C. (1938). The phytoplankton of the River Shannon and Lough Derg. *Proc. Ir. Acad.* **45B**, 85–124.

STOCKNER, J. G. and LUND, J. W. G. (1970). Live algae in post-glacial lake deposits. *Limnol. Oceanogr.* **15**, 41–58.

SWALE, E. M. F. (1959a). Some uncommon Chlorophyceae from the River Lee. *Br. phycol. Bull.* **1**, 71–72.

SWALE, E. M. F. (1959b). The algal flora of the River Lee. 1. Introduction and Rhodo-phyceae. *Essex Nat.* **30**, 173–174.

SWALE, E. M. F. (1962). Notes on some algae from the Reddish Canal. *Br. phycol. Bull.* **2**, 174–175.

SWALE, E. M. F. (1964). A study of the phytoplankton of a calcareous river. *J. Ecol.* **52**, 433–446.

SWALE, E. M. F. and BELCHER, J. H. (1964). The Algal flora of the River Lee. 3. Vol-vocales and Chlorococcales. *Essex Nat.* **31**, 193–199.

TALLING, J. F. (1951). The element of chance in pond populations. *Naturalist, Hull* **1951**, 157–170.

TAYLOR, F. J. (1954). Notes on the phytoplankton of Saddington Reservoir, Leicester-shire. *Naturalist, Hull* **1954**, 141–148.

WAILES, G. H. (1939). The plankton of Lake Windermere, England. *Ann. Mag. nat. Hist.*, ser. 2 **3**, 401–414.

WANLESS, T. W. (1964). "Ponds." *Northumberland Durham Naturalists' Trust Newsletter*, p. 10.

WEISS, F. E. and MURRAY, H. (1909). On the occurrence and distribution of some alien aquatic plants in the Reddish Canal. *Mem. proc. Manchester lit. phil. Soc.* **53**, 1–8.

WEST, G. S. (1909). The Peridineae of Sutton Park, Warwickshire. *New Phytol.* **8**, 181–196.

WEST, W. (1890). A contribution to the freshwater algae of North Wales. *J. R. microsc. Soc.* **10**, 277–306.

WEST, W. (1912). Clare Island Survey. Pt. 16. Freshwater algae. *Proc. R. Ir. Acad.* **31**, 1–62.

WEST, G. S. and FRITSCH, F. E. (1932). "A Treatise on the British Freshwater Algae". Revised edition. Cambridge University Press, London.

WEST, W. and WEST, G. S. (1900–01). The algal-flora of Yorkshire. *Trans. Yorks. Nat. Union (Bot.)* **5**, 1–239.

WEST, W. and WEST, G. S. (1902). A contribution to the freshwater algae of the North of Ireland. *Trans. R. Ir. Acad.* **23B**, 1–101.

WEST, W. and WEST, G. S. (1904–12). "A Monograph of the British Desmidiaceae", 4 vols. Ray Society, London.

WEST, W. and WEST, G. S. (1909). The British freshwater phytoplankton, with special reference to the desmid-plankton and the distribution of British desmids. *Proc. R. Soc. Lond.* **81**, 165–206.

WHEELER, B. D. and WHITTON, B. A. (1971). Ecology of Hell Kettles. 1. Terrestrial and sub-aquatic vegetation. *Vasculum* **56**(3), 25–37.

WHITTON, B. A. (1973). Freshwater plankton. *In* "The Biology of Blue-green Algae". (N. G. Carr and B. A. Whitton, eds), 353–367. Blackwell, Oxford.

WHITTON, B. A. and DALPRA, M. (1968). Floristic changes in the River Tees. *Hydro-biologia* **32**, 545–550.

WILLIAMS, E. G. (1951). Plankton organisms of the River Dee, near Chester. *Proc. Chester Soc. nat. Sci. Lit. Art* **4**, 143–152.

WOOD, R. B. and GIBSON, C. E. (1973). Eutrophication and Lough Neagh. *Wat. Res.* **7**, 173–187.

WOODHEAD, N. (1937). Studies in the flora of Anglesey and Caernarvonshire lakes. *NWest. Nat.* **12**, 160–171.

WOODHEAD, N. and TWEED, R. D. (1938). Some freshwater algae in Southern Hampshire. *Proc. Bournemouth nat. Sci. Soc.* **37**, 50–67.

WOODHEAD, N. and TWEED, R. D. (1947). Some algal floras of high altitude in Snowdonia. *NWest. Nat.* **22**, 34–42.

WOODHEAD, N. and TWEED, R. D. (1953). Notes on freshwater algae in Yorkshire (v.c. 63), Cheshire and Salop. *NWest. Nat.* **24**, 266–271.

WOODHEAD, N. and TWEED, R. D. (1954–55). The freshwater algae of Anglesey and Caernarvonshire. *NWest. Nat.* **25**, 85–122, 255–296, 392–435, 564–601; **26**, 76–101, 210–228.

9 | Freshwater Invertebrates

T. T. MACAN

Freshwater Biological Association, Ambleside, Westmorland

Abstract: Rivers have been altered for drainage, navigation and power and by waste disposal. Lakes have been enriched or made into reservoirs. These changes have altered the fauna, but existing knowledge is well documented. Accordingly attention is directed here to changes due to invaders from abroad. One snail and one crustacean have spread rapidly, though little is known about their method of transport or of the vacant niche that they appear to have found. Ten other animals have spread less rapidly. Three are associated with water warmed either incidentally by industry or deliberately for the cultivation of tropical plants. There seems to be a connection between water with a higher salt content than is usual and four species. What factor favours the remainder is not known. One species is known from one locality only and two have arrived but failed to establish themselves permanently. Several are thought to have travelled from America with plants for aquarists, but there is no firm evidence.

INTRODUCTION

Mollusca have been a popular, well studied group for a long time, and reliable records of some species at least are available from the early years of the nineteenth century. Among the insects the Coleoptera, and to a lesser extent the Odonata, have enjoyed a similar popularity. Apart from these, the freshwater fauna was not well known, even at the beginning of the present century, and reliable records from the last century are few. For example, to the Heteroptera, by no means a neglected group, Dr G. A. Walton, a student at the time, added 5 species, some 8% of the total, to the British list between 1936 and 1939. The adults of other insect orders lack the variety of form and colour that attract the collector, and furthermore are often on the wing for a few weeks only. Complete keys to their immature stages, if prepared at all, are generally only two or three decades old and consequently records for these groups do not go back far.

Systematics Association Special Volume No. 6, "The Changing Flora and Fauna of Britain", edited by D. L. Hawksworth, 1974, pp. 143–155. Academic Press, London and New York.

Groups outside the class Insecta, apart from the molluscs already mentioned, and to a lesser extent some of the Crustacea, have not been studied thoroughly. Between the wars British material was identified by means of French or German keys, and doubt often remained about which species occurred in Britain and which did not. It was, for example, impossible to make reliable statements about the flatworms until Professor T. B. Reynoldson started serious ecological work on them after the war. He found it necessary to embark on a critical review of the taxonomy, and the eventual upshot (Reynoldson, 1967) was a key in which for the first time a reliable check-list of the British species and means of distinguishing them became available in one publication.

Man has been dredging and embanking rivers both to promote navigation and to get rid of excess water for centuries, but his operations have probably become more devastating biologically as mechanical apparatus has been brought into use. Damming to provide power probably reached a maximum in the Middle Ages, but modern dams are much larger. Rivers have doubtless been used to dispose of waste for a very long time but sufficient waste to affect fauna and flora severely is a modern development. Waste matter and also fertilizer from farmland enriches both rivers and lakes. Natural bodies of standing water have also been converted into reservoirs, and the resulting increase in the rise and fall of the water level has altered fauna and flora. On the credit side is the creation of new lakes by means of dams to provide power and water. Freshwater organisms that inhabit small bodies of water have probably also benefited from human excavation for a variety of purposes.

The effects of all this activity on the biota, though far from fully explored, have been well documented, notably by Hynes (1960, 1970). The first book is devoted exclusively to the subject; the second is concerned with all aspects of running water. Macan (1970) describes the effect of enrichment on the Lake District lakes. Recent papers that have come to the writer's notice are Collins (1971), Learner et al. (1971) and Hawkes and Davies (1971). There would be slight value in repeating all the information here. The effects of all this interference are reversible. It is not possible to judge whether any species have been lost or gained as a result of it.

The changes to be discussed here are due to immigrants but it is also worth noting two within what are believed to have been established populations. Among the Coleoptera, for example, Balfour-Browne (1940, p. 93) records that *Ilybius subaeneus* Erichson was known only from south-east England in the 1860s and 70s. In 1937 it was common in Nottinghamshire, in 1930 in Durham, both counties where previously workers had been active and had not recorded it. In 1933 Balfour-Browne himself took it near Forfar in a small loch from which he had been collecting since 1908. There is no explanation of this spread, nor of the present-day restriction to Scotland of *Acilius canaliculatus* Nicolai which was known in south-east England in the last century. The Coleoptera are one of the very few groups which have been collected long enough intensively enough for this kind of change to be revealed.

The Odonata provide an example of a possible loss. Corbet *et al.* (1960) record that a colony of *Coenagrion scitulum* (Rambur) in a low-lying corner of Essex was watched from 1946 till 1953, in which year flooding by the exceptionally high tide overwhelmed it. They believe that this species may now be extinct in the British Isles.

Thienemann (1950) devotes a whole section of his long book on the European freshwater fauna to what he calls "Neuste Einwanderer" (most recent immigrants). In it he refers to eighteen species, of which four have in some unknown way succeeded in crossing the barriers between zoogeographical regions, one has been deliberately introduced, and the rest have apparently extended their range inland from the Ponto-Caspian or Mediterranean centres, or from coastal brackish waters. There is dispute about some of the species in the last two categories, and some workers argue from the finding of subfossil shells that they have always been present and have merely become more abundant. Even an increase in number is hard to prove, for during the last century observers have become much more numerous too.

Not all the species mentioned by Thienemann have reached the British Isles, and a few that have established themselves there have not extended their range to the continent. Sixteen species have probably or certainly arrived in these islands only recently and they may be grouped in a variety of ways of which three are: (1) position in the animal kingdom; (2) country of origin; (3) degree of success.

The first is the most popular but perhaps the least useful and, for present purposes, the third has been chosen. The sixteen immigrants can be graded as: (1) outstandingly successful—3 species; (2) moderately successful—10 species; (3) one foothold only—1 species; (4) failures—2 species. The ten moderately successful species may be subdivided into three whose occurrence is related to temperature, four related to salinity and three of which nothing definite can be said at present.

OUTSTANDINGLY SUCCESSFUL SPECIES

The three outstandingly successful immigrants are *Potamopyrgus jenkinsi* (Smith), *Crangonyx pseudogracilis* Bousfield and *Aphanomyces astaci* Schikora, which causes crayfish disease.

A recent, well documented account of *P. jenkinsi* is given by Fretter and Graham (1962), and the very latest reference is probably that by Calow (1973). The policy here is to quote only the later papers, in which full reference to earlier work may be found. Fretter and Graham conclude that its origin is still unsettled, one school maintaining, partly on subfossil evidence, that it is a mutant of a brackish-water species, the other that it was transported somehow from the antipodes. It was described in 1889 from brackish marshes bordering the Thames estuary and first taken in 1893 in freshwater, after which it colonized most of England and Wales in about thirty years, though it is still unrecorded from much of highland Scotland. T. Warwick (*in litt.*) informs me that it now occurs in the Shetland Islands, from where it was not recorded in the Conchological Society's 1951 census, and that in 1965 he found it abundantly in many places on Barra in the Outer Hebrides, whereas in 1936 he took it there in one or two places only. There are only speculations about the reasons for its success.

C. pseudogracilis was first recorded in England by Crawford (1937), whereupon Tattersall published that a specimen from a London suburb had been sent to him some years previously, but that he had refrained from mentioning it because the emergence of an American amphipod from an English tap sounded so improbable. As it is widespread in North America and unknown in continental Europe (Straškraba, 1967), its status as an immigrant has not been questioned. The most popular suggestion is that it crossed the Atlantic in a consignment of one of the water plants that are in demand to decorate and aerate aquaria. In 1955 Hynes published a map showing almost continuous distribution in that part of central England which can be reached by way of the canals and inland navigable waterways. As several writers have pointed out, new records may reflect awareness on the part of collectors that a new species, this one easily mistaken for a small *Gammarus* by the unalerted, has been discovered, rather than reflecting a recent arrival. However, Hynes was able to follow its advance into new territory in the Llangollen area, and, when it was first taken in Windermere, it had colonized only half that lake. A few years later it was abundant everywhere in it and was found in some others as well (Macan, 1970). In 1959 Warwick recorded it at Grangemouth in ponds used for seasoning timber, and he suggested that this first record for Scotland was a fresh invasion, the animal having survived on timber shipped from North America. Again there is no reliable information about how this immigrant traverses land barriers or how it competes successfully with the native fauna.

In 1876 a disease that almost annihilated the crayfish population appeared in France and spread rapidly. Later it moved eastward and gradually travelled round the Baltic Sea to Sweden (Vivier, 1965). The lack of resistance by the crayfish suggests that it was an organism that had not attacked them before. It was shown (Schäperclaus, 1954) to be the fungus *Aphanomyces astaci*, an organism that has been studied more recently by Unestam and Weiss (1970). Its arrival in and spread through England is less well documented, but Duffield (1933, 1936) records alternating abundance and extreme scarcity of *Astacus* which suggests an epidemic disease. What effect this sudden disappearance of a large and abundant animal has on the rest of the community is not known. Crayfish are of commercial importance on the continent and therefore fluctuations in numbers are noticed immediately. This would not be true of most freshwater organisms, and one may wonder whether others are affected by epidemics.

MODERATELY SUCCESSFUL SPECIES

Physa acuta Drap. According to Frömming (1956) this species was confined to south and west Europe until early in this century when it started to extend its range eastwards across northern Europe. This extension of range was facilitated by the transport of cultivated aquatic plants, particularly those which, coming from the tropics, must be kept warm in winter. Frömming asserts that this snail is now to be found in every botanic garden between Amsterdam and Moscow, and that it spreads outwards from these centres but is annihilated by severe winters. Dr M. P. Kerney, recorder of the Conchological Society, has been kind enough to supply the latest information about this genus in a letter. He writes that there may be three species, two of them American, in Britain in addition to the native *P. fontinalis*, but that the taxonomy is in a confused state. Langford (1971, 1972) records *Physa acuta* in the Rivers Trent, Ouse and Witham and in the cooling-tower ponds in four power stations in the midlands.

Branchiura sowerbyi Beddard, a large unmistakable oligochaete, is a south-east-Asian species that was described from specimens from the *Victoria regia* tanks in Regent's Park. Later it was recorded in similar conditions in Dublin, Kew and Oxford. Mann (1958, 1965) found it abundantly in the River Thames below Earley power station, and records that it is the only animal that occurs below but not above the warm effluent. However, it is apparently able to survive in unheated conditions in Britain, for P. M. Jenkin informed Mann in a letter that she had observed a colony for five years in the Kennet and Avon Canal and had recently found another in the River Avon at Bradford. Aston (1968) recorded it in a warmed effluent at Coventry and in two more unheated localities, the River Otter in Devon and the Long Water at Hampton Court.

Planorbis dilatatus (Gould) is an American species that was first found in

F

Britain in canals in the Manchester area. Dance (1970) records it from L. Trawsfynydd, which is warmed by a power station, and he gives an account of its early history. Up to 1919 it was abundant in canals around Manchester, generally in water warmed by cotton mills. It was not found in this region between that year and 1969, though how much time was spent looking for it is not clear. In 1919 it was recorded also in the River Tame at Dukinfield and there is a doubtful later record from the Hebrides. Fryer (1954) found it in the canal near Huddersfield. Two other species mentioned in this article are recorded from this canal (Fryer, 1950, 1952): *Crangonyx pseudogracilis* and *Orchestia cavimana*. It is the only place from which three newcomers have been recorded and the only peculiarity which might account for this appears to be the proximity of the home of an eminent naturalist. This point emphasizes how widely scattered good collectors are and how incomplete the records for some species may be.

All three species in this group, *Physa acuta*, *Branchiura sowerbyi*, and *Planorbis dilatatus*, appear to be favoured by a temperature above that normal for Britain, and have benefited from industrial use of water as a cooling medium as well as from the cultivation of plants that require warm water. They may well have arrived originally with these plants and have almost certainly been distributed locally with them. Aston (1968) found that immature *Branchiura* grow fastest between 25° and 30°C, a range rarely reached in natural waters in Britain, but this appears to be the only work on thermal requirements that has been done. Each of the species has been recorded in water that is not warmed, which has led to the suggestion that selection is producing a strain with a lower optimum. Another possibility is that in the places in question some adverse ecological factor was bearing less heavily than usual on the population, which was able in consequence to survive in the unfavourably low temperature. One may remark that few measurements of temperature are recorded in the literature and the possibility of some natural warming cannot be left out of account.

Corophium curvispinum G.O. Sars var. *devium* Wundsch is thought to have spread from the Caspian and Black Sea regions, probably along canals, since about 1900. It was described as a new species in Germany in 1919 but the name was later demoted to the rank of variety. It was first recorded at Tewkesbury by Crawford (1935). Moon (1970) recorded it on the other side of England in the Grand Union Canal and mentioned a record at Stourport-on-Severn in 1962. Professor Moon has kindly supplied further information in a letter. He has found the animal, often in considerable numbers, in long stretches of the Grand Union Canal in Northamptonshire, Leicestershire and Buckingham-shire; commonly in many parts of the Coventry Canal; in some parts of the

Oxford Canal and in the Worcester Canal in the Penkridge–Stafford area. T. E. Langford reports findings from some of these localities. It occurs also in Cheshire (D. G. Holland, in preparation).

Dreissena polymorpha (Pallas) is also believed by some to have originated in the Caspian and Black Sea regions but there are others who maintain that it has been present in northern Europe for a long time. It seems to be well established that, whatever its history before then, it increased greatly in abundance in northern and western Europe early in the last century. It was recorded in the London docks in 1824, in the Humber in 1831–33, and in the Forth–Clyde canal in 1834. Kerney and Morton (1970) give a map of its present-day distribution which coincides closely with the canal region, with outposts at Exeter and in the Forth–Clyde area.

Cordylophora lacustris Allman is recorded by Thienemann (1950) as a marine and brackish-water species that has been invading fresh water in a few places. I have found no recent references to its spread.

Gammarus tigrinus Sexton, a native of the eastern seaboard of America, first recorded in Britain in 1939, appears to thrive not only in dilute sea water but in saline water when the proportions of the main ions are not as in the sea. Its occurrence was outlined by Hynes (1955) and I have found no further reference to it apart from two records by Langford, but detailed information about its occurrence in parts of Cheshire will be available when a study by D. G. Holland is published. In England it is recorded in several localities in the drainage area of the Bristol Avon and in Cheshire, both inland and in estuarine marshes. All these places are distinctly saline except one, and that contains water with an unusual salt content. In Ireland, in contrast, *G. tigrinus* occurs in fresh water in Lough Neagh and the River Bann. *Gammarus pulex* was unknown in Ireland until introduced recently at a few places, and Hynes suggests that the distribution of *G. tigrinus* can be explained on the assumption that it is more tolerant of a high concentration of salt than *G. pulex* but cannot compete with it in fresh water.

In 1957 Schmitz (1960) introduced it into saline stretches in the River Werra in Germany successfully, as was evident from collections made two years later. At about the same time, some dozens of specimens were released into the Ijsselmeer but there is doubt whether this small number could have produced the enormous population found a few years later. *G. tigrinus* appears to have replaced *G. duebeni* in much of the lake except in the most saline regions. *G. pulex* is confined to the least saline regions and *G. tigrinus* does not penetrate fresh water here, but it does elsewhere in places to which it appears to have spread from the Ijsselmeer (Nijssen and Stock, 1966). Advance was rapid in

1965 and 1967 but slower in 1967 (Dennert *et al.*, 1968). Nijssen and Stock mention that it occurs in shallow water over a sandy bottom, the kind of conditions in which they have not seen the other Dutch species. Hynes (1955) also mentions that the places where it has been found in England "are much more pondlike than most *Gammarus* habitats." *Gammarus tigrinus* is unlike the other invaders from brackish water in that it has been able to occupy an apparently empty niche in inland waters with a high salinity. The other three have spread from docks and estuaries into rivers and canals. This has been taking place during a period of industrial development and a steady increase in the amount of waste matter discharged into waterways. Sewage raises the salinity of water and may provide a source of food for filter feeders. There is, therefore, a possibility that increasing impurity of our waters has opened them up to these originally brackish-water species.

Finally, among the moderately successful, come an amphipod and two planarians not obviously associated with any single ecological factor. *Orchestia cavimana* Heller: the status of this amphipod as a freshwater animal is doubtful since it lives in rubbish at the water's edge, much as the similar but more familiar sand-hopper inhabits piles of seaweed on the sea-shore. However, it can live in water and sometimes takes refuge there when disturbed. Its habitat is probably one which is not often searched, and therefore it is difficult to disentangle new records that represent an advance by the species from those due to increased activity and awareness of naturalists. To add to this uncertainty, *O. cavimana*, if it is a genuine newcomer, has travelled only from the Mediterranean region. Curry *et al.* (1972) give a history of its occurrence in Britain. It was first found in 1942 beside the Thames at Richmond, and in 1946 and 1947 it was taken further upstream at Oxford. At about the same time it was recorded from the River Yare in Norfolk, and in 1950 Fryer found it in the canal at Huddersfield. Curry *et al.* present fresh records from an artificial lake, a river and a canal in Cheshire. It is therefore to be expected in canals further south; this is confirmed (*in litt.*) by Moon, who has specimens from the Grand Union Canal at Slapton, Buckinghamshire, and by Langford (*in litt.*) who suspects that it may be much more widespread in the Trent area than his single finding at Ratcliffe-on-Soar, Nottinghamshire indicates. Holland (in preparation) has found it in a number of places in Cheshire, which together with his new records of *Crangonyx pseudogracilis* and *Gammarus tigrinus* indicates yet again that known distribution may reflect distribution of collectors.

Dugesia tigrina (Girard) is an American planarian first recorded in Europe at the beginning of the century (Dahm, 1955). Reynoldson (1956) showed that earlier records of *Dugesia gonocephala* (now *subtentaculata*) were due to wrong determinations, and that the species was probably *D. tigrina*. He recorded it for certain in two lakes at Hellifield, in western Yorkshire, in the River Thames at Pangbourne and in the River Wye at Symonds Yat. In his key (1967) he mentioned the Greater London area and the Norfolk Broads. Pickavance (1971) studied a dense population in Cole Mere. Dahm (1955) recorded a report of it in Malham Tarn. R. J. Spittle, of the Devon River Authority, informs me that there is a flourishing colony in the River Axe.

Pickavance (1971) has investigated the diet and concludes that *D. tigrina* is able to compete with indigenous species because it is quicker to take advantage of what is available.

Planaria torva (O. F. Müller) is widespread in Finland and its distribution in Britain, mainly in the vicinity of ports and canals, suggests that it may be a recent arrival that probably travelled on timber. This information was kindly conveyed by Professor T. B. Reynoldson in a letter, from which it is also evident that investigation of this species is in progress at the moment.

ONE FOOTHOLD ONLY

Asellus communis Say is placed by itself in a group characterized by occurrence in one place only. It is an American species (Williams, 1972) found in Bolam Lake, Northumberland, by Sutcliffe (1972) in 1962.

FAILURES

Finally there are two species which may be classed as failures. *Lymnaea catascopium*, an American snail, was found in 1929 in a warm engine-pond in a timber yard in Leith, and in 1943 Kevan gave an account of the wide variation in shell form in this immigrant population. It was evidently well established in this pond but it did not spread beyond it, and recently M. P. Kerney informed me that the pond has been filled in and the species is almost certainly extinct.

Eriocheir sinensis Milne-Edwards, the Mitten Crab, probably made the journey from its homeland, China, to the River Weser in the ballast tanks of a steamship. Its spread round the Baltic and down the Channel and Atlantic coasts was one of the most spectacular invasions that fresh water has seen. It did cross the channel, for a single specimen was taken on the screen of the intake to a Thamesside power station, but apparently failed to establish a foothold. It has spread to the southern rivers of France but Hoestlandt (1959), who gave a map showing where it had reached by various dates, believed that it had attained regions where the temperature was beginning to be too high for it.

GENERAL OBSERVATIONS

It should perhaps be mentioned that Fryer (1965, 1968, 1969ab) has added to the British list four species of copepod parasites on fish. There are strong indications that one is a recent immigrant and all may be, but all occur in Europe and they have been too little studied in the past for anything definite to be known about their distribution.

It is noteworthy that there are no insects in this list or in that of Thienemann. Perhaps their powers of flight have enabled them to occupy every niche more thoroughly than animals that cannot transport themselves, and newcomers have faced insurmountable competition. It is worth a passing comment that insects tend to predominate in unproductive waters, the other groups in productive waters, and the trend during the last century and longer has been towards enrichment of inland waters.

It is disappointing that it has not been possible to do much more than present a catalogue of facts. How new arrivals have travelled will probably always be a matter of speculation. Aquarists' plants, tropical water plants and timber have all been mentioned as vehicles more than once; that they are is highly probable but hard to prove incontrovertibly. Canals have certainly facilitated spread, though *Potamopyrgus jenkinsi* and *Crangonyx pseudogracilis*, having spread rapidly far beyond the regions served by them, show that they are not essential. The enrichment of inland waters by our ever-increasing volume of waste may possibly have benefited some immigrants.

A successful species must have good powers of dispersal and must find a niche that is not fully occupied or which it can occupy more successfully than the present incumbent. The students of the planarians have made some study of this, but nothing is known about members of other groups. The newcomers have been found, the other members of the community have sometimes been recorded too, but why an immigrant has been able to establish itself in a community that has been evolving and adapting to the conditions for a long time cannot be explained. Indeed, we cannot be confident that the range of some of the species is particularly well known; it will have been evident from what has been written that a handful of enthusiasts have made a number of new records in the area that they can cover. Collecting and recording have not been regarded as serious science for some decades now, but it is to be hoped this will change in the future in view of the growing interest in the environment.

ACKNOWLEDGEMENTS

That the compilation of this account has been greatly facilitated by correspondents who have given freely of their time and information will have been evident in the text, and my sincere thanks go to Mr D. G. Holland, Dr M. P. Kerney, Mr T. E. Langford, Professor H. P. Moon, Professor T. B. Reynoldson, Mr R. J. Spittle and Dr T. Warwick. I also acknowledge gratefully assistance from three colleagues, Dr G. Fryer, F.R.S., Dr. D. W. Sutcliffe and Dr G. Willoughby.

REFERENCES

ASTON, R. J. (1968). The effect of temperature on the life cycle, growth and fecundity of *Branchiura sowerbyi* (Oligochaeta: Tubificidae). *J. Zool. Lond.* **154**, 29–40.

BALFOUR-BROWNE, F. (1940). "British Water Beetles". Ray Society, London.

COLLINS, J. M. (1971). The Ephemeroptera of the River Bela, Westmorland, *Freshwat. Biol.* **1**, 405–409.

CALOW, P. (1973). *Potamopyrgus jenkinsi* Smith, at Malham, with particular reference to its invasion ecology. *Naturalist, Hull,* **1973**, 29–32.

CORBET, P. S., LONGFIELD, C. and MOORE, N. W. (1960). "Dragonflies". Collins (*New Naturalist*, **41**), London.

CRAWFORD, G. I. (1935). *Corophium curvispinum* G.O. Sars, var *devium*, Wundsch in England. *Nature, Lond.* **136**, 685.

CRAWFORD, G. I. (1937). An amphipod, *Eucrangonyx gracilis* S.I. Smith, new to Britain. *Nature, Lond.* **139**, 327.

CURRY, A., GRAYSON, R. F. and MILLIGAN, T. D. (1972). New British records of the semi-terrestrial amphipod *Orchestia cavimana. Freshwat. Biol.* **2**, 55–57.

DAHM, A. G. (1955). *Dugesia tigrina* (Girard) an American immigrant into European waters, *Verh. int. Ver. Limnol.* **12**, 554–561.

DANCE, S. P. (1970). Trumpet Ram's-Horn Snail in North Wales. *Nature in Wales* **12**, 10–14.

DENNERT, H. G., DENNERT, A. L. and STOCK, J. H. (1968). Range extension in 1967 of the alien amphipod *Gammarus tigrinus* Sexton, 1939, in the Netherlands. *Bull. zool. Mus. Univ. Amsterdam* **1**, 79–81.

DUFFIELD, J. E. (1933). Fluctuations in numbers among freshwater crayfish, *Potamobius pallipes* Lereboullet. *J. Anim. Ecol.* **2**, 184–195.

DUFFIELD, J. E. (1936). Fluctuations in numbers of crayfish, *J. Anim. Ecol.* **5**, 396.

FRETTER, V. and GRAHAM, A. (1962). "British Prosobranch Molluscs". Ray Society, London.

FRÖMMING, E. (1956). "Biologie der mitteleuropäischen Süsswasserschnecken". Duncker and Humblot, Berlin.

FRYER, G. (1950). A Yorkshire record of the amphipod *Orchestia bottae* (M. Edw.). *Naturalist, Hull* **1950**, 148.

FRYER, G. (1952). The amphipod *Eucrangonyx gracilis* (S. I. Smith) and its occurrence in Yorkshire. *Naturalist, Hull* **1952**, 65–66.

FRYER, G. (1954). The Trumpet Ramshorn Snail *Menetus* (*Micromenetus*) *dilatatus* (Gould) east of the Pennines, *Naturalist, Hull* **1954**, 86.

FRYER, G. (1965). The parasitic copepod *Tracheliastes polycolpus* Nordmann in some Yorkshire rivers: the first British records. *Naturalist, Hull* **1965**, 51–56.

FRYER, G. (1968). The parasitic copepod *Lernaea cyprinacea* L. in Britain. *J. nat. Hist.* **2**, 531–533.

FRYER, G. (1969a). The parasitic copepod *Ergasilus sieboldi* Nordmann new to Britain. *Naturalist, Hull* **1969**, 49–51.

FRYER, G. (1969b). The parasitic copepods *Achtheres percarum* Nordmann and *Salmincola gordoni* Gurney in Yorkshire. *Naturalist, Hull* **1969**, 77–81.

HAWKES, H. A. and DAVIES, L. J. (1971). Some effects of organic enrichment on benthic invertebrate communities in stream riffles. *Symp. Br. ecol. Soc.* **11**, 271–293.

HOESTLANDT, H. (1959). Répartition actuelle du crabe chinois (*Eriocheir sinensis* H. Milne Edwards) en France. *Bull. Franc. Piscic.* **194**, 5–13.

HYNES, H. B. N. (1955). Distribution of some freshwater Amphipoda in Britain. *Verh. int. Ver. Limnol.* **12**, 620–628.

HYNES, H. B. N. (1960). "The Biology of Polluted Waters". Liverpool University Press, Liverpool.

HYNES, H. B. N. (1970). "The Ecology of Running Waters". Liverpool University Press, Liverpool.

KERNEY, M. P. and MORTON, B. S. (1970). The distribution of *Dreissena polymorpha* (Pallas) in Britain. *J. Conch.* **27**, 97–100.

KEVAN, D. K. McE. (1943). Study of an introduced North American freshwater mollusc, *Stagnicola catascopium* (Say). *Proc. R. Soc. Edinb.*, **B**, **41**, 430–461.

LANGFORD, T. E. (1971). "The biological assessment of thermal effects in some British rivers". *CERL Symp. Freshwater Biology and Electric Power generation.* Leatherhead [Ref. RD/L/M 312].

LANGFORD, T. E. (1972). A comparative assessment of thermal effects in some British and N. American rivers. *In* "River Ecology and Man" (Oglesby, R., Carlson, C. and McCann, J. eds) pp. 319–351. Academic Press, New York.

LEARNER, M. A., WILLIAMS, R., HARCUP, M. and HUGHER, B. D. (1971). A survey of the macro-fauna of the River Cynon, a polluted tributary of the River Taff (South Wales). *Freshwat. Biol.* **1**, 339–367.

MACAN, T. T. (1970). "Biological studies of the English lakes". Longmans, London.

MANN, K. H. (1958). Occurrence of an exotic oligochaete *Branchiura sowerbyi* Beddard, 1892, in the River Thames. *Nature, Lond.* **182**, 732.

MANN, K. H. (1965). Heated effluents and their effects on the invertebrate fauna of rivers. *Proc. Soc. Wat. Ttmt.* **14**, 45–53.

MOON, H. P. (1970). *Corophium curvispinum* (Amphipoda) recorded again in the British Isles, *Nature, Lond.* **226**, 976.

NIJSSEN, H. and STOCK, J. H. (1966). The amphipod, *Gammarus tigrinus* Sexton, 1939, introduced in the Netherlands (Crustacea). *Beaufortia* **13**, 197–206.

PICKAVANCE, J. R. (1971). The diet of the immigrant planarian *Dugesia tigrina* (Girard) 1. Feeding in the laboratory 2. Food in the wild and comparison with some British species. *J. Anim. Ecol.* **40**, 623–650.

REYNOLDSON, T. B. (1956). The occurrence in Britain of the American triclad *Dugesia tigrina* (Girard) and the status of *D. gonocephala* (Dugès). *Ann. Mag. nat. Hist.*, ser. 12, **9**, 102–105.

REYNOLDSON, T. B. (1967). A Key to the British species of freshwater triclads. *Sci. Publ. Freshwat. biol. Ass.* **23**, 1–28.

SCHÄPERCLAUS, W. (1954). "Fischkrankheiten". Akademie Verlag, Berlin.

SCHMITZ, W. (1960). Die Einbürgerung von *Gammarus tigrinus* Sexton auf dem europäischen Kontinent. *Archs. Hydrobiol.* **57**, 223–225.

STRAŠKRABA, M. (1967). Amphipoda. *In* "Limnofauna Europaea" (J. Illies, ed.), pp. 202–209. Fischer, Stuttgart.

SUTCLIFFE, D. W. (1972). Notes on the chemistry and fauna of water-bodies in Northumberland with special emphasis on the distribution of *Gammarus pulex* (L.), *G. lacustris* Sars and *Asellus communis* Say (new to Britain). *Trans. nat. Hist. Soc. Northumb.* **17**, 222–248.

THIENEMANN, A. (1950). Verbreitungsgeschichte der Süsswassertierwelt Europas. *Die Binnengewässer* **18**, i–xvi, 1–809.

UNESTAM, T. and WEISS, D. W. (1970). The host–parasite relationship between freshwater crayfish and the crayfish disease fungus *Aphanomyces astaci*: response to infection by a susceptible and a resistant species. *J. gen. Microbiol.* **60**, 77–90.

VIVIER, P. (1965). La "peste", un facteur de régulation des populations d'écrivisses (*Astacus*). *Mitt. int. Ver. Limnol.* **13**, 49–62.

WARWICK, T. (1959). *Crangonyx pseudogracilis* Bousfield, an introduced freshwater amphipod, new to Scotland. *Glasgow Nat.* **18**, 109–110.

WILLIAMS, W. D. (1972). Occurrence in Britain of *Asellus communis* Say, 1818, a North American freshwater isopod. *Crustaceana, Suppl.* **3**, 134–138.

10 | Changes in the Freshwater Fish Fauna of Britain

A. WHEELER

Department of Zoology, British Museum (Natural History), London

Abstract: Freshwater fishes illustrate as much as any other group of animals the effects of man's interference with the biota and environment of the British Isles. At least twenty species have been introduced within historical time to England and elsewhere (not all successfully). Redistribution of native species has been a continuous feature for over a century and threatens to increase under the guise of "fishery management". The Irish fauna has been greatly enriched by the introduction of native English and exotic species.

In addition eutrophication and pollution of rivers and inland waters have produced considerable changes in the fauna, in extreme cases leading to local extinction of fish life. The need to control and manage water resources has produced other changes in the fish fauna. Abstraction and drainage practices often result in greatly altered aquatic environments, while the construction of water storage reservoirs and modification of natural lakes have both added to, and altered aquatic habitats.

INTRODUCTION

The freshwater fishes of the British Isles show as much as any animal group the effects of man's influence on the biota and the aquatic environment. Pollution of inland waters and the drainage of wetlands are two obvious factors, but the interference with the environment has been so wide-ranging that many more subtle factors have to be considered. In addition, human activity has been responsible for considerable redistribution of native species within the British Isles, as well as the introduction of exotic species, both forms of biological enrichment which have masked much of the original distribution of British fishes. So widespread has been the man-made alterations to the ichthyofauna that it is difficult to detect any change which might be due to natural causes,

Systematics Association Special Volume No. 6, "The Changing Flora and Fauna of Britain", edited by D. L. Hawksworth, 1974, pp. 157–178. Academic Press, London and New York.

such as climatic change. This paper discusses some of the changes in the distribution and abundance of fishes in the British Isles as a whole caused by varied alterations to the aquatic environment, and briefly considers the biological implications of future management of water resources in Great Britain.

The nomenclature follows Wheeler (1969).

FACTORS CONTROLLING THE DISTRIBUTION OF FRESHWATER FISHES

Discussion of the changes in the status of freshwater fishes is not meaningful without some statement of their original distribution. For biogeographical studies, freshwater fishes can be divided into two categories based on their tolerance to salinity. Some species, such as the salmon (*Salmo salar*), allis shad (*Alosa alosa*), twaite shad (*A. fallax*), eel (*Anguilla anguilla*), and stickleback (*Gasterosteus aculeatus*) are euryhaline, for they can tolerate a high degree of salinity. The first four are migratory species which spend a significant part of their lives at sea, the anadromous shads and salmon spawning in fresh water, while the catadromous eel returns to the sea to spawn. The stickleback on the other hand is not migratory, and is well known as a freshwater fish in southern England but can be found in estuaries and salt marsh pools, while in Scottish waters it is a common intertidal fish living in full-salinity marine habitats as well as in fresh water. Clearly in considering the distribution of these euryhaline species the important factor is accessibility of freshwater habitats from the sea, and it is not surprising that euryhaline species are the most widely distributed in the British Isles. A special case must be made for the relict populations of charr (*Salvelinus alpinus*), and whitefishes (*Coregonus* spp.) which are found in the English Lake District, and many Scottish, Welsh and Irish lakes. Given climatic conditions of some severity, these are also anadromous fishes. Migratory charr today range as far south as Oslo Fjord, Norway (*ca*. 59° N) and are regularly found off southern Iceland. Whitefishes are migratory in the low-salinity areas of the Baltic Sea, and in the Arctic Ocean rivers of the Northern Hemisphere. Both groups are believed to owe their presence in the lakes of the British Isles to immigration during the immediately post-glacial period when temperatures were still depressed, and to have become subsequently isolated in the various lakes or their catchments.

Most of the remaining freshwater fishes are stenohaline, intolerant of any degree of salinity. The original distribution of the native primary freshwater fishes is of considerable interest. Some, such as the roach (*Rutilus rutilus*), bream (*Abramis brama*), perch (*Perca fluviatilis*) and pike (*Esox lucius*) are widely distributed today no doubt as a result of man's influence coupled with their considerable adaptability. Others, such as the silver bream (*Blicca bjoerkna*), barbel (*Barbus barbus*), bleak (*Alburnus alburnus*), ruffe (*Gymnocephalus cernua*), and spined loach (*Cobitis taenia*), are, or were until recently, confined to the eastern rivers of England. In general terms, all of the primary freshwater fishes are found in the rivers of eastern England from Yorkshire to the Thames, and the number of species declines to the west or the north of this area. The extreme of this is seen in Ireland, as in most biogeographical discussions, where no primary freshwater fish is native.

The reason advanced for this concentration of the fish fauna in eastern England and its sparsity elsewhere is that colonization took place during a period of post-glacial depression of sea level which resulted in the existence of land between the British Isles and continental Europe (Charlesworth, 1957). The European rivers, notably the Rhine, would during this period have been joined to the eastward flowing rivers of England and by this means a substantial part of the freshwater fish fauna would have been transferred. The eastern Scottish rivers would not have been so connected for they discharge into the deeper northern North Sea basin, the floor of which would not have been exposed during the period of depressed sea level.

To summarize, the native freshwater fish fauna of the British Isles is thus composed of three main elements:

(1) obligate freshwater fishes which entered during a connection with the rivers of western Europe;

(2) euryhaline fishes which entered rivers from the sea during the period since the last glaciation;

(3) euryhaline fishes which are relicts of migratory fishes landlocked in inland waters since the early post-glacial period.

DIRECT ALTERATIONS TO THE FISH FAUNA

This relatively comprehensible original distribution of Britain's fishes has been grossly obscured by direct human agency. Exotic species have been introduced, native species have been redistributed within Great Britain and introduced to Ireland, and the natural river systems joined with one another by canals, thus facilitating the spread of several species.

1. Introduction of Exotic Species

Wheeler and Maitland (1973) recently discussed the history of the introduction of exotic fishes into the British Isles. They cited literary evidence for the introduction of nineteen species within the last century. Two additions should be made to their list: the goldfish (*Carassius auratus*), and the carp (*Cyprinus carpio*) which was introduced in early historical times. Of these species several have failed to establish themselves and are not considered further here.

Introduced species which have become established, even in only one locality, include the rainbow trout (*Salmo gairdneri*), brook charr (*Salvelinus fontinalis*), goldfish, carp, bitterling (*Rhodeus sericeus*), wels (*Silurus glanis*), zander (*Stizostedion lucioperca*), and the centrarchids *Ambloplites rupestris*, *Lepomis gibbosus*, and *Micropterus salmoides*. In addition, a population of the tropical North and West African cichlid *Tilapia zillii* has survived for ten years in the Church Street Canal, St Helens, Lancashire, where the temperature is elevated by the discharge of warmed cooling water from a large glass works. Two local populations of the tropical South American guppy (*Poecilia reticulata*) have been recorded, one in the St Helens canal, the other in a comparable habitat in the River Lee, Hackney, east London, where the river water is locally warmed by the discharge of cooling water from an electricity generating station. The continued survival of these species in both habitats is clearly contingent upon the continued warming of the water. The River Lee population of guppies may already be extinct as the local generating station has ceased working and decrease of pollution in the river at this point will allow native fishes to extend their range with resulting predation on and competition with this exotic species.

Both these exotic species are presumed to owe their introduction to the disposal of surplus fish kept in aquaria. The occurrence of the bitterling in Lancashire is believed to be due to similar release, for this species is kept in cold water aquaria. The detailed history of its establishment is given by Wheeler and Maitland (1973), and it is evident that owing to dispersal by anglers who used them for live bait for predatory fishes, and to naturalists who aided its spread by releasing them in fresh water, the bitterling became locally common in certain contained waters in Lancashire and Cheshire. Unused live bait fish were even released by anglers in the Lake District in Esthwaite Water, Rydal Water and Grasmere. Populations of this species have existed in Lancashire since the 1920s (Ellison and Chubb, 1963). It has more recently been found in the Praes branch of the Shropshire Union Canal, a population which probably results from a fresh liberation of aquarium specimens.

The bitterling is a widely distributed European fish being found from central France eastwards; it is not found in Scandinavia nor south of the Alps. Another introduced European species which is, however, found in the Baltic basin is the wels. This has been imported into Britain on a number of occasions in the past century, most successfully in 1880 when 70 fish were put into the Duke of Bedford's lakes at Woburn. A further introduction (although it is not known from whence the specimens were obtained) was made to the Tring reservoirs in Hertfordshire. Populations became established at both sites, but because of their relative inaccessibility the wels did not spread greatly outside the original sites. Since 1947 the Woburn lakes have supplied a number of specimens to angling clubs in the Bedfordshire and Buckinghamshire area, and they have been stocked in a number of isolated waters. Despite this the wels does not seem to have expanded its range dramatically, possibly because it has not so far been introduced to a river system of suitable size.

Another central European species which has been introduced is the zander or pike-perch. The early history of its release in England paralleled that of the wels: it was first established in 1878 at Woburn, and restocked in 1910. In the Woburn lakes it was well established but did not spread until 1947 and 1950 when a number of fish were netted and later introduced into lakes in the Bedfordshire area, again without any major subsequent spreading. In 1963, however, some 97 specimens obtained from Woburn were introduced into the Relief Channel of the Great Ouse. Within five years many zander were being caught by anglers along the Channel, and it became clear that the original stock had bred very successfully. The Relief Channel communicates through its head and tail sluices with the River Ouse and indirectly to the Rivers Little Ouse and Wissey. In January 1969 the first report of zander in the River Wissey was made, and during that year they were taken by anglers in the River Cam, the Old Bedford River, and in the River Nene. Following its introduction to this river system the zander had clearly made a dramatic expansion in its range, and there now seems little doubt that in time it will spread throughout the lowland rivers of England at least.

The zander is a special case because the controversy which followed its introduction to the Relief Channel brought it to the attention of the angling public who in general regard large (it grows to 5 kg in weight) predatory fishes very favourably. Its present availability over a wide area of East Anglia means that it is now accessible for unauthorized (and illegal in most areas) redistribution. The prediction that it would quickly be carried by anglers to other river systems made by Wheeler and Maitland (1973: 61), has been justified, for the angling press (*Anglers' Mail*, 14 February 1973), reports a 2 kg zander caught below Tewkesbury Weir in the River Severn, and an earlier report of two small specimens seen to be released in the River Lee in Hertfordshire has been noted.

Several North American species have been introduced at various times, most of them with the object of improving angling sport. The three centrarchids listed earlier have established local populations in one or more isolated waters, but have not extended their range greatly. This is in marked contrast to the situation in continental Europe where the large-mouth black bass (*Micropterus salmoides*) and the pumpkinseed (*Lepomis gibbosus*) are locally well established (Vooren, 1972). Most of the reports of very successful naturalization are in southern Europe or central Europe so that their relative lack of success in the British Isles may be due to climatic factors. More successful North American species are the salmonids, the brook charr and the rainbow trout. The former has had a long history of introduction and appears to have established reproducing populations in about eleven lakes mostly in mountainous regions of Great Britain. It requires very specific habitats, appears to be often unable to establish itself in the presence of brown trout in the same water, and frequently hybridizes with that species producing sterile offspring (Vooren, 1972). Such hybrids have been found occurring in Wise Een Tarn in the English Lake District (Frost, 1968). Possibly the gene loss as a result of hybridization is one factor in its failure to become well established in the same waters as trout occur.

The rainbow trout, which is native to the Pacific coastal drainage of North America from Alaska to Mexico, has also been introduced in many waters. In general the number of successful introductions where self-maintaining populations have become established is few. Worthington (1941) in a survey of the status of this species in Britain reported only 14 waters in the south of England and Ireland where spawning took place. Maitland (1972) has published a map of the recorded distribution of this species which shows that it is found in well over a hundred waters, but in the great majority of these its presence is a result of stocking from hatchery bred fish. Natural reproduction continues to be rare.

The carp is a long established exotic species in the British Isles. It is a native of the eastern Mediterranean and Black Sea basins which has been introduced to most of temperate Europe and many other parts of the world. It is said to have been first mentioned as an English fish in Dame Juliana Berner's "Boke of St Albans", published in 1496 (Regan, 1911), and its introduction may well have taken place some time in the fifteenth century. In Ireland it was introduced between 1634 and 1640 (Went, 1950). Although it is a moderately common fish in England and parts of Wales and Ireland it is confined by its preference for slow-running or still waters to the lowlands. In many lakes even in southern England water temperatures are only marginally suitable for successful spawning, and the year class structure of some populations is such that it suggests that adequate survival of fry only occurs every few years. Many of the recorded carp populations undoubtedly owe their existence to stocking by angling interests.

The general conclusion from this discussion of introduced exotic fishes seems to be that of the many species which have been liberated in Britain, few have been successfully established. Of those that have established themselves several in isolated, often private, waters have not spread, but others when released into waterways or rivers have spread, in one case dramatically.

2. Redistribution of Native Species

Comparison of the maps of recorded distribution produced by Maitland (1972) with the original distribution presumed for these species (p. 159) shows that many have extended their range considerably. Such species as perch, pike, roach, and bream are widespread throughout England, all are found in Ireland (although the roach is relatively sparsely distributed), and all except the bream are found in Scotland, where the pike and perch are well distributed. All four species are highly thought of by anglers and there seems little doubt that they have been deliberately transported in attempts to diversify the fishing, or from some other angling-connected motive. Several examples can be cited to show how swiftly some fishes can establish themselves and expand when introduced to a river system. The zander in the Great Ouse has already been mentioned. Roach and dace were introduced to Ireland in recent times. The roach is found in the Foyle River system and the Erne due to the release of fish stocked in a pond in Co. Tyrone. They were first noticed in 1957 when they were locally common, now they are abundant in both river systems. Roach are also found in the Co. Cork Blackwater, where an angler released them with dace in 1889; the dace especially in this river is now abundant.

A well documented history of introduction can be constructed for the barbel, which until the 1890s was confined to the rivers of eastern England flowing into the North Sea. In 1896 barbel were transferred to the Hampshire Stour, from whence they colonized the Avon. This species was introduced in 1950 to the River Severn, and from thence to the Warwickshire Avon, and in 1955 and 1962 to the Bristol Avon. In all these rivers barbel soon expanded their range and became abundant, and recently some were reintroduced to eastern rivers.

Even small fishes which are not directly of interest to anglers appear to have been widely distributed by man. Only introduction can account for the presence of stone loach (*Noemacheilus barbatulus*), gudgeon (*Gobio gobio*), and minnow (*Phoxinus phoxinus*) in Ireland. Introduction again can only account for the presence of these species and the bullhead (*Cottus gobio*) in isolated catchments in the north and west of Great Britain. The evidence for this is circumstantial, for stockings and transportations of such little fishes have mostly gone unrecorded, but the past climatic and geological history of the British Isles leaves no other solution to account for their present distribution. One or two recorded cases can, however, be cited as supporting evidence. Orkin (1957) recorded that minnows had been released in northern Scottish rivers to "provide an additional source of food for brown trout", while the Cornish naturalist Jonathan Couch (1789–1870) recorded in his manuscript notes that he had introduced the stone loach to the small stream that runs through the village of Polperro in Cornwall.

A recent development in the practice of redistribution of native species has been the importing of stocks of fishes from continental Europe. Several cases have been reported within the last year involving the transfer of roach from Belgium to the Thames system, and from Denmark to the Rivers Welland and Nene. Such imports have to be licensed by the Ministry of Agriculture, Fisheries and Food and are subject to certification as being free of disease. However, it is relevant to question the wisdom of such imports and the validity of the controls. Inspection can only be made on a sample of the fishes concerned, many parasitic infections are internal with no obvious external symptoms, and as in both cited cases the fish were not quarantined but simply released into the river, the risk of introducing new parasitic infection if not diseases would seem to be high. In addition, unless careful inspection of the fishes is made there is a very real possibility of introducing exotic species with the stock of native species. J. Leeming of the Kent River Authority (*in litt.*, 23 February 1973) reported taking an ide (*Leuciscus idus*) from a consignment of roach imported into his area; other cases of stocks purchased on the continent contaminated by other species have come to my notice.

These cases are but examples of the many redistributions of native fishes that have occurred. Many others have gone unrecorded, some have taken place unofficially, but the transportation of native angling species has been continuous for many many years in lowland areas and is especially significant in waters close to urban areas.

3. The Canal System and its Effects on the Fish Fauna

Amongst the acts of man which have affected the present day distribution of our primary freshwater fishes the building of canals had a profound effect. Canal construction on a grand scale did not begin until after 1761 when Francis, 3rd Duke of Bridgewater successfully completed the Bridgewater Canal from Worsley to Manchester, and from Worsley to Hollin ferry on the River Mersey (Hadfield, 1959). Canals such as the Duke's "Cut" or the earlier canal the Sankey Brook (now the St Helens canal), can have made little impact on the distribution of aquatic organisms for they were simply an extension of a river system, but the later ambitious schemes were to link all the major

English rivers into a web of connecting waterways. Thus, the Trent and Mersey canal was completed around 1766; by 1790 the Thames navigation was connected to it by the Oxford Canal, and a year earlier the Thames was connected to the lower Severn by the Thames and Severn Canal. The Thames was in 1810 connected to the Bristol Avon by the Kennet and Avon Canal, and via the Wiltshire and Berkshire Canal, while to the south the Thames system was extended to Chichester and Littlehampton on the Sussex coast through the Wey and Arun Canal (1816). Finally, the Grand Junction Canal started in 1810 in its course joined the River Lea, the Thames, the River Nene, through the Welford Branch to the River Trent, or via Warwick connecting with the Avon, and through a series of small canals in the Birmingham area the River Severn, and the Shropshire Union to the River Dee and the River Mersey.

At its peak the nation-wide inland waterway system extended to some 4000 miles of navigable canals or rivers. Its impact on the aquatic fauna of England was great for it provided a potential means of distribution throughout the country. Navigable locks are no barrier to fish movement, not even to anadromous fish (Nichols and Louder, 1970), indeed the very act of transference of water through a lock is certain to aid the dispersal of even relatively immobile organisms. Not surprisingly, the fishes most common in canals are those lowland river species, such as the roach, bream, perch, pike, ruffe, and to a lesser extent carp, and bleak. These species would have had the greatest opportunity to colonize canals, because with few exceptions the canals connected to the rivers in their lowland reaches. Moreover, these fishes are clearly well adapted to life in still or slow-flowing, eutrophic waterways, and spawn in quiet backwaters on submerged vegetation such as many canals provided even when in regular use. The present wide distribution of the bleak and ruffe (neither species likely to be transported for their angling potential and both of which were originally inhabitants of eastern rivers only), owes much to the construction of canals.

The effects of canals on the distribution of our fishes is seen to its fullest extent in England. While canals were constructed in Wales, Scotland, and Ireland, it was on a less grandiose scale, and their construction had less far-reaching effects on fish distribution for there was a greater degree of uniformity in the fauna of the river systems brought into connection, However, new waterways represented an opportunity for angling interests to stock with fishes, and in addition to natural invasion of canals by fishes their dispersal was no doubt aided by man. One such case concerns the Brecon and Usk Canal in South Wales where roach and dace were introduced soon after its construction, and from the canal later became widespread in the River Usk itself (R. Millichamp, personal communication).

The importance of canals in hastening the distribution of fishes is emphasized by the role they have played in the rapid spread in England of exotic aquatic invertebrates, such as *Corophium curvispinum* (Moon, 1970), and of *Crangonyx pseudogracilis* (Hynes, 1955) both of which are now widespread (see also Chapter 9).

4. The Effects of Introductions and Redistribution

Considering the widespread nature of the redistribution of native and the introduction of exotic fishes in the British Isles, considerable alteration to aquatic ecosystems must have taken place. It seems, however, that little research effort has been directed towards assessing the extent of these changes. Vooren (1972) has recently discussed the consequences of the introduction of exotic species into Europe and the Netherlands especially, and Lachner *et al.* (1970) for North American introductions where the problem has received serious study. Vooren's conclusions that several successful introductions, particularly those of piscivorous species, have produced unexpected and detrimental side-effects deserve further discussion in relation to observations in the British Isles.

The pike is now well distributed in Ireland following its introduction to that country. Its spread has everywhere been at the expense of the stocks of salmon and trout, which are native to that country. In an attempt to improve the salmonid stocks, pike are now being netted, and their spawning grounds treated with piscicidal products (Went, 1957; Kennedy, 1969). Shafi and Maitland (1971) showed that in Loch Lomond pike preyed very heavily on *Salmo* sp. and the endemic whitefish in the loch. In Lake Windermere pike also occur, owing to introduction at some time in the past. The programme of control of pike here has resulted in an increase in the number of charr and trout in the lake (Frost and Kipling, 1967). It can be assumed that pike introduced in other waters in the western and northern parts of the British Isles have had similar effects on the stocks of naturally occurring fishes. This is, of course, a direct relationship between predator and prey but there seems to be little information available on the consequences of the reduction of the prey-species populations. It can, however, be inferred from Macan's (1965) experimental introduction of trout into a moorland fishpond from which fish had been removed some years before, that the presence of trout has a considerable effect on the population size and distribution of the invertebrates. In this tarn, tadpoles, *Notonecta*, and dytiscid beetles all disappeared from open water and were only found in the shallow water at the margins after the reintroduction of trout.

Cyprinid fishes which in general feed on invertebrates and plants have a less direct effect on other fishes than do piscivorous predators. The barbel, which has been released in the primarily salmonid rivers of the west of England may compete in a minor way for food, but is more likely to interfere with the spawning redds of the salmonids. No research has been published to show the extent of competition between this cyprinid and the salmonids, and the barbel is now too well established for it to be controlled in the Severn and other rivers.

In Ireland the dace in the Cork Blackwater has become so abundant that it competes with the native trout for food and space (Healy, 1956). Roach in the same river likewise offer some competition. The minnows introduced to Scottish rivers as a potential food for trout (Orkin, 1957) will have acted as competitors with the young trout as much as food for the adults. Minnows and stone loach in the River Endrick, Scotland, have been shown to have similar food requirements to salmon and trout, and Maitland (1965) concluded that "it is probable that under certain conditions at least there is competition for food among them". Both minnow and stone loach must have been introduced to this river at some time.

Exact assessment of competition between two or more species of fish in the natural situation is not easy because of the complexity of most biological systems, but the examples quoted suggest that introduction of alien species to river systems in the British Isles has resulted in many cases of competition with and predation on fishes native to the system and must be presumed to have produced considerable changes in the fauna.

ALTERATIONS TO THE ECOSYSTEM

A major factor in the changes in the fish fauna of the British Isles lies in the conflict between the various uses of water by man and the requirements of fish. A number of alterations have been brought about directly as a result of water use by the human population and others have resulted indirectly.

1. Pollution of the Water

The problems of the many forms of pollution of fresh water have received extensive treatment although much of the early literature was concerned with the public health aspects of polluted waters. Within the second half of the present century a considerable change of emphasis has taken place and many studies of the effects of pollution on the aquatic ecosystem have been made. Many of these have been summarized by Jones (1964) and by Hynes (1960). Pollution as used here is a catch-all term for the addition in the course of water use of any substance, including heat, which alters the chemical or physical condition of the water.

Amongst the many pollutants that have affected fish life is the addition of untreated or inadequately treated sewage or sewage effluent from domestic sources. The effects of this form of pollution are essentially deoxygenating, the organic material in the sewage being broken down by the micro-organisms present in the water at the expense of the dissolved oxygen. In mild cases the dissolved oxygen may be reduced below the optimum required for the least tolerant species; in severe cases the depletion of oxygen is such that whole reaches of the river about the discharge point are anaerobic and no fishes and very few other animals can survive. In addition to domestic sewage other organic effluents have a similar effect; milk washings and slaughterhouse refuse have both been implicated in the past, but with the present concern for cleaner rivers, discharges of this type are becoming fewer and may eventually be negligible in the British Isles.

A well-known example of this kind of pollution concerned the tidal parts of the River Thames especially where it flows through metropolitan London. This part of the river once supported abundant fish-life which provided a living for many fishermen in riverside communities, and before the days of fast goods transport an important source of protein for the inhabitants of London. With the introduction of the water closet in the mid-1800s, the products of which were discharged directly or indirectly to the Thames, and the great increase in the metropolitan population which followed the Industrial Revolution, the river became deoxygenated and the fish-life extinguished. This situation continued, with short-lived improvements, until relatively recently, when following extensive rebuilding of sewage treatment works by the Greater London Council, and determined control of other forms of pollution by the controlling agency, the Port of London Authority, a substantial improvement has taken place. Today it is salinity that controls the further spread of marine and freshwater species, and euryhaline species penetrate from the sea into freshwater conditions (Wheeler, 1970). More than sixty species of fish have been captured between 1968 and 1972 in the area where no fish could be found in 1957–58.

Although the major polluting load to the tidal Thames was from oxygen-reducing effluents, other pollutants played a considerable part in worsening the condition of the water. One such was the effluent discharged by early town gas-producing plants which discharged a liquid waste so potent that even eels placed into it were killed in seconds. Toxic effluents are produced by many industries and although at one time they were commonly discharged into rivers, today most are either piped for treatment at sewage treatment works or are subject to special disposal arrangements.

Pollution by toxic heavy metals has, however, been a long-standing problem. Lead mines in Cardiganshire are believed to date back to Roman times (Jones, 1964), but it was not until the nineteenth century when the industry became greatly developed that severe pollution problems were encountered. Then the ore was ground very fine and washed with water (often dammed headwater streams close to the mine site) to extract the metal. The washings which contained substantial quantities of very finely divided metals were collected in settling pits and dumped on land, and from here were leached into the river systems. In Cardiganshire the lead-polluted rivers were studied by Carpenter (1924), who was able to show that these rivers were fishless owing to the toxic effects of the lead (and possibly zinc), not because of suspended matter choking the fishes' gills. The persistent effects of pollution by heavy metals was later demonstrated by Jones (1958) who detected 0·6 ppm of zinc in the Cardiganshire river Ystwyth 35 years after active working of the mines was abandoned. Mining for the ores of heavy metals was essentially confined to the western parts of Great Britain from Cornwall to Scotland. While such mining was practised many rivers and inland waters were severely affected, including such large areas as Coniston Lake in the Lake District which suffered from effluents of the local copper mines (Watson, 1925, p. 94).

The effects of metallic pollutions in extreme cases are the complete extinction of all fish-life in the river below the source of the pollution, which was often in the headwaters of the river. The best known example of this is the Cardiganshire River Rheidol, where at one time no fish could be found in the main river. Sticklebacks were recorded in 1925, some seven years after the mine workings ceased, but in 1948 the fish fauna was described as decidedly poor, and it was not until 1950 or 1952 that salmon returned to the river (Jones and Howells, 1969). Thanks to restocking by local fishery interests a substantial salmon population is now present in the river.

Thermal pollution of fresh waters is of some significance in this discussion and has received considerable attention in the U.S.A. (see Parker and Krenkel, 1969). Many industries use river water in cooling processes, returning it warmed above ambient to the river. The effects of warmed water on fish-life vary with the rise in temperature encountered and also with climatic effects and the condition of the water. During the warm months there may be complete avoidance of the warm water areas by some species (the stenothermic salmonids especially); in cold seasons fishes are known to have congregated within the warm plume of water; the only cases of temperature-induced mortality are where fish have been trapped in areas where the temperature has risen sharply.

The effect of thermal pollution in the River Trent which is already polluted with domestic and industrial effluents, has been recently studied by Alabaster *et al.* (1971). In this river where dissolved oxygen levels were already low it was suspected that raising the temperature would reduce the available oxygen and increase the toxicity of poisons in the water. However, it was established that even though water temperatures rose below the power-station effluent discharge, the effects on the water were not serious to non-salmonid fishes. Where cooling-towers were used to cool the turbine water these workers even reported a slight improvement in dissolved oxygen levels and less ammonia below the point of discharge. This suggests that, subject to adequate control, the discharge of heated effluents into non-salmonid rivers does little harm to the native fish. When the river water used for cooling is passed through cooling-towers it may even have a beneficial effect in polluted rivers. The discharge of heated effluents does, however, have an effect on the invertebrates present in the river (Langford, 1971), and may therefore have an indirect effect on the fish.

The elevation of temperature in rivers has more effect on migratory fishes, especially the stenothermal salmonids. In the case of the anadromous American shad (*Alosa sapidissima*), Leggett and Whitney (1972) have shown that water temperature is the controlling factor in their oceanic migration, and that the peak of spawning runs into rivers at various latitudes in both the Atlantic and Pacific takes place when water temperatures are near 18·5° C. In this case it seems likely that elevated water temperatures in rivers might prove to be a decisive factor in the timing of the spawning migration and affect the success of spawning. Similar data do not seem to be available for the European shads (*Alosa fallax* and *A. alosa*) the status of which is seriously worsened, possibly through the effects of general estuarine pollution. But there is reason to suspect that thermal pollution is a causative factor in their decline and may deter spawning runs of shad from becoming established in rivers which become less polluted.

In the case of salmonid fishes considerable research has been undertaken into the effects of warming of river water, but in the British Isles thermal polluting loads have rarely been located on salmon and trout rivers except in the estuarine reaches where greater mixing and volume of the water reduces the adverse effects. As a consequence no inhibition of migration by salmonids which was entirely attributable to thermal causes appears to have been observed in Britain. Evidence from Canadian rivers, however, suggests that migration of Atlantic salmon may be inhibited in exceptionally warm seasons by increased water temperatures due to discharge of heated effluents (Alabaster *et al.*, 1971). However, the elevation of river water temperatures above ambient may present problems to the migratory salmonids in conditions of low river flow and in polluted waters which would not affect non-salmonid fishes. Such conditions could be materially worsened by severe droughts and heavy water abstraction.

2. Abstraction and Drainage Practices

Abstraction of water to supply drinking water for the human population and for irrigation has led in the present century to considerable changes in river flow. As the consents to abstract water are governed by requirements to maintain agreed minimum flows, major rivers are not dramatically affected. The flow in tributary streams, however, can be seriously reduced with considerable changes consequently taking place in the fish fauna. This problem is most acute in south-east England, in which area some streams virtually dry out in places following prolonged drought, and many are greatly reduced in depth and flow seasonally. The changes that this has brought about in the fish fauna do not appear to have been systematically studied. The present abundance of minnows in parts of rivers like the Roding (Essex) and Lee (Hertfordshire), is possibly due to the present shallowness of the water favouring this species rather than the larger cyprinids or potential predators like the trout, pike, chub and perch.

The importance of river flow to the upstream movement by migratory fishes has received some attention. Both salmon and sea trout move upstream during periods of higher than seasonally normal flow. There appears to be no particular median flow above which migration will not proceed, even allowing for seasonal variation, but the causative factor appears to be the short-term rise in flow or freshets, with the accompanying changes in temperature and dissolved oxygen (Alabaster, 1970). It is clearly important in rivers where flows may be reduced by abstraction to make provision for such conditions to be maintained and thus to preserve the normal pattern of upstream fish migration. The possibility of such interruptions to migratory fishes in Ireland has been discussed by McGrath (1960) with special reference to dams built in rivers. He gave details of one case in which salmon were induced to ascend the River Liffey by an artificial freshet produced by pumping water from a balancing reservoir at Leixlip, only to be killed by the silt and deoxygenated water from the final flow off the bottom of the reservoir which had acted as a settling pool to the tributary streams.

The clearing of the bed of natural watercourses to permit the rapid run-off of flood waters is normal practice for river engineers; its effects on fish-life in the river have been serious but little appreciated. The effects of drainage schemes in various parts of the British Isles seem to have been accepted as, rather than proven to be, detrimental especially to salmonid fishes. Toner *et al.* (1964) have studied the effect on salmon of the arterial drainage works on the River Moy in Ireland. Their work was primarily concerned with dredging operations on a large tributary and they found that the recovery of potential food organisms was slow although within two years chironomids and ephemeropterans had recovered but trichopteran larvae were still scarce. The food habits of young fish had changed considerably as a result of the change in the fauna. Runs of spawning adults and the number of redds made had not changed considerably, but survival of the fry in natural redds had declined (mortality being as high as 93% in the spawning season following drainage work).

The effects of drainage on non-salmonid fishes is likely to be as severe as these observations suggest it is on salmonids. Most of the cyprinids and the pike spawn on submerged vegetation and as the banks and bed of rivers are left bare by dredging, the survival of fry in the year or years following is likely to be low. The disturbance and disappearance of adequate weed cover for the young, as well as drainage of former habitats, is suggested as a possible cause for the depletion in numbers of the burbot (Marlborough, 1970).

3. Hydro-electric Schemes and Dammed Lakes

The development of rivers as sources of power for hydro-electric schemes can have a considerable effect on the local population of migratory fishes. Scottish rivers, in particular, have been so developed and a good deal of attention has been devoted to the problems arising especially for salmon; this topic has recently been discussed by Pyefinch (1966). The development of a hydro-electric scheme on a river can affect it in several ways. It can alter the contour of the river, at its most extreme changing it to a series of steps with impoundments at each point, as is seen in some Swedish rivers where virtually the whole head of the river has been utilized. Secondly, it imposes a uniformity of flow over the year which dampens down the freshets so important to the return of migrating salmonids. Thirdly, it imposes a physical barrier across the path of the migrating fish, and although fish-passes can be, and are, built into the development some interference is still inevitable. In addition, unless effective safeguards are installed there may be significant losses of immature fishes during their passage downstream, for the major flow of the river is through the turbine intake which has to be specially screened to prevent fishes being swept through the turbines. Other hazards to salmonids presented by the development of rivers for hydro-electric schemes are that the water held in impoundments is a suitable habitat for predatory fishes and birds, while migrating fishes held overlong in impoundments because of the regulated flow are frequently prone to fungal attacks (McGrath, 1960). Increased predation can represent a very serious loss, as Mills (1964) showed in the River Bran, on the Conon system in the Scottish highlands, where only about 0·05% of planted salmon fry survived migration below the impoundments of the river, partly because of the predatory activities of pike, trout, and birds, chiefly goosanders (*Mergus merganser*).

These considerations suggest that the development of hydro-electric schemes can have a considerable effect on the fish-life of the river. They tend to favour non-migratory fishes at the expense of the anadromous species, of which the salmon is the most important. However, the fact that salmon still run in such rivers indicates that hydro-electric development is not necessarily devastating in its effects.

The effects of dammed lakes on the fish fauna is a topic closely associated with hydro-electric schemes, although many upland lakes are also dammed as storage reservoirs for drinking water, or for flood prevention. The essential result of these uses is that the level of the water is made to vary over a greater range than is natural, with consequent changes in the composition of the littoral fauna. The extent of these changes has been shown by Hynes (1961) to be considerable, from the stranding and thus mortality of small fishes and aquatic invertebrates, to changes in abundance and dominance amongst the invertebrates. Profound alterations to the littoral fauna of Llyn Tegid (Lake Bala) have followed since its level changes began in 1955 (Hynes, 1961), and Hunt and Jones (1972a) have demonstrated a considerable increase in the numbers of profundal animals since level fluctuation began. The effects on the fish fauna of Llyn Tegid of these changes in potential food organisms has not been recorded but for Llyn Celyn, another regulated Welsh lake, Hunt and Jones (1972b) have discussed this topic. Llyn Celyn was formed by damming a river and its level fluctuates considerably, the littoral fauna is sparse, and many of those groups that normally form the food of trout are absent. The food of trout in this lake is mainly chironomids and aerial insects; the growth of trout is poor, by comparison with that of fishes in nearby lakes. Although many regulated lakes are in highland areas and are thus principally populated by salmonid fishes, the cyprinid species are equally affected by the impoverishment of the littoral zone. Most cyprinids spawn on submerged vegetation and their fry remain in the vegetation for a considerable period after hatching. It follows from this that any diminution of the littoral vegetation will directly affect the breeding success of most non-salmonid fishes.

There seems, therefore, little doubt that the fish fauna of natural lakes in which the level fluctuates will be adversely affected by the resulting sparsity of life in the littoral zone. Even stocking with trout as in the case of Llyn Celyn resulted in a poor fishery, with trout of inferior size to nearby natural populations.

POTENTIAL FUTURE CHANGES TO THE FISH FAUNA

Much of the observed change in the British freshwater fish fauna has arisen from use of aquatic resources. The increase in human population, and increasing industrial use of water, has led to concern for future water supplies especially in the lowland areas, and it is pertinent to consider what implications for fishes lie in the various solutions to water shortage.

Water storage barrages have been much discussed especially for Morecambe Bay, the Solway Firth, and the Wash, in which proposals have been advanced to dam an area of the sea into which fresh water will flow to form a reservoir. The biological implications of such barrages have been discussed by Gilson (1966) who drew attention to the importance of depth and turbidity and their effect on the productivity of such lakes. Gilson's conclusion (for the Morecambe Bay and Solway Firth barrages) was that good trout populations would be initially established but that these might be replaced by populations of cyprinids, with pike, and perch as predators. In the Wash barrage this change would be speedy and the predators would be supplemented by the zander which would be particularly suited to this type of water and thrive as it has in the Ijsselmeer in Holland (Willemsen, 1969).

The success of these barrage lakes as fish habitats will, however, lie in their management, for if the water level is regulated then the littoral flora and fauna will be impoverished, and the fish fauna accordingly diminished with declining food resources and breeding areas. Fishery management will also no doubt be attempted in order to maintain angling-preferred species, such as trout, in favour of cyprinids, but these lowland waters will inevitably become eutrophic due to run-off of fertilizers from the land and sewage effluents discharged to the feeder rivers and the non-salmonid species will speedily dominate the trout. This process will probably proceed very quickly in spite of attempts to manage the fish populations.

Another proposal to alleviate the water shortage of the south-east of England is the transference of water from water-rich areas to impoverished areas. This is already being carried out by the transference of water from the Great Ouse system to the Essex River Authority area by pipeline. The possibility that aquatic organisms including young fishes may be transferred in this way is remote as the water is screened at intake and chlorinated, but it nevertheless exists. More extensive schemes may be pursued but in terms of fish-life they will run from faunistically poorer areas (such as the north-west of England) to the faunistically richer areas of the midlands and south, so there may be no damaging interchange of organisms. Nevertheless, such schemes imply a regulation of flow from donor rivers or impoundments and this may have deleterious effects on their fish fauna.

The increase in angling as a recreation may lead to further changes in the freshwater fish fauna. Fishery management is increasingly tending towards the maintenance of artificial populations of fishes. At one extreme there are continued efforts to stock waters with trout, removing at the same time non-salmonid species; such fisheries have so far been rarely successful as long-term projects. In other cases it may be assumed that stocking with native species, mostly cyprinids, will continue either from stock reared in fish farms, or from fish imported from Europe. The hazards represented in the latter case have already been mentioned; there seems to be every reason to expect that the number and extent of such transferences will increase in the future, and as the motive is essentially short-term improvement of fishing, the consequences on the native fish and invertebrate fauna may well be serious.

The increased demand for sport fishing will also inevitably lead to pressure for the introduction of other exotic sport fishes, like the American large-mouth bass (*Micropterus salmoides*), and others. This species, which has been widely established in Europe and has a high reputation as an anglers' fish, is a predator on a wide range of invertebrates and fishes at all times of its life. Its biology in France has been studied by Wutz-Arlet (1952) who shows that it is a direct competitor with trout, and also that as a piscivore its effect on a mixed fishery could be serious. If pleas for the introduction of such species by anglers or angling interests are allowed the likely consequences for the native fauna will be serious.

SUMMARY

The freshwater fish fauna of the British Isles can be seen to be essentially what man has made of it. Many native species have been redistributed outside their original range, and have colonized new areas by means of canals joining river systems. Many new habitats, such as canals, reservoirs and lakes, have been constructed and stocked with fishes. Exotic species have been introduced, but with the exception of the zander, carp and the rainbow trout, few have established themselves and the rainbow trout is only marginally successful.

Aquatic habitats have been greatly altered. Many of the lowland areas have been drained with consequent impoverishment of the fish fauna (the near-extinction of the burbot may be mainly due to drainage of the Fens). River-flows and lake levels have been controlled for a number of reasons with changes in the abundance of certain fishes. Pollution by addition of deoxygenating or toxic substances, and surplus heat has changed other rivers resulting in a local impoverishment of fish life. Eutrophication of lakes and rivers has equally affected the fish fauna.

Overall the effects of two millenia of activity by man in Britain can be seen in restricted distribution and scarcity of some species (especially salmonids) and the wider distribution and increased abundance of others. Migratory fishes, such as the sturgeon, shads, salmon, and sea trout have suffered most from pollution and the development of rivers. The salmonids and whitefishes have declined as a result of eutrophication, predation and possibly competition with non-salmonid native species introduced to lakes and rivers. On the other hand, the cyprinids, perch and pike have become more widely distributed, and have been favoured by many of the developments such as impoundments and eutrophication of our rivers and the construction of new aquatic habitats.

Proposals for the future management of water resources in Britain suggest that these trends will continue.

REFERENCES

ALABASTER, J. S. (1970). River flow and upstream movement and catch of migratory salmonids. *J. Fish. Biol.* **2**, 1–13.

ALABASTER, J. S., GARLAND, J. H. N. and HART, I. C. (1971). Fisheries, cooling-water discharges and sewage and industrial wastes. *Symposium on Freshwater Biology and Electrical Power Generation.* CERL RD/L/M/312, 3–8.

CARPENTER, K. E. (1924). A study of the fauna of rivers polluted by lead mining in the Aberystwyth district of Cardiganshire. *Ann. appl. Biol.* **11**, 1–23.

CHARLESWORTH, J. K. (1957). "The Quaternary Era", 2 vols. Arnold, London.

ELLISON, N. F. and CHUBB, J. C. (1962). The marine and freshwater fishes. *Fauna of Lancashire and Cheshire.* **41**, 1–34.

FROST, W. E. (1968). Fish. *Freshwat. biol. Ass. Ann. Rep.* **36**, 42.

FROST, W. E. and KIPLING, C. (1967). A study of reproduction, early life, weight–length relationship and growth of pike, *Esox lucius* L., in Windermere. *J. anim. Ecol.* **36**, 651–693.

GILSON, H. C. (1966). The biological implications of the proposed barrages across Morecambe Bay and the Solway Firth. in *Man-Made Lakes. Sym. Inst. Biol.* **15**, 129–137.

HADFIELD, C. (1959). "British Canals, An Illustrated History". Phoenix House, London.

HEALY, A. (1956). Roach and dace in the Cork Blackwater. *Rep. Sea Inld. Fish. Eire*, 3–14.

HUNT, P. C. and JONES, J. W. (1972a). The profundal fauna of Llyn Tegid, North Wales. *J. zool.* **168**, 9–49.

HUNT, P. C. and JONES, J. W. (1972b). The littoral fauna of Llyn Celyn, North Wales. *J. Fish. Biol.* **4**, 321–331.

HYNES, H. B. N. (1955). Biogeography and palaeolimnology. Distribution of some freshwater Amphipoda in Britain. *Verh. int. Verein. Limnol.* **12**, 620–628.

HYNES, H. B. N. (1960). "The Biology of Polluted Waters". Liverpool University Press.

HYNES, H. B. N. (1961). The effect of water-level fluctuations on littoral fauna. *Verh. int. Verein. Limnol.* **14**, 652–656.

JONES, A. N. and HOWELLS, W. R. (1969). Recovery of the River Rheidol. *Effluent and Water Treatment J.* November 1969: 6 pp.

JONES, J. R. E. (1958). A further study of the zinc-polluted river Ystwyth. *J. anim. Ecol.* **27**, 1–14.

JONES, J. R. E. (1964). "Fish and River Pollution". Butterworths, London.

KENNEDY, M. (1969). Irish pike investigations. I. spawning and early life history. *Irish Fish. Invest., ser.* A (*Freshwater*), **5**, 1–33.

LACHNER, E. A., ROBINS, C. R. and COURTENAY, N. R. (1970). Exotic fishes and other aquatic organism introduced into North America. *Smithsonian Contr. Zool.* **59**, 1–29.

LANGFORD, T. E. (1971). The biological assessment of thermal effects in some British rivers. *Symposium on Freshwater Biology and Electrical Power Generation.* CERL RD/L/M/ 312, 9–39.

LEGGETT, W. C. and WHITNEY, R. R. (1972). Water temperature and the migration of American shad. *U.S. Fish. Wildl. Serv. Fish. Bull.* **70**, 659–670.

MACAN, T. T. (1965). The effect of the introduction of *Salmo trutta* into a moorland fishpond. *Mitt. Internat. Verein. Limnol.* **13**, 194–199.

McGRATH, C. J. (1960). Dams as barriers or deterrents to the migration of fish. *In* "Soil and Water Conservation", Vol. 4, 81–92. I.U.C.N., Brussels.

MAITLAND, P. S. (1965). The feeding relationships of salmon, trout, minnows, stone-loach and three-spined stickleback in the River Endrick, Scotland. *J. Anim. Ecol.* **34**, 109–133.

MAITLAND, P. S. (1972). A key to the freshwater fishes of the British Isles. *Sci. Publ. Freswat. biol. Ass.* **27**, 1–139.

MARLBOROUGH, D. (1970). The status of the burbot *Lota lota* (L.) (Gadidae) in Britain. *J. Fish Biol.* **2**, 217–222.

MILLS, D. H. (1964). The ecology of the young stages of the Atlantic salmon in the River Bran, Ross-shire. *Freshwat. Salm. Fish. Res.* **32**, 1–58.

MOON, H. P. (1970). *Corophium curvispinum* (Amphipoda) recorded again in the British Isles. *Nature, Lond.* **226**, 976.

NICHOLS, P. R. and LOUDER, D. E. (1970). Upstream passage of anadromous fish through navigation locks and use of the stream for spawning and nursery habitat Cape Fear River, N.C., 1962–66. *U.S. Fish & Wildlife Service, Circular* 352, 1–12.

ORKIN, P. A. (ed.) (1957). "Freshwater fishes" (By O. Schindler). London.

PARKER, F. L. and KRENKEL, P. A. (1969). "Thermal Pollution: Status of the Art". Report No. 3 Dept. of Environmental and Water Resources Engineering, Vanderbilt University, Tennessee. Chapter III, 65 pp.

PYEFINCH, K. A. (1966). Hydro-electric schemes in Scotland. Biological problems and effects on salmonid fisheries. *In* "Man-Made Lakes" *Symp. Inst. Biol.* **15**, 139–147.

REGAN, C. T. (1911). "The Freshwater Fishes of the British Isles". Methuen, London.

SHAFI, MUHAMMED and MAITLAND, P. S. (1971). Comparative aspects of the biology of pike, *Esox lucius* L., in two Scottish lochs. *Proc. R. Soc. Edinb.* B, **71**(1), 41–60.

TONER, E. D., O'RIORDAN, A. and TWOMEY, E. (1964). The effects of arterial drainage works on the salmon stock of a tributary of the River Moy. *Ir. Fish. Invest. ser.* A, **1**, 36–55.

VOOREN, C. M. (1972). Ecological aspects of the introduction of fish species into natural habitats in Europe, with special reference to the Netherlands. *J. Fish. Biol.* **4**, 565–583.

WATSON, J. (1925). "The English Lake District Fisheries". 2nd edition. Foulis, London and Edinburgh.

WENT, A. E. J. (1950). Notes on the introduction of some freshwater fish into Ireland. *J. Dept. Agric. Dublin* **47**, 119–124.

WENT, A. E. J. (1957). The pike in Ireland. *Ir. Nat. J.* **12**, 177–182.

WHEELER, A. (1969). "The Fishes of the British Isles and North-west Europe". Macmillan, London.

WHEELER, A. (1970). Fish return to the Thames. *Sci. J.* **6**(11), 28–32.

WHEELER, A. and MAITLAND, P. S. (1973). The scarcer freshwater fishes of the British Isles I. Introduced species. *J. Fish. Biol.* **5**, 49–68.

WILLEMSEN, J. (1969). Food and growth of pike-perch in Holland. *Proc. 4th British Coarse Fish Conference*, 72–78.

WORTHINGTON, E. B. (1941). Rainbows. A report on attempts to acclimatize rainbow trout in Britain. *Salm. Trout Mag.* **100**, 241–20; **101**, 62–99.

11 | The Distribution of Mammals in Historic Times

G. B. CORBET

Department of Zoology, British Museum (Natural History), London

Abstract: No species of mammal has become extinct in Britain since the wolf was exterminated during the eighteenth century. Of the 56 species now present on the British mainland 14 have been introduced. Only one of these, the grey squirrel, has spread at the expense of a native species. Most recorded changes in distribution can be attributed to the direct influence of man, through persecution or through artificial extension of range. Deer were at first greatly reduced by overhunting and destruction of forest but have recovered by reintroduction and through reafforestation. The ranges of wild cat, polecat and pine marten were severely reduced during the nineteenth century by persecution in the interests of game preservation but the last thirty years have seen substantial recoveries.

At present the only species in serious danger of extinction are the greater horseshoe and mouse-eared bats. Other declining species are the other bats, otter and dormouse although as yet there has been only marginal loss of range by any of these. The remaining species are either holding their position or are increasing.

INTRODUCTION

If we leave aside the wholly marine cetaceans, and the vagrant seals and bats that have never been known to breed here, the present mammalian fauna of the British mainland amounts to 56 species, made up as shown in Table I. In addition there are two shrews and one rodent confined to small islands and believed to be introduced. Of these 56 species, 31 occur also in Ireland and there are no additional species in Ireland that do not also occur in Britain. For a check-list of British species see Corbet (1969a).

Historical information on some identifiable species of mammals extends back to before the Norman conquest and it is therefore convenient to begin this

Systematics Association Special Volume No. 6, "The Changing Flora and Fauna of Britain", edited by D. L. Hawksworth, 1974, pp. 179–202. Academic Press, London and New York.

G

survey in the Roman period. However apart from some significant extinctions and introductions it is only during the course of the nineteenth century that sufficient data became available to enable any more subtle changes to be described.

For the insectivores, bats, lagomorphs and rodents the volumes of Barrett-Hamilton and Hinton (1910–21) provide detailed accounts of the history of these species, although it is curious that in spite of the otherwise exhaustive treatment no attempt was made to produce maps showing distribution. For the deer a detailed account of their complex history (with maps) was provided by

TABLE I. Composition of the mammal fauna of the British mainland since Roman times (excluding vagrants)

	Extant		Extinct	
	Indigenous	Introduced	Indigenous	Introduced
Marsupialia	—	1	—	—
Insectivora	5	—	—	—
Chiroptera	15	—	—	—
Carnivora	8	1	2	—
Pinnipedia	2	—	—	—
Artiodactyla	2	5	3[a]	—
Lagomorpha	2	1	—	—
Rodentia	8	6	1	1
Totals	42	14	6	1
	56		7	

[a] Two of these may have become extinct before Roman times.

Whitehead (1964), but for the carnivores, including many of the most interesting species from this point of view, there is no comparable synthesis, although a considerable and very scattered literature exists. The cetaceans, although only marginally relevant to this survey, seem worth including since an efficient system for reporting strandings has operated since 1913 and provides much useful data on changes in abundance of certain species (Fraser, in press).

The changing distribution of certain species has been surveyed in considerable detail mainly for economic reasons, e.g. the squirrels (Shorten, 1957; Lloyd, 1962), the black rat (Bentley, 1959, 1964), the coypu (Laurie, 1946; Davis, 1963; Norris, 1967), the mink (Thomson, 1968) and the seals (Bonner, 1972).

In 1965 the Mammal Society inaugurated a programme in collaboration with the Biological Records Centre of the Nature Conservancy, to prepare distribution maps of all species. A very provisional set of maps has been published (Corbet, 1971) and the survey is still continuing. A more detailed survey of deer is being undertaken by the British Deer Society.

The great majority of species for which changes in distribution and status can be clearly demonstrated are those that have been directly influenced by man. These influences include persecution by hunting for food, fur or sport; persecution because of damage to crops, livestock or sporting interests; deliberate introduction for sport or amenity; and accidental introduction by escape from captivity or by unseen importation. With the remaining species very few cases of alteration in gross range can be reported, either because the species are likely always to have been ubiquitous, as with the shrews and voles, or because records from the past are too few to indicate significant changes, as with most bats.

In reviewing the subject the extinct and introduced species will be dealt with first, followed by a systematic survey of the extant, indigenous species (including in the last category any whose indigenous status is in doubt).

EXTINCT SPECIES

There is no evidence that any mammal has become extinct in Britain during the last 2000 years other than six large species that have undoubtedly been exterminated by direct overhunting; and one species, the musk-rat, that was introduced and then successfully and deliberately exterminated as a pest (discussed under introduced species below). By going back beyond the Roman period we could extend the list of species that were almost certainly exterminated by hunting to include the elk (*Alces alces*) and the giant deer (*Megaceros giganteus*); and even the mammoth (*Elephas primigenius*) and woolly rhino (*Coelodonta antiquitatis*) if, as I am inclined, one subscribes to the theory that man, rather than the elements, was responsible for the loss of these species (Martin and Wright, 1967).

The history in Britain of the six species detailed below was described by Harting (1880) and very little of significance can be added to his account except to emphasize how little we know for certain. The information given below is from Harting unless otherwise stated. Since the middle of the eighteenth century when the wolf was finally exterminated no further species have been lost to Britain, although some, for example the pine marten, have come rather close to it.

1. *Wolf* (Canis lupus)

The wolf probably became extinct in England about 1500 but survived in Scotland until about 1740 and in Ireland until perhaps 1770. It is an adaptable species with an enormous range from the Arctic coasts to Arabia and there is no reason to doubt that the wolf would have survived well in Britain given freedom from persecution and the continued existence of deer on which they could prey.

2. *Brown bear* (Ursus arctos)

The date of extinction of the brown bear is much more conjectural and is confused by the practice of bear baiting and the undoubted importation of captive bears for that purpose. Harting considered that they were probably extinct by the tenth century, but it is possible that they survived in the wilder parts of the country to a later date. Bears were hunted for sport, for food and for fur and were caught alive for baiting.

3. *Beaver* (Castor fiber)

Literary references suggest the presence of beaver in Wales in the tenth and twelfth centuries but it is probable that by then they were already rare and that they became extinct soon after that time. Deforestation of the banks of lowland rivers must have played a part in the extinction but there can be little doubt that hunting for fur was the principal cause.

4. *Wild boar* (Sus scrofa)

The wild boar survived in England until some time in the seventeenth century. Subsequently several attempts at reintroduction were made but mostly in enclosed parks and no general recolonization took place. The wild boar is a species native to broad-leaved forest and can be considered to have been near the northern edge of its range in Britain.

5. *Reindeer* (Rangifer tarandus)

Following Harting the reindeer is generally quoted as having survived in Scotland until the twelfth century. This however is based on a single reference in the Orkney Saga to the earls of Orkney hunting them in Caithness and must be considered very dubious. Subfossil material does not provide any clear evidence of reindeer since the end of the Pleistocene, in early post-glacial times. On grounds of habitat it is not improbable that reindeer could have survived in Scotland although it is curious that another tundra species, the Norway lemming (*Lemmus lemmus*), should also have become extinct in Britain since the late glacial period in spite of the survival of apparently suitable habitats in the Scottish Highlands.

The reindeer recently reintroduced in the Cairngorms are domesticated animals in a carefully managed herd and cannot be considered wild or even feral.

6. *Urus* (Bos primigenius)

There is no reason to believe that domestication of wild cattle took place in Britain but the history of the wild species is bedevilled by confusion with feral individuals of domestic cattle. Harting was noncommittal about this species, concluding that although it was likely to have become extinct in lowland England before the Roman period it may have survived much later in Scotland. There is no archaeological evidence of its survival beyond the Bronze Age.

INTRODUCED SPECIES

The fifteen introduced species (Table I) include some that were deliberately liberated (e.g. grey squirrel); some that escaped from close captivity (e.g. mink, coypu); some that were liberated mainly in enclosed parks and subsequently escaped (e.g. deer); and those like the rats and house mouse that came unwanted and unseen.

During the nineteenth century many attempts were made to introduce exotic mammals (Fitter, 1959) but none resulted in a successful colonization outside parks until the end of the century when grey squirrel and fat dormouse were established. In the systematic list below, colonizations that were only temporary and local are excluded—the only species listed that has subsequently become extinct is the musk-rat.

Three species (the white-toothed shrews and the Orkney/Guernsey field vole) are treated as introduced on circumstantial evidence. But one should not ignore the possibility that other species generally considered indigenous could have been introduced by man. Of the species included here as indigenous the yellow-necked mouse (*Apodemus flavicollis*) in particular could conceivably have been introduced.

More detailed accounts of the history of most of these species were given by Fitter (1959) and of the deer by Whitehead (1964).

1. *Red-necked wallaby* (Macropus rufogriseus)

This species, from Tasmania and eastern Australia, has frequently been liberated within enclosed parks and zoos and breeds well under these conditions. At least two feral populations now exist as a result of escapes from such park colonies. The better documented is the one in the Peak District (mainly north Derbyshire) which became feral about 1940 but has spread slowly, with several setbacks in hard winters, having been recorded in about seven 10 km squares by 1970 (Yalden and Hosey, 1971).

Another feral population appears to exist in central Sussex, where at least three enclosed colonies have been maintained, but it is known only from casual sightings and road casualties, with for example five records in 1969 (Page, 1970). This is a species that is likely to survive but with a slow rate of expansion due to its liability to high mortality in winter.

2. *White-toothed shrews* (Crocidura *spp.*)

Three species of white-toothed shrews are found in continental Europe but none on the British mainland. One of the continental species, *C. russula*, occurs on Alderney, Guernsey and Herm in the Channel Isles; another, *C. suaveolens*, occurs on Jersey and Sark in the Channel Isles and most of the Scilly Isles. Considering the overall distribution and ecology of these species, the impoverished fauna of these islands and their recent geological history, it seems very probable that these populations have all arisen from introduction by man (Corbet, 1961, 1969b). They were first recorded on the Channel Isles (Alderney) in 1898 and on the Scilly Isles in 1914 but could easily have been overlooked prior to these dates.

3. *American mink* (Mustela vison)

This is the most recent introduction that can be considered well established. Mink farms increased greatly after 1945 and by 1957 escaped mink were breeding in the wild. Feral animals at first spread widely in Devon but the very rapid colonization of a very large part of the country from Cornwall to Inverness and Ireland has been due to expansion from a considerable number of separate centres (Thompson, 1968; Dean and O'Gorman, 1969).

It might be considered that the mink has spread rapidly in a niche that is doubly vacant: the European mink (*Mustela lutreola*) has apparently never occurred in Britain, and the polecat (*M. putorius*) has been exterminated from all but Wales and the immediately adjacent parts of England. However since these species coexist in Europe, the polecat being much less associated with water than the mink, there is no reason to believe that the polecat will significantly stop the expansion of mink in Wales and already the two species coexist in Pembrokeshire, Shropshire and Herefordshire (Corbet, 1971).

4. *Fallow deer* (Dama dama)

The status of the fallow deer as an indigenous or introduced species in Britain has for a long time been in dispute. The suggestion that it might be indigenous usually stems from the presence of this or related species in Pleistocene deposits in England. But these are now all recognized as interglacial and there are no post-glacial records of fallow deer prior to the Roman period. In particular its absence amongst the abundant deer remains at the Mesolithic site of Star Carr, Yorkshire, seems significant (Fraser and King, 1954). According to Whitehead (1964) historical references to fallow in England go back to about A.D. 238 and it is generally agreed that they were well established by Roman times.

Throughout most of their recorded history in Britain fallow have been primarily park deer but wild populations arose from the break-up of some parks during the Civil Wars in the seventeenth century and again during the present century, especially during World War II. They are now very widespread outside parks in southern England but more local elsewhere.

5. *Sika* (Cervus nippon)

This eastern Asiatic species has been widely kept in parks since the 1860s and feral populations in Dorset and the New Forest date from 1896 and 1904 respectively. In these areas they have expanded locally and are now well established. Elsewhere scattered feral colonies have become established in several parts of Scotland and Ireland but they remain rather local and expansion has been slow.

6. *Muntjac* (Muntiacus reevesi)

The British population of this small oriental deer, now widespread from Somerset to Norfolk and Lincolnshire, has arisen mainly from a single centre of escape from captivity, at Woburn in Bedfordshire, early this century. Much of the expansion has been very recent, since 1940, and is probably still continuing. Dansie (1970) attributed this success to their unobtrusive habits and use of recent afforestation, and pointed out that the population is unusually healthy due to lack of specific internal parasites and to moderate selection by dogs and traffic.

7. *Chinese water deer* (Hydropotes inermis)

This species has a similar history to the muntjac except that there has been little spread beyond Bedfordshire. It is possible that independent feral populations may still survive from escapes in Shropshire and Hampshire. In its native range it is a species of open marshlands but like all deer in this country other than the red, it uses woodland for cover although remaining a grazer.

8. *Feral goat* (Capra)

Domestic goats have been kept in Britain since the Neolithic period. Feral populations of long standing now exist in many upland parts of the country, e.g. in North Wales, the Cheviots, Galloway, the Scottish Highlands and Ireland; and especially on small islands. Although extremely elusive to the casual observer, these herds can be controlled and in a recent survey Whitehead (1972) recorded many temporary colonizations in various parts of the country. Apart from a few park herds goats are not normally tolerated other than on land of minimal economic value such as mountain and sea-cliff. The distribution is and probably always has been controlled purely by the extent to which landowners have tolerated them.

9. *Rabbit* (Oryctolagus cuniculus)

The history of the rabbit in Britain has been documented in detail by Fitter (1959), Thompson and Worden (1956) and Sheail (1971). It was introduced in the late twelfth century as an animal to be kept in confinement and bred for fur and meat. Although enclosed rabbit warrens spread very quickly throughout at least the lowland parts of the country it is uncertain when rabbits became feral on any large scale. But it is likely that it was not until the nineteenth century that it really became abundant and it was during that period that it first spread widely throughout the Scottish Highlands (Barrett-Hamilton and Hinton, 1910–21). Fitter (1959) very reasonably attributed the progressive increase in abundance during the hundred years up to 1954 to the massive destruction of predators (avian and mammalian) and to the general improvement in agriculture. The introduction of myxomatosis in 1954 and 1955 decimated the population throughout the country but probably did not reduce the range except on a very local scale. Recovery has been slow and in most areas densities are still far below the level prior to 1954. Thompson and Worden (1956) provided a detailed account of the spread of myxomatosis and its effects.

The history of the rabbit in Ireland is likely to have been very similar to that in Britain.

10. *Grey squirrel* (Sciurus carolinensis)

The first recorded introduction of the American grey squirrel took place in 1876 in Cheshire and between then and 1929 many separate introductions were made in many parts of the country from southern England to the central lowlands of Scotland and in Ireland. Many were successful and there has been a steady colonization of the country at the expense of the native red squirrel (*Sciurus vulgaris*). This has been documented in detail by Shorten (1957) and Lloyd (1962). Spread has been fast in south-eastern England where woods are not only abundant but tend to be linked by well wooded hedgerows. In the north and west the discrete nature of the woods and scarcity of hedgerows has prevented such a rapid extension of range.

There has been much debate about the mechanism whereby the red squirrel has been replaced by the grey, and indeed whether the changes in the two species really have been correlated. Throughout its range the red squirrel is more closely associated with coniferous than with deciduous forest. Before the introduction of the grey squirrel the red had an erratic history in Britain (see below) but was generally abundant at the end of the nineteenth century. There seems no doubt that in the first decades of this century the red squirrel suffered a severe decline even in areas unaffected by grey squirrels and this has been attributed to epidemic disease, especially of coccidiosis (Shorten, 1954). But the more recent spread of the grey and decline of the red squirrel have been much more closely correlated (Lloyd, 1962; Corbet, 1971; also further, unpublished maps prepared by Lloyd). This can be seen especially in Norfolk, Cornwall, north-west Wales and Lincolnshire where there have been significant replacements of red by grey since 1959.

It is still not possible to identify the factors involved but it seems probabl that it concerns competition for food during the hardest part of the winter. In southern England the grey has succeeded in replacing the red even in coniferous woodland; in the north the isolation of many of the new coniferous forests may be the red squirrel's only defence against similar replacement.

11. *Orkney and Guernsey voles* (Microtus arvalis)

Microtus arvalis is one of the commonest species of vole in much of continental Europe but in Britain is found only on the Orkney Islands and Guernsey, in each case to the exclusion of *M. agrestis*, the species found throughout the British mainland. There is no historical evidence for the introduction of these island populations but there are strong reasons for suspecting such introduction (Corbet, 1961). It has been on Orkney at least since 1805 (Barrett-Hamilton and Hinton, 1910–21).

12. *House mouse* (Mus musculus)

The house mouse was probably the first mammal to be introduced to Britain by human agency. None of the very early historical references to mice allows clear identification of species and information on its date of introduction can therefore only come from subfossil finds. Many of these are difficult to interpret because of the possibility of intrusion through burrows into older strata, but one recent record, hitherto unpublished, seems to confirm its presence in the pre-Roman Iron Age. This is based on the rostrum of two animals and one mandible, identified by the author and now in the British Museum (Natural History), from a site at Gussage All Saints, Dorset. They were in a sealed layer with other small mammals and with no sign of subsequent disturbance.

The house mouse is now ubiquitous in Britain wherever there is human activity, but there is no information available relating to the history of its spread throughout the country in general.

13. Black rat (Rattus rattus)

Introduction of this rat in Britain is thought to have occurred about the end of the twelfth century and they were well established during the following century. It seems to have declined very rapidly during the eighteenth century on the introduction of the brown rat (R. norvegicus), being already scarce by 1783 and almost extinct by 1890 (Fitter, 1959). Throughout the present century it has been well established only in ports and on some small islands such as Lundy. Surveys in 1956 and 1961 showed a considerable contraction of range between these dates even at the ports (Bentley, 1959, 1964), and the reduction is probably continuing, mainly due to better protection of edible merchandise rather than to direct control of the rats.

14. Brown rat (Rattus norvegicus)

The arrival of the brown rat in Britain about 1720 was part of a general colonization of western Europe. In Britain it spread very rapidly and had colonized most of the country, replacing the black rat, by the end of the eighteenth century although it appears not to have reached some parts of the Scottish Highlands until the middle of the nineteenth century (Matheson, 1962). The brown rat not only replaced the black in buildings and towns but it colonized the agricultural countryside to a much greater extent than the black ever did. It has been one of our most abundant mammals ever since and the constant warfare against it has had little significant effect on its overall status.

15. Fat dormouse (Glis glis)

This dormouse was deliberately released at Tring, Hertfordshire, in 1902. It quickly became well established locally but its spread has been slow and it has not been recorded more than about 30 km from its origin. Its history was reviewed in detail by Thompson (1953). Although in some ways resembling the grey squirrel, it is probably more completely arboreal and less inclined to colonize across open ground. It has a wide distribution in continental Europe. Its absence in Britain can be considered an accident of our insularity and there is no reason to doubt that a much larger area of England could be ecologically suitable.

16. *Coypu* (Myocastor coypus)

The South American coypu was introduced to Britain to be bred for its fur about 1929 and during the 1930s many escapes were recorded. However, feral animals became well established only in East Anglia and in the Thames valley near Slough. They continued spreading until about 1962 since when an intensive campaign of control has progressively confined them to the marshy areas of eastern Norfolk (Norris, 1967). Although strictly confined to waterways, their feeding extended to adjacent fields and serious damage was done to crops of sugar beet in particular.

17. *Musk rat* (Ondatra zibethicus)

This North American relative of the voles was introduced for fur farming at the same time as the coypu and like it escaped and became established in several areas in the 1930s. It occupied quite large tracts in central Scotland and the Severn valley but an extensive trapping campaign succeeded in exterminating it by 1937. Like the coypu it is confined to the vicinity of water but its extermination was facilitated by the fact that, unlike the coypu, it did not colonize areas of marsh like the Norfolk Broads. Its history in Britain was described by Warwick (1934, 1940).

INDIGENOUS, EXTANT SPECIES

1. *Hedgehog* (Erinaceus europaeus) *and Mole* (Talpa europaea)

Both of these species are now ubiquitous on the mainland and may always have been so. However there are accounts that suggest, though not very convincingly, that colonization of the extremities of the Scottish Highlands occurred only during the nineteenth century (Barrett-Hamilton and Hinton, 1910–21).

2. *Shrews* (*Soricidae*)

The three shrews found on the British mainland—common shrew (*Sorex araneus*), pygmy shrew (*S. minutus*) and water shrew (*Neomys fodiens*)—can be considered ubiquitous with the exception that the water shrew may be rather local in the Scottish Highlands. They are almost certainly indigenous and there is no evidence for any long-term changes in either range or abundance. The pygmy shrew is very widespread on the smaller islands, including the Outer Hebrides, the common and water shrews less so although present on many of the Inner Hebrides. It is probable that many of these populations owe their origin to introduction by man (Corbet, 1961) but there is neither historical nor subfossil evidence to indicate if and when this took place.

3. Bats (Chiroptera)

Bats are by far the least known of our mammals. In the second edition of his "History of British Quadrupeds", Bell (1874) included fourteen species as known from Britain, twelve of them presumed residents and two known from single specimens and possibly vagrant. This remained the situation until within the last twenty years when there has been a resurgence of interest in bats. As a result two species, the whiskered bat (*Myotis mystacinus*) and the long-eared (*Plecotus auritus*) have been found to consist of pairs of sibling species (Stebbings, *in litt.*; Corbet, 1964). One of the possible vagrants, the mouse-eared bat (*Myotis myotis*), has been shown to be resident since at least 1957 but in very small numbers, whilst a further species, *Pipistrellus nathusii*, has been recorded in Dorset from a single specimen (Stebbings, 1970).

Knowledge of distribution, past or present, is too scanty to reveal any clear-cut changes in range with the single exception of the greater horseshoe bat (*Rhinolophus ferrumequinum*) (Fig. 1). At the end of the nineteenth century this species was authentically recorded from at least seven localities in Kent and two in the Isle of Wight. The absence of any subsequent records from these areas seems significant, especially in the case of Kent. In the main part of its range in south-western England Racey and Stebbings (1972) have estimated, on the basis of mark and recapture data, that the total British population may be as low as 400 whereas estimates made in 1955 indicated that colonies were then at least five times the present size.

The greater horseshoe bat is a colonial species roosting mainly in caves. Other species are less easy to count (and less vulnerable to disturbance), but there are many indications that populations are declining, owing to pesticides, deliberate extermination and loss of roosting sites (Racey and Stebbings, 1972). Many species are heavily dependent upon man-made sites for roosting. Our most abundant species for example, the pipistrelle (*Pipistrellus pipistrellus*), normally uses buildings for the summer nursing colonies which may number several hundred females, while many species use old mine workings, railway tunnels, etc. for roosting, especially in winter. There are no indications to what extent the original utilization of these sites resulted in extension of range or simply changes in habits.

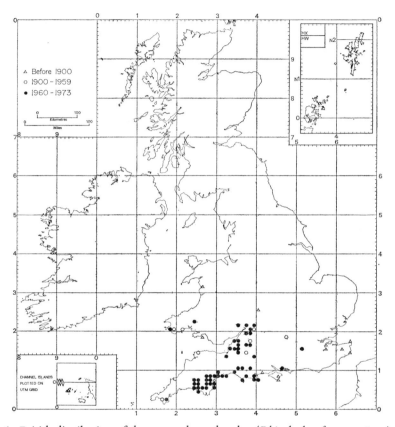

Fɪɢ. 1. British distribution of the greater horseshoe bat (*Rhinolophus ferrumequinum*).

4. *Fox* (Vulpes vulpes)

Locally the fox has had a complex history, being widely persecuted as vermin as well as being the object of a highly specialized hunt which led not only to its protection but even at times to its importation. But there is no reason to believe that in terms of its overall range it has ever been different from what it is now— very nearly ubiquitous. In very recent years it has been increasingly penetrating suburban and urban areas; its colonization of the London suburbs has been described in detail by Teagle (1967). On the other hand the loss of covert and hedgerow has reduced its numbers in some arable areas, for example in East Anglia.

5. Pine marten (Martes martes), Polecat (Mustela putorius) and Wild cat (Felis silvestris)

These three species of carnivore have had a very similar history although differing in detail. From a consideration of their habitat requirements there is no reason to doubt that they were originally ubiquitous on the British mainland wherever there was woodland, with the pine marten present additionally in Ireland.

It is likely that already by 1800 the marten and the cat were restricted by destruction of forest, especially in the lowlands; the polecat is less dependent

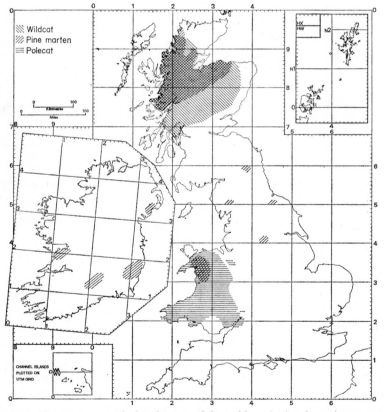

FIG. 2. Probable current British distribution of the wild cat (*Felis silvestris*), pine marten (*Martes martes*) and polecat (*Mustela putorius*). In each case this represents an expansion beyond the minimum range reached earlier this century.

upon forest and can survive in fairly intensively used agricultural land if direct persecution is not heavy. During the course of the nineteenth century all three species were heavily persecuted, especially in the interests of game preservation but also for their threat to poultry. By the middle of the century the wild cat had been virtually exterminated throughout the lowlands and that situation had been reached by the end of the century in the case of the marten and polecat. It is likely that all three reached their lowest ebb sometime between 1910 and 1940, but in different areas. The marten survived in the extreme north-western Highlands of Scotland, in the Lake District, in Wales and in Ireland; the polecat only in central Wales; and the cat only in the Scottish Highlands but more widespread than the marten (Fig. 2).

All three have been expanding their ranges in recent years, certainly since 1940. The marten has expanded especially in the Scottish Highlands although it is still less widespread there than the wild cat. There is less definite evidence of expansion from the other parts of the range. A number of isolated individuals and even breeding populations at widely separated parts in Yorkshire and elsewhere are difficult to interpret—are they overlooked survivors, long distance immigrants, introductions or escaped pets? The polecat has been increasing rapidly in recent years and is now found abundantly throughout Wales and most of the adjacent English counties. Walton (1968) demonstrated a considerable expansion of range between 1962 and 1967. The wild cat has not shown any general expansion beyond the Scottish Highland boundary but it has expanded considerably within the Highlands (Jenkins, 1962).

In each of these cases the reasons for the expansion are generally believed to be the same: expansion of woodland, especially in the form of state forests, and relaxation of direct persecution, especially in the state forests where their value for the control of rodents has generally overridden considerations of game preservation.

6. *Stoat* (Mustela erminea) *and* *Weasel* (Mustela nivalis)

In spite of heavy persecution by gamekeepers, both of these species have remained ubiquitous. They are versatile with respect to habitat, requiring only a minimum of cover along with the small rodents that such cover normally provides.

7. *Badger* (Meles meles)

Although badgers normally make their setts in woodland, small coverts and thick hedgerows will suffice and much of their feeding is done on adjacent open ground. The species has therefore not suffered so severely as some others from destruction of woodland and has likewise survived persecution by gamekeepers and farmers without substantial loss of range. It remains almost ubiquitous in both Britain and Ireland. For details of its history see Neal (1948).

8. Otter (Lutra lutra)

Until recently the otter has remained widespread in Britain and Ireland in spite of persecution by fishing interests and by the hunt. But there is evidence of a considerable decline in lowland rivers in recent years, possibly correlated with increasing recreational pressure as well as with industrial pollution and pesticide poisoning (Anon., 1969).

9. Common seal (Phoca vitulina) and Grey seal (Halichoerus grypus)

Both species of seals that live and breed around the British coasts have been subjected to exploitation for probably thousands of years. Recently this has consisted mainly of the killing of pups for fur. Common seal pups do not remain on land for more than a few hours after birth and the breeding populations are more dispersed. They have therefore suffered less from over hunting and there are no reasons to believe that gross changes in range or abundance have taken place. Bonner (1972) reported that at present about 1800–2000 pups are taken each year in Britain.

The grey seal has shown more marked changes in both range and abundance. They tend to form large breeding colonies and the pups remain ashore and vulnerable for several weeks. Legislation to control the taking of grey seals was enacted in 1914 and since then there has been a steady increase in numbers— from about 4000–5000 in 1928 to about 36 000 at present (Bonner, 1972). Depopulation of small islands has been followed by the establishment of seal colonies. St Kilda for example had no breeding seals before it was evacuated in 1930 but now produces about a hundred pups annually. The breeding behaviour of the grey seal allows the taking of more accurate censuses than in the case of any other of our mammals, by counting pups. The expansion of the breeding colony on the Farne Islands, Northumberland has been closely documented in this way, the count rising from 751 pups in 1956, to 1956 pups in 1970 (Bonner and Hickling, 1971).

10. Hares (Lepus)

Our two hares, the brown hare (Lepus capensis) and the mountain hare (L. timidus), constitute our only example of a pair of closely related species with mutually exclusive ranges within the British mainland. Both have been grossly affected by man but the original distribution of the mountain hare can be assumed to have been Ireland and all the higher ground of the Scottish Highlands; of the brown hare the remainder of the mainland. The mountain hare has been introduced beyond the Highlands, beginning in the 1830s when it was established in southern Scotland (Peebles, etc.) with introductions to the Pennines and to North Wales later in the nineteenth century (Barrett-Hamilton and Hinton, 1910–21). These populations survive although the species has apparently never become very widespread in Wales. There has been a complex history of introduction and extinction of both species on the smaller islands.

11. Red squirrel (Sciurus vulgaris)

The recent decline of the red squirrel has already been mentioned in dealing with the grey squirrel, but before the introduction of the grey towards the end of the nineteenth century the red squirrel had a history of instability. Over a long period its range must have been progressively reduced as a result of deforestation and in Scotland this resulted in complete extinction of the native stock by 1840. But by that time the species had been reintroduced in the developing plantations in a number of the areas from which it had first disappeared, e.g. in Midlothian in 1772 and in Perthshire in 1793. From these and other centres they had recolonized the entire county by the end of the nineteenth century (Shorten, 1954). Most of these introductions were of animals from England or Wales but some Scandinavian animals were released in Perthshire and perhaps elsewhere.

In England reafforestation throughout the nineteenth century, especially with conifers, was in time to save the species from a similar fate and during the second half of the century they increased considerably and were widespread and abundant by the 1890s when the grey squirrel began to increase.

12. Voles and wood mice

Three species of small rodents, the bank vole (*Clethrionomys glareolus*), the field vole (*Microtus agrestis*) and the wood mouse (*Apodemus sylvaticus*) are at present ubiquitous in Britain and there is no reason to suspect any major changes. On a local scale the destruction of woodland would only have eliminated the bank vole and wood mouse where no bushes or hedgerows remained, as in large areas of sheep pasture in southern Scotland. The field vole is notorious for its periodic fluctuations in numbers and in 1891–93 they reached plague proportions in southern Scotland. It is not normally abundant on heavily grazed ground but the exclusion of rabbits and other grazers associated with the establishment of forest plantations provides ideal conditions and very high densities may develop under these circumstances.

The bank vole was unknown in Ireland until 1964 when it was found in Co. Kerry. Since suitable habitat is abundant throughout Ireland it seems very probable that this represented a recent, although undocumented, introduction. No evidence of spread has yet been obtained but surveys conducted in 1967 and 1969 have provided a good basis for future monitoring (Fairley and O'Donnell, 1970).

The yellow-necked mouse (*Apodemus flavicollis*) has a very curious distribution in south-east England and the Severn Valley (Corbet, 1971), but since it was only clearly recognized as distinct from the wood mouse in 1894 it is not possible to say whether the range has been changing. The peculiar distribution could perhaps be interpreted as the result of slow and incomplete recolonization after deforestation had reached its peak, but the possibility of accidental introduction by man should not be ruled out.

The water vole (*Arvicola terrestris*) is as widespread as the small voles but is likely to be scarce and local in the Scottish Highlands. There is no evidence of any gross changes in its status.

13. Harvest mouse (Micromys minutus)

The harvest mouse now appears to be confined to the south-eastern half of England, with two outliers in North Wales and Cheshire (Corbet, 1971). It is generally stated that it formerly occurred in the north of England and eastern Scotland and quoted as an example of gross reduction of range due to the mechanization of agriculture. But there is virtually no evidence for such a reduction. It has been widely confused with the wood mouse (*Apodemus sylvaticus*) which is frequently found in harvest fields. Barrett-Hamilton and Hinton (1910–21) detail a number of probable harvest mouse nests found in Scotland last century but it seems that no actual specimen survives.

This species survives in hedgerows, wood margins and on waste ground with a dense growth of tall annual weeds and is not dependent upon cereal crops.

14. Dormouse (Muscardinus avellanarius)

The dormouse depends upon woodland with a rich shrub layer and it has almost certainly suffered a decline during the course of this century as covert woodland has been destroyed. It has survived well wherever there is suitable habitat in southern England but there are reasons for believing that there has been a significant reduction of range in England north of the Humber. Nineteenth-century records show it to have been distributed north at least to the Tyne but no recent (post-1960) records have been received from north of the Humber (Fig. 3).

15. *Red deer* (Cervus elaphus) *and Roe deer* (Capreolus capreolus)

The history of our native deer has been one of progressive reduction as woodland was destroyed until by the end of the eighteenth century they had been virtually eliminated from the lowlands, except for some red deer in parks.

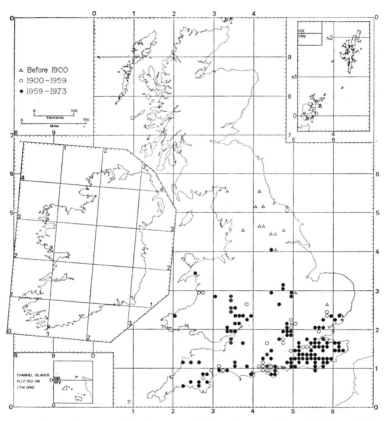

Fig. 3. British distribution of the dormouse (*Muscardinus avellanarius*). (Pre-1960 records are not shown for all areas but none are available from further north than shown here).

(The roe has never been kept in parks to any great extent.) In England red deer survived in Exmoor and the Lake District, the roe only in the north.

Reintroduction of roe took place in southern England and in East Anglia at various times throughout the nineteenth century and the species is now widely distributed in the southern counties from Cornwall to Sussex and in East Anglia (see Page, 1962, for a concise account with maps).

There have been many park herds of red deer in England and some of these have given rise to small feral populations during the course of this century (Whitehead, 1964).

16. Cetaceans

A system for recording strandings of whales, dolphins and porpoises on the British coasts, with the help of the Coastguard service, was begun by the British Museum (Natural History) in 1913 and continues to provide valuable data, the total number of identifications to 1966 being 1549. As one would expect the figures reflect the decline of the large whales, with for example four blue whales (*Balaenoptera musculus*) between 1913 and 1923 but none thereafter; and 31 common rorquals (*B. physalis*) between 1913 and 1947 but only four between 1948 and 1966. But it has also shown changes in frequency of the smaller species: common dolphins (*Delphinus delphis*), a southern species, were second in order of frequency (after porpoise) in the period 1913–47, with 116 strandings, but had fallen to seventh place in the period 1948–66, with only 19 strandings. Species that are more frequent in the recent period are pilot whale (*Globicephala melaena*) with 46 strandings between 1948 and 1966, and Sowerby's beaked whale (*Mesoplodon bidens*) with 10 (Fraser, in press).

CONCLUSIONS

The foregoing data may be summarized in chronological sequence. Prior to 1800 the most significant changes were associated with the progressive reduction in forest. With respect to some of the large mammals subjected to hunting pressure, the loss and fragmentation of forest facilitated their extermination. Urus and reindeer may have become extinct before the Roman period; brown bear and beaver probably between the ninth and twelfth centuries; wild boar in the seventeenth and wolf in the eighteenth century. By 1800 the fallow deer had been introduced but was mainly in parks; the house mouse was probably widespread and the brown rat had already gone a long way towards replacing the black. The species most affected by deforestation without becoming extinct were the red and roe deer, which by then had virtually disappeared from the lowlands; the red squirrel, already extinct in most of Scotland and scarce in England; and probably also the pine marten which was hunted for fur in the surviving woodland.

From our point of view the nineteenth century can be categorized above all as the era of game preservation. Several factors led to this—the loss of large game already described; the emergence, in larger numbers than ever before, of a new class that could afford to hunt; and the mobility provided by the railways. A second important factor, the gradual reafforestation which began at the end of the eighteenth century, was closely linked with game preservation. The results were felt by all groups of mammals. Firstly the species hunted were not allowed to diminish further and were reintroduced to areas from which they had been lost. The survival of hares, foxes and otters was thereby assured; roe deer were reintroduced to southern England; mountain hares were introduced in southern Scotland and England beyond their original range. Predators, on the other hand, were ruthlessly persecuted. Wild cat, pine marten and polecat were rapidly reduced in range whilst the equally severe persecution of predatory birds significantly reduced the predation pressure on rabbits, rodents and shrews. The fox, badger and otter were in an equivocal position, being hunted for sport but at the same time persecuted for their supposed or real depredations on other game. Persecution of stoats, weasels, hedgehogs and rats was massive but had no more than local and temporary effects on their numbers or range. The red squirrel increased enormously during the nineteenth century. The demands of the fur trade were met largely by importation and it was only towards the end of the century that red squirrels had to be controlled because of damage to forestry.

Significant introductions during the nineteenth century were few in spite of a great interest in the subject. Many exotic deer, however, were introduced to parks, only later to add to our wild fauna. Towards the end of the century introduced grey squirrels began to spread.

The twentieth century has been an age of introduction, largely unintentional. The grey squirrel has continued to expand its range at the expense of the red. Other introduced species seem to have succeeded in spreading without any corresponding reduction in native species. Amongst the rodents the fat dormouse has spread only slowly and has done only a little damage (to conifers); coypu and musk-rat proved too damaging and were successfully controlled and exterminated respectively. Neglect of deer parks, especially during the two wars, allowed exotic deer to escape, and accelerating reafforestation has allowed them and the native species to expand. Amongst the carnivores the mink has made an extremely rapid colonization of the country, much more rapid than the recovery of the very similar native polecat.

The polecat, wild cat and marten were saved from extinction by a number of factors. Persecution by gamekeepers was greatly reduced during the two wars because of lack of man-power and has since been reduced because of a growing realization, backed by ecological studies, that such predators have little effect on the available "crop" of game. The recovery of these species has also been greatly facilitated by the development of state forests.

Coming to the developments of the last two decades, several additional factors, although not new, are rapidly becoming more significant. The decline of the otter, in contrast to other carnivores which either remain abundant or are expanding, can probably be attributed to the fact that three increasingly important adverse factors all happen to affect it—industrial pollution (in this case of rivers), pollution through their food by agricultural chemicals, and the pressures on the countryside of an increasingly large, mobile and affluent population. The other species that appear to be declining, the bats, have also suffered from the last two factors. But many bats are on the extreme edge of their range in Britain and are therefore especially vulnerable to any adverse change of environment, however slight.

Apart from the increasing use of chemicals, other agricultural changes have been significant but usually local in their effect. The removal of hedges and small coverts, for example, destroys necessary habitat for many small mammals as well as fox and badger.

On the national scale however we can conclude that most species are at present either holding their position or expanding. Only the greater horseshoe bat and the mouse-eared bat are in any immediate danger of extinction in Britain (and the latter has probably never been otherwise); whilst only the other bats, the otter and the dormouse show any sign of declining, although so far this has had only a marginal influence on their overall range in the country.

REFERENCES

ANON. (1969). The otter in Britain. Oryx 10, 16–22.

BARRETT-HAMILTON, G. E. H. and HINTON, M. A. C. (1910–21). "A History of British Mammals". Gurney and Jackson, London.

BELL, T. (1874). "A History of British Quadrupeds", 2nd edition. van Voorst, London.

BENTLEY, E. W. (1959). The distribution and status of Rattus rattus L. in the United Kingdom in 1951 and 1956. J. Anim. Ecol. 28, 299–308.

BENTLEY, E. W. (1964). A further loss of ground by Rattus rattus L. in the United Kingdom during 1956–61. J. Anim. Ecol. 33, 371–373.

BONNER, W. N. (1972). The grey and common seal in European waters. Oceanogr. Mar. Biol. Ann. Rev. 10, 461–507.

BONNER, W. N. and HICKLING, G. (1971). The grey seals of the Farne Islands. Trans. nat. Hist. Soc. Northumb. 17, 141–162.

CORBET, G. B. (1961). Origin of the British insular races of small mammals and of the Lusitanian fauna. Nature, Lond. 191, 1037–1040.

CORBET, G. B. (1964). The grey long-eared bat Plecotus austriacus in England and the Channel Islands. Proc. zool. Soc. Lond. 143, 511–515.

Corbet, G. B. (1969a). "The Identification of British Mammals." British Museum (Natural History), London.

Corbet, G. B. (1969b). The geological significance of the present distribution of mammals in Britain. *Bull. Mammal Soc.* **31**, 14–16.

Corbet, G. B. (1971). Provisional distribution maps of British mammals. *Mammal Rev.* **1**, 95–142.

Dansie, O. (1970). "Muntjac (*Muntiacus* sp.)". [Publ. no. 2]. British Deer Society.

Davis, R. A. (1963). Feral coypu in Britain. *Ann. appl. Biol.* **5**, 345–348.

Deane, C. D. and O'Gorman, F. (1969). The spread of feral mink in Ireland. *Ir. Nat. J.* **16**, 198–202.

Fairley, J. S. and O'Donnell, T. (1970). The distribution of the bank vole *Clethrionomys glareolus* in south-west Ireland. *J. Zool. Lond.* **161**, 273–276.

Fitter, R. S. R. (1959). "The Ark in our Midst". Collins, London.

Fraser, F. C. (in press). "Report on cetaceans stranded on the British coasts from 1948 to 1966." British Museum (Natural History), London.

Fraser, F. C. and King, J. (1954). Faunal remains. *In* "Excavations at Star Carr, an early mesolithic site at Seamer, near Scarborough, Yorkshire" (J. G. D. Clark, ed.), pp. 70–95. Cambridge University Press.

Harting, J. E. (1880). "British animals Extinct within Historic Times". Trübner, London.

Jenkins, D. (1962). The present status of the wild cat (*Felis silvestris*) in Scotland. *Scot. Nat.* **70**, 126–138.

Laurie, E. M. O. (1946). The coypu (*Myocastor coypus*) in Great Britain. *J. Anim. Ecol.* **15**, 22–34.

Lloyd, H. G. (1962). The distribution of squirrels in England and Wales, 1959. *J. Anim. Ecol.* **31**, 157–166.

Martin, P. S. and Wright, H. E. (eds) (1967). "Pleistocene Extinctions: the Search for a cause". Yale University Press, Newhaven and London.

Matheson, C. (1962). "The Brown Rat" (Animals of Britain, **16**). Sunday Times Publications, London.

Neal, E. (1948). "The Badger". Collins, London.

Norris, J. D. (1967). A campaign against feral coypu (*Myocastor coypus* Molina) in Great Britain. *J. appl. Ecol.* **4**, 191–199.

Page, F. J. T. (1962). "Roe Deer" (Animals of Britain, **12**). Sunday Times Publications, London.

Page, F. J. T. (1970). "The Sussex Mammal Report, 1969". Sussex Naturalists Trust, Henfield.

Racey, P. A. and Stebbings, R. E. (1972). Bats in Britain—a status report. *Oryx* **11**, 319–327.

Sheail, J. (1971). "Rabbits and their History". David and Charles, Newton Abbot.

Shorten, M. (1954). "Squirrels". Collins, London.

Shorten, M. (1957). Squirrels in England, Wales and Scotland, 1955. *J. Anim. Ecol.* **26**, 287–294.

Stebbings, R. E. (1970). A bat new to Britain, *Pipistrellus nathusii*, with notes on its identification and distribution in Europe. *J. Zool. Lond.* **161**, 282–286.

TEAGLE, W. G. (1967). The fox in the London suburbs. *London Nat.* **46**, 44–68.

THOMPSON, H. V. (1953). The edible dormouse (*Glis glis* L.) in England, 1902–51. *Proc. zool. Soc. Lond.* **122**, 1017–1024.

THOMPSON, H. V. (1968). British wild mink. *Ann. appl. Biol.* **61**, 345–349.

THOMPSON, H. V. and WARDEN, A. N. (1956). "The Rabbit". Collins, London.

WALTON, K. C. (1968). The distribution of the polecat, *Putorius putorius*, in Great Britain, 1963–67. *J. Zool. Lond.* **155**, 237–240.

WARWICK, T. (1934). The distribution of the musk rat (*Fiber zibethicus*) in the British Isles. *J. Anim. Ecol.* **3**, 250–267.

WARWICK, T. (1940). A contribution to the ecology of the musk rat (*Ondatra zibethica*) in the British Isles. *Proc. zool. Soc. Lond.* (A) **110**, 165–201.

WHITEHEAD, G. K. (1964). "The Deer of Great Britain and Ireland". Routledge and Kegan Paul, London.

WHITEHEAD, G. K. (1972). "The Wild Goats of Great Britain and Ireland". David and Charles, Newton Abbot.

YALDEN, D. W. and HOSEY, G. R. (1971). Feral wallabies in the Peak District. *J. Zool. Lond.* **165**, 513–520.

12 | The Changing Status of Breeding Birds in Britain and Ireland

J. T. R. SHARROCK

Ornithological Atlas, British Trust for Ornithology, Tring

Abstract: The number of species breeding regularly in Britain and Ireland was fairly static from 1800 to 1949, varying from 178 to 181 in each decade, but since 1950 there has been a net gain (attributed largely to increased protection) of about five species per decade, leading to a current total of at least 192.

At least 129 of the nearly 200 species which have bred regularly at some time since 1800 have shown marked changes in range and/or numbers. The proportion of these expanding their ranges or increasing in numbers (successful) has increased dramatically, while the proportion contracting their ranges or decreasing in numbers (failing) has decreased. Changes in excess of the average were exhibited by species persecuted by man (more failing and fewer successful until 1920) and breeding in wetland habitats (recovering more quickly from a disastrous situation, by the second half of the nineteenth century).

The differences between species breeding in the north and south of Britain are not clear cut as many factors are operating. On the whole, northern species have been more successful than southern species at all times since 1800, except in the second half of the nineteenth century, and the difference has been most marked since 1960.

INTRODUCTION

There have been dramatic changes in the avifauna of Britain and Ireland during the past 172 years, these affecting residents, summer visitors, winter visitors and passage migrants alike. The third and fourth of these groups are not considered in this paper, which is confined to an analysis of the changes which have taken place in the breeding birds.

Anyone considering the changes in status among breeding birds in Britain and Ireland owes an enormous debt to J. L. F. Parslow, whose extensive review of the literature (Parslow, 1967–68, 1973) forms the essential background to any further analysis. This paper relies largely on his work and also that of the

Systematics Association Special Volume No. 6, "The Changing Flora and Fauna of Britain", edited by D. L. Hawksworth, 1974, pp. 203–220. Academic Press, London and New York.

British Ornithologists' Union (1971), Alexander and Lack (1944), Witherby *et al.* (1940–41) and Kinnear (1935). Parslow (*op. cit.*) has reviewed the changes which are known to have taken place in the status of each individual species and it is, therefore, unnecessary to repeat these data here. This paper is concerned with the detection of any underlying patterns.

The English and scientific names in this paper follow Hudson (1971); authorities for scientific names may be found in British Ornithologists' Union (1952, 1971). Scientific names are given in Table I and, therefore, are not repeated in the text.

QUALITATIVE CHANGE

The number of species breeding regularly in Britain and Ireland during 1800–1949 (excluding introductions and reintroductions made since 1800) varied from 178 to 181 in each decade and over this 150 years there was no significant change—losses due to extinctions being matched by gains due to colonizations or recolonizations (Fig. 1). In this period, 11 species became extinct in Britain

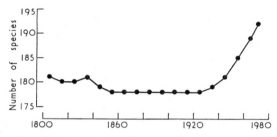

FIG. 1. Number of species (other than those introduced or reintroduced since 1800) breeding regularly in Britain and Ireland in each decade from 1800 to 1972.

and Ireland, but two later returned, and nine colonized, giving no overall change. Since 1950, however, only one species has become extinct, four have recolonized and at least eight have colonized. It is, of course, difficult to judge which of the species breeding in Britain for the first time in recent years constitute firm colonists and which are merely a temporary part of the avifauna. Nevertheless, it is clear that there has recently been a net gain of nearly five species per decade. This is probably largely due to increased protection, both of species and of habitats, and much of the credit must be due to the Royal Society for the Protection of Birds and similar bodies. In recent years much effort has been devoted to preserving habitats, reserves being set up with the immediate aim of preventing extinctions and promoting the firm establishment of colonists and recolonists.

More species are now breeding regularly in Britain and Ireland than at any time since ornithological recording began. The current total of regular breeding species is at least 192 and a total of over 200 species (naturally occurring or introduced prior to 1800) were proved to breed in Britain and Ireland during 1968–72 in the course of the British Trust for Ornithology/Irish Wildbird Conservancy Atlas Project.

QUANTITATIVE CHANGES

Of the nearly 200 species known to have bred regularly in Britain and Ireland at some time since 1800, at least 129 have shown marked changes in status (range and/or numbers). Parslow (*op. cit.*) noted that marked changes occurred in 117 out of 187 species during 1800–1940 and 115 out of 186 species during 1940–67. As he also noted, it is probable that significant changes occurred in even more species than this, since it is difficult to detect changes in numbers (even large ones) of the commonest species. This is likely to be particularly true in the early part of the period, when observers were far fewer than at present.

The data are far too imprecise for numerical comparisons to be made. It is usually possible, however, to categorize each species as (i) increasing in numbers or expanding its range (hereafter referred to as "successful"), (ii) decreasing in numbers or contracting its range (hereafter referred to as "failing"), or (iii) remaining unchanged. The nineteenth-century information is often too vague to do this satisfactorily other than for the two halves of the century. Since 1900 the data are far more amenable to such treatment but short-term effects (e.g. severe winters) can only be eliminated by dealing with periods in excess of one decade. For the purpose of analysis, therefore, the past 172 years have been divided into six unequal periods: 1800–49, 1850–99, 1900–19, 1920–39, 1940–59 and 1960–72. The status changes of the 129 species are shown for each of these six periods in Table I.

TABLE I. Summary of status changes in 129 species during six periods from 1800 to 1972

Species	1800–49	1850–99	1900–19	1920–39	1940–59	1960–72
Great Crested Grebe PWS (*Podiceps cristatus*)	−	+	+ +	+ +	+ +	+
Slavonian Grebe WN (*P. auritus*)			+	+	+	+
Black-necked Grebe W (*P. nigricollis*)			+	+ +	−	0

TABLE 1—*continued*

Species	1800–49	1850–99	1900–19	1920–39	1940–59	1960–72
Fulmar N (*Fulmarus glacialis*)	0	+	+	++	++	+
Gannet P (*Sula bassana*)	0	0	+	+	+	+
Bittern PWS (*Botaurus stellaris*)	—		+	0	+	0
Wigeon WN (*Anas penelope*)	+	++	+	+	+	0
Pintail W (*A. acuta*)		+	0	0	+	0
Shoveler W (*A. clypeata*)	0	0	++	+	+	0
Tufted Duck W (*Aythya fuligula*)		++	++	+	+	+
Pochard W (*A. ferina*)	0	+	+	+	+	+
Common Scoter WN (*Melanitta nigra*)	0	0	0	+	0	0
Eider N (*Somateria mollissima*)	—	+	+	+	+	+
Red-breasted Merganser N (*Mergus serrator*)	0	+	++	+	+	+
Goosander N (*M. merganser*)		++	+	+	+	0
Shelduck (*Tadorna tadorna*)	0	0	+	+	+	+
Golden Eagle PN (*Aquila chrysaetos*)	—	—	—	+	0	0
Buzzard P (*Buteo buteo*)	—	—	—	+	+	+
Sparrowhawk P (*Accipiter nisus*)	—	—	—	+	+−	−+
Red Kite PS (*Milvus milvus*)	—	+	0	0	+	+
White-tailed Eagle P (*Haliaeetus albicilla*)	—	—	—			
Honey Buzzard P (*Pernis apivorus*)	—	—	—	0	0	+
Marsh Harrier PWS (*Circus aeruginosus*)	—	—	—	0	+	—
Hen Harrier PN (*C. cyaneus*)	—	—	—	—	+	+

Species	1800–49	1850–99	1900–19	1920–39	1940–59	1960–72
Montagu's Harrier PS (*C. pygargus*)	0	0	0	0	+	−
Osprey PN (*Pandion haliaetus*)	−	−	−		+	+
Hobby PS (*Falco subbuteo*)	−	−	0	0	0	0
Peregrine PN (*F. peregrinus*)	−	−	0	0	+−	0
Merlin N (*F. columbarius*)	−	−	−	−	−	−
Kestrel (*F. tinnunculus*)	0	0	0	0	0	−
Red Grouse N (*Lagopus lagopus*)	0	0	0	0	−	−
Black Grouse N (*Lyrurus tetrix*)	−	−	−	−	−	+
Partridge (*Perdix perdix*)	0	0	−	−	−	−
Quail S (*Coturnix coturnix*)	−	−	−	−	+	+
Spotted Crake W (*Porzana porzana*)	−	−	0	0	0	0
Corncrake (*Crex crex*)	0	−	−	−	−	−
Great Bustard PS (*Otis tarda*)	−					
Oystercatcher (*Haemotopus ostralegus*)	−	−	+	+	+	+
Lapwing (*Vanellus vanellus*)	0	−	0	−	−	−
Ringed Plover (*Charadrius hiaticula*)	0	0	0	−	−	−
Little Ringed Plover S (*C. dubius*)					+	+
Kentish Plover PS (*C. alexandrinus*)	−	−	−	−	−	
Golden Plover N (*Pluvialis apricaria*)	0	0	−	−	−	−
Dotterel PN (*Eudromias morinellus*)	0	−	−	−	0	0
Snipe W (*Gallinago gallinago*)	−	+	+	0	−	−
Woodcock (*Scolopax rusticola*)	+	+	0	0	0	+

TABLE 1—*continued*

Species	1800–49	1850–99	1900–19	1920–39	1940–59	1960–72
Curlew N	0	0	+	+	+	+
(*Numenius arquata*)						
Whimbrel N	0	—	—	0	+	+
(*N. phaeopus*)						
Black-tailed Godwit WS	—				+	+
(*Limosa limosa*)						
Wood Sandpiper WN						+
(*Tringa glareola*)						
Redshank W	—	+	+	+	0	0
(*T. totanus*)						
Ruff PWS	—					+
(*Philomachus pugnax*)						
Avocet PWS	—				+	+
(*Recurvirostra avosetta*)						
Red-necked Phalarope PWN	—	—	0	0	0	0
(*Phalaropus lobatus*)						
Stone Curlew S	—	—	—	+	—	—
(*Burhinus oedicnemus*)						
Great Skua PN	—	—	0	+	+	+
(*Stercorarius skua*)						
Great Black-backed Gull	—	0	+	+	+	+
(*Larus marinus*)						
Herring Gull	0	0	+	+	++	++
(*L. argentatus*)						
Common Gull N	+	+	+	+	+	0
(*L. canus*)						
Black-headed Gull PWN	—	—	+	+	+	+
(*L. ridibundus*)						
Kittiwake P	—	—	+	+	+	+
(*Rissa tridactyla*)						
Roseate Tern	—	0	+	0	+	0
(*Sterna dougallii*)						
Little Tern	—	—	+	+	—	—
(*S. albifrons*)						
Sandwich Tern	—	—	0	0	+	+
(*S. sandvicensis*)						
Great Auk P	—					
(*Pinguinus impennis*)						
Guillemot P	0	0	0	0	—	—
(*Uria aalge*)						
Puffin P	0	0	0	—	—	—
(*Fratercula arctica*)						

Species	1800–49	1850–99	1900–19	1920–39	1940–59	1960–72
Stock Dove S (*Columba oenas*)	0	+	+	+	0	−
Woodpigeon P (*C. palumbus*)	+	+	+	+	+	+
Turtle Dove S (*Streptopelia turtur*)	+	+	+	0	0	0
Collared Dove (*S. decaocto*)					++	++
Cuckoo (*Cuculus canorus*)	0	0	0	0	−	−
Barn Owl P (*Tyto alba*)	0	0	0	0	−	−
Snowy Owl N (*Nyctea scandiaca*)						+
Tawny Owl P (*Strix aluco*)	−	−	+	+	+	0
Long-eared Owl P (*Asio otus*)	0	0	−	−	−	−
Nightjar S (*Caprimulgus europaeus*)	0	0	0	−	−	−
Green Woodpecker S (*Picus viridis*)	0	0	0	+	+	+
Great Spotted Woodpecker S (*Dendrocopos major*)	−	+	+	+	+	0
Wryneck S (*Jynx torquilla*)	0	−	−	−	−	−
Woodlark S (*Lullula arborea*)	0	−	0	+	+	−
Golden Oriole S (*Oriolus oriolus*)		+			+	+
Raven P (*Corvus corax*)	−	−	0	+	+	+
Rook (*C. frugilegus*)	0	0	0	+	+	0
Jackdaw (*C. monedula*)	0	0	+	+	+	0
Magpie PS (*Pica pica*)	0	0	0	+	+	+
Jay PS (*Garrulus glandarius*)	0	0	0	+	+	+
Chough (*Pyrrhocorax pyrrhocorax*)	−	−	−	−	−	0
Coal Tit (*Parus ater*)	0	0	0	0	+	+

TABLE 1—*continued*

Species	1800–49	1850–99	1900–19	1920–39	1940–59	1960–72
Crested Tit N	−	−	0	0	+	+
(*P. cristatus*)						
Nuthatch S	0	0	0	0	+	+
(*Sitta europaea*)						
Bearded Tit WS	−	−	0	0	+	++
(*Panurus biarmicus*)						
Mistle Thrush	++	+	+	+	+	+
(*Turdus viscivorus*)						
Fieldfare N						+
(*T. pilaris*)						
Song Thrush	0	0	0	0	−	−
(*T. philomelos*)						
Redwing N				+	+	++
(*T. iliacus*)						
Ring Ouzel N	0	0	−	−	−	0
(*T. torquatus*)						
Blackbird	0	+	+	+	+	+
(*T. merula*)						
Wheatear N	0	0	−	−	−	−
(*Oenanthe oenanthe*)						
Stonechat	0	0	0	−	−	−
(*Saxicola torquata*)						
Whinchat N	0	0	0	−	−	−
(*S. rubetra*)						
Redstart	0	0	0	−	+	+
(*Phoenicurus phoenicurus*)						
Black Redstart S					+	−
(*P. ochruros*)						
Nightingale S	0	0	0	0	0	−
(*Luscinia megarhynchos*)						
Grasshopper Warbler S	0	0	0	0	0	+
(*Locustella naevia*)						
Savi's Warbler WS	−					+
(*L. luscinioides*)						
Marsh Warbler WS	0	0	0	0	−	−
(*Acrocephalus palustris*)						
Blackcap S	0	0	0	0	0	+
(*Sylvia atricapilla*)						
Dartford Warbler S	−	−	−	−	−	−
(*S. undata*)						
Chiffchaff S	0	+	0	0	+	+
(*Phylloscopus collybita*)						

Species	1800–49	1850–99	1900–19	1920–39	1940–59	1960–72
Goldcrest (*Regulus regulus*)	+	+	+	+	+	+
Firecrest S (*R. ignicapillus*)						+
Pied Flycatcher S (*Ficedula hypoleuca*)	0	+	+	+	+	0
Grey Wagtail (*Motacilla cinerea*)	0	0	+	+	+	0
Red-backed Shrike PS (*Lanius collurio*)	0	0	–	–	–	–
Starling (*Sturnus vulgaris*)	–	+	+	+	+	+
Hawfinch S (*Coccothraustes coccothraustes*)	+	+	+	0	0	0
Greenfinch (*Carduelis chloris*)	0	0	+	+	0	0
Goldfinch PS (*C. carduelis*)	–	–	+	+	+	+
Siskin N (*C. spinus*)	0	0	+	+	+	+
Redpoll N (*Acanthis flammea*)	0	0	+	–	+	++
Serin S (*Serinus serinus*)						+
Bullfinch (*Pyrrhula pyrrhula*)	0	0	0	0	+	+
Chaffinch (*Fringilla coelebs*)	0	0	0	0	–	–
Corn Bunting (*Emberiza calandra*)	0	0	0	0	–	–
Cirl Bunting S (*E. cirlus*)	0	0	0	–	–	–
Reed Bunting (*E. schoeniclus*)	0	0	0	0	+	+
House Sparrow (*Passer domesticus*)	+	+	+	+	0	0
Tree Sparrow (*P. montanus*)	0	0	0	0	–	+

+ = increasing in numbers or expanding its range (successful). – = decreasing in numbers or contracting its range (failing). 0 = remaining unchanged or unknown. ++ = greatly increasing in numbers or greatly expanding its range (not treated separately from + in analysis). The code letters alongside English names indicate, P = persecuted species, W = wetland species, N = northern species, S = southern species.

It must be stressed that the designations as successful (+), failing (−) or showing no significant change or unknown (0) in Table I, while always based on published evidence, are personal assessments. No doubt other analysers would occasionally come to different conclusions in individual cases, depending to an extent on the emphasis placed upon local changes (which are usually ignored in Table I), but the overall trends are acknowledged facts and minor discrepancies due to differing personal opinions would probably balance. Table I is intended only as a basis for further analysis; the fortunes of an individual species are far better traced by referring to Parslow's work (*op. cit.*).

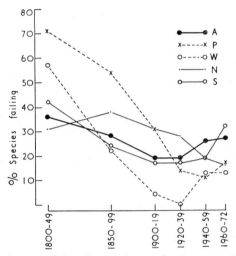

FIG. 2. Percentages of species failing (see text) in each of six periods during 1800–1972. A=all species showing changes (n=129); P=persecuted species (n=35); W= wetland species (n=23); N=northern species (n=32); S=southern species (n=41). The groups are defined in the text and the species in each group are indicated in Table I.

Summing, separately, the successful and failing species in each period produces crude assessments of the changes since 1800 (Figs 2–3). The proportion of successful species has increased steadily and dramatically from 7% in the first half of the nineteenth century to about 50% in the period since 1940. There has been a reciprocal but less dramatic decrease in the number of failing species, from 36% in the early nineteenth century to 19% in the first four decades of this century and then a slight increase to just over 25% since 1940. Improvements in observation and recording over this period may have emphasized these changes (since the percentages of species in which some change has been noted have increased steadily—43%, 48%, 52%, 58%, 78% and 76% in the six periods) but there is no reason to suppose that the appearance or increase of a species is more likely to have been noted and recorded than the disappearance or decline. If only those species in which changes were noted in each period are taken into account, those failing have fallen from 84% in the early nineteenth century, to 58% in the late nineteenth century and have remained relatively steady at 35% during the whole twentieth century (the figures for successful species self-evidently being 16%, 42% and 65%). The general patterns, therefore, are probably genuine. It is not necessary to assume this, however, for valid comparisons to be made between different sections of the avifauna.

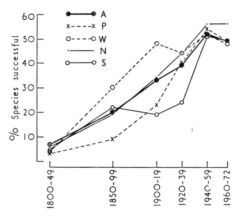

FIG. 3. Percentages of species successful (see text) in each of six periods during 1800–1972. A=all species showing changes (n=129); P=persecuted species (n=35); W=wetland species (n=23); N=northern species (n=32); S=southern species (n=41). The groups are defined in the text and the species in each group are indicated in Table I.

The direct and indirect effects of man upon the avifauna are best represented by selecting two groups—persecuted species and wetland species (wetlands being habitats greatly affected by man and having a distinct avifauna). Persecuted species include those commercially exploited (e.g. Great Crested Grebe —feathers for millinery), collected for food (e.g. Gannet), eggs collected for food (e.g. Black-headed Gull), persecuted by egg-collectors (e.g. Red-backed Shrike), destroyed in the name of game-preservation (e.g. Golden Eagle), shot for taxidermy (e.g. White-tailed Eagle), destroyed to protect crops (e.g. Woodpigeon) and trapped as cage-birds (e.g. Goldfinch). Some species would fall into several of these categories. Wetland species are those which breed in wet habitats, from moorland dubh-lochs to reed-beds and large lakes, but excluding riparian species. Once again, the selection of these species (indicated by P and W in Table I) is a personal one.

Taking the persecuted species, the decline in the number of failing species (Fig. 2) has been very much more marked than the average for all species. No less than 71% of the persecuted species could be defined as failing in the early nineteenth century compared with about 15% since 1920. The increase in successful species was below average, however, until 1920 (Fig. 3). This pattern is to be expected, for as living standards have improved and bird protection laws have been strengthened, there has been a steady reduction in pressure on species which were formerly persecuted.

The pattern in wetland species is similar to that in persecuted species, but the recovery since the early nineteenth century was more rapid. 57% were failing and only 4% successful in the early nineteenth century but even by the second half of the nineteenth century only 21% were failing and 30% successful (in both cases "better" than the average for all species). The main period of wetland destruction was prior to and in the early nineteenth century and the improvement in the position of wetland species reflects a recovery from a disastrous situation. The westward expansions of the ranges of a number of waterfowl influence this pattern but the development of artificial waters (e.g. flooded gravel-pits and reservoirs) has clearly helped to compensate for the loss of natural wetlands.

It is less simple to carry out comparisons within other habitats since, for example, few species are *confined* to coniferous woodland or deciduous woodland or heathland and, indeed, these habitats are seldom distinct within Britain. There are also only a small number of species confined to other distinct habitats (e.g. only 14 of the 129 species are strictly coastal), providing samples that are too small.

Many of the changes in range and status of birds have been attributed to climatic changes. At the simplest level one might expect an inverse relationship between species breeding in the north of Britain and those breeding in the south. Species were divided, therefore, into "northern" and "southern". Northern species were defined as those in which the bulk of the population is currently north of a line from the Humber to the Mersey *or* which have a range which has contracted northwards or has spread southwards. Southern species were defined as those in which the bulk of the population is currently south of a line from the Humber to the Mersey *or* which have a range which has contracted southwards or has spread northwards.

An average proportion of northern species were successful until 1920, since when they have slightly exceeded the average for other species (Fig. 3).

An average proportion of southern species were successful from 1800 to 1899 and since 1940 but during the first four decades of this century 14–15% fewer southern species were successful compared with all species (Fig. 3).

A below-average proportion of northern species were failing in the early nineteenth century and again since 1940, but an above-average proportion were failing from 1850 to 1939 (Fig. 2). Failing southern species have conformed more closely to the average throughout the period, but an above-average number were failing in the early nineteenth century and since 1960 (Fig. 2).

TABLE II. Trends within northern, southern, persecuted and wetland species during six periods since 1800

Group	Status	1800–49	1850–99	1900–19	1920–39	1940–59	1960–72
Northern	Failing	+5	−10	−12	−9	+7	+11
$n = 32$	Successful	−1	−1	+1	+5	+4	+7
Southern	Failing	−6	+4	+2	+2	+7	−5
$n = 41$	Successful	−2	+2	−14	−15	−1	0
Persecuted	Failing	−35	−26	−12	+5	+15	+10
$n = 35$	Successful	−4	−11	−10	+1	+2	0
Wetland	Failing	−21	+6	+15	+19	+13	+14
$n = 23$	Successful	−3	+10	+15	+5	0	−1

+ = below average number of failing species or above average number of successful species; − = above average number of failing species or below average number of successful species; 0 = average number. The numerals are percentage difference from average for 129 species.

Thus, there has been a confused pattern in the proportions of successful and failing northern and southern species over the past 172 years (summarized in Table II). This is not surprising, for no single factor is operating. Even if one ignores the effects of cold winters on resident species, habitat changes and the effects of human activity, one species may be adversely affected by a series of wet springs (killing nestlings), another by cold springs (repressing spring migration into an area at the north of its range) and a third by wet or cold summers (reducing its insect food supply). Thus, one southern species may be successful at the same time that another southern species with the same distribution is failing; a successful northern species may be extending its range southwards while a successful southern species is extending its range northwards. Many such specific examples could be quoted.

Goosanders colonized Scotland in the 1870s, spread to south-west Scotland in the 1940s, colonized Northumberland in 1941 and Cumberland in 1950.

The range is still expanding and the first Welsh breeding record was in 1972
(Fig. 4). Green Woodpeckers were confined to England and Wales until 1951,
when they bred for the first time in Scotland; they are currently still spreading
in Scotland (Fig. 5). Woodlarks were in a phase of expansion from 1920 to

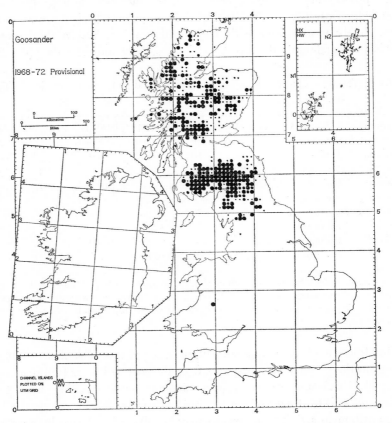

FIG. 4. A northern species which has spread southwards. Provisional map of distribution
of Goosander (*Mergus merganser*) during 1968–72 from preliminary results of the
British Trust for Ornithology/Irish Wildbird Conservancy Atlas Project. The
small, medium and large dots signify, respectively, possible, probable and
confirmed breeding.

1951 and during 1950–53 were breeding in about 40 English and Welsh counties.
During 1968–72 breeding was proved in only 15 counties and probably occurred
in only another nine (Fig. 6), having become extinct in at least 16 counties in
the intervening 14 years. The current withdrawal southwards of southern

British species like Woodlark, Red-backed Shrike and Wryneck is occurring simultaneously with colonization by south European species such as Savi's Warbler and Cetti's Warbler and in few cases has it been possible to do more than guess at the controlling factors. The Wryneck provides an interesting case

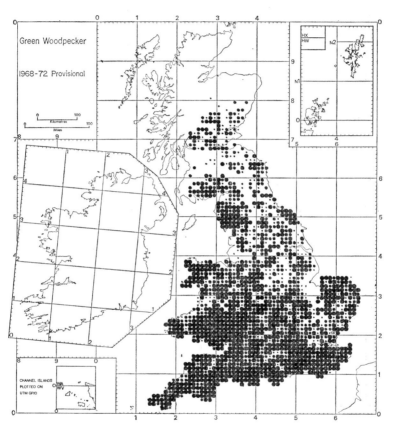

FIG. 5. A southern species which has spread northwards. Provisional map of distribution of Green Woodpecker (*Picus viridis*) during 1968–72 from preliminary results of the British Trust for Ornithology/Irish Wildbird Conservancy Atlas Project. The small, medium and large dots signify, respectively, possible, probable and confirmed breeding.

for it has shown a very marked contraction in the south. In the nineteenth century it bred in every English and Welsh county except Cornwall and Northumberland. In 1958 the total British population was estimated to be only 100–200 pairs and by 1966 only 25–30 pairs in the extreme south-east of

England (Monk, 1963; Peal, 1968). Spring records in Scotland increased from about 1950, however, and breeding has been proved and a number of summering birds located—probably representing a colonization from Scandinavia (Burton *et al.*, 1970). Similarly, there is a recent case of probable breeding of

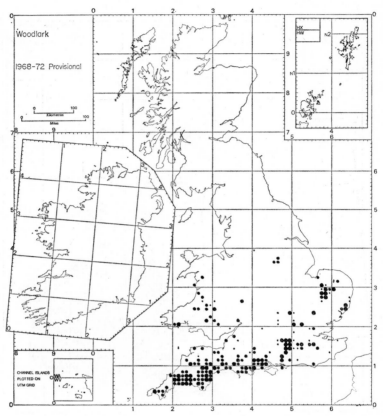

Fig. 6. A southern species which has contracted southwards. Provisional map of distribution of Woodlark (*Lullula arborea*) during 1968–72 from preliminary results of the British Trust for Ornithology/Irish Wildbird Conservancy Atlas Project. The small, medium and large dots signify, respectively, possible, probable and confirmed breeding.

Red-backed Shrike in Orkney despite the decline of the southern English population from about 253 pairs in 1960 to less than 90 pairs in 1971 (Bibby, 1973).

Despite the anomalies, however, it is possible to deduce some broad trends from Table II.

CONCLUSIONS

Compared with the average, persecuted species were declining in 1800–1919 but have increased since 1920. Wetland species were declining in 1800–49 but have been increasing since. Northern species were declining markedly in 1850–1939 but were increasing marginally in 1800–49 and markedly since 1940. Southern species were declining in 1800–49, 1900–39 (markedly) and since 1960 but were increasing (marginally) in 1850–99 and 1940–59. The changes in persecuted and wetland species have been very noticeable but the pattern in northern and southern species has been less clear-cut. The only case of a coincident large increase in successful species and large decrease in failing species (compared with the average) concerns northern species since 1960. Comparing only the northern and southern species, the only period which appears to have been more advantageous to southern species was the second half of the nineteenth century; a higher proportion of northern species have been successful and(or) a lower proportion failing (to give a balance in favour of northern species) in all of the other five periods analysed here.

The B.T.O./I.W.C. Atlas Project has now provided a base-line for distribution studies, and repeat surveys at intervals will result in distribution changes being precisely known for distinct periods of time. Thus, it may be possible in the future to discover the links between external factors and range changes which at present, apart from the effects of man, appear to be very subtle.

ACKNOWLEDGEMENTS

I am most grateful to Mr K. Williamson for many helpful comments, to Dr D. L. Hawksworth for drawing my attention to Kinnear (1935), to the British Trust for Ornithology for allowing me to include breeding distribution maps prior to the publication of "The Atlas of breeding birds in Britain and Ireland" and to Mr J. L. F. Parslow and Mr T. Poyser (author and publisher, respectively) for loaning me the page-proofs of the forthcoming book "Breeding Birds of Britain and Ireland".

REFERENCES

ALEXANDER, W. B. and LACK, D. (1944). Changes in status among British breeding birds. *Brit. Birds* **38**, 42–45, 62–69, 82–88.

BIBBY, C. J. (1973). The Red-backed Shrike: a vanishing British bird. *Bird Study* **20**, 103–110.

BRITISH ORNITHOLOGISTS' UNION (1952). "Check-list of the Birds of Great Britain and Ireland." British Ornithologists' Union, London.

BRITISH ORNITHOLOGISTS' UNION (1971). "The Status of Birds in Britain and Ireland." British Ornithologists' Union, Oxford and Edinburgh.

BURTON, H., LLOYD EVANS, T. and WEIR, D. N. (1970). Wrynecks breeding in Scotland. *Scott. Birds* **6**, 154–156.

HUDSON, R. (1971). "A Species List of British and Irish Birds." British Trust for Ornithology, Tring.

KINNEAR, W. B. (1935). Changes in the British fauna and flora in the past fifty years. (2) Birds. *Proc. Linn. Soc. Lond.* **148**, 35–39.

MONK, J. F. (1963). The past and present status of the Wryneck in the British Isles. *Bird Study* **10**, 112–132.

PARSLOW, J. L. F. (1967–68). Changes in status among breeding birds in Britain and Ireland. *Brit. Birds* **60**, 2–47, 97–123, 177–202, 261–285, 396–404, 495–508, **61**, 49–64, 241–255.

PARSLOW, J. L. F. (1973). "Breeding Birds of Britain and Ireland: a Historical Survey." T. and A. D. Poyser, Berkhamsted,

PEAL, R. E. F. (1968). The distribution of the Wryneck in the British Isles, 1964–1966. *Bird Study* **15**, 111–126.

WITHERBY, H. F., JOURDAIN, F. C. R., TICEHURST, N. F. and TUCKER, B. W. (1940–41). "The Handbook of British Birds." H. F. and G. Witherby, London.

13 | New Bird Species Admitted to the British and Irish Lists Since 1800

K. WILLIAMSON

British Trust for Ornithology, Tring

Abstract: The decadal average of new bird species added to the British and Irish lists since 1800 is 13. During the first half of the nineteenth century an above-average rate was maintained, but a number of these species later proved to be passage migrants or winter visitors. Additions from 1870 to the end of the century, and again from 1910–49, were few, but 50 new species were added between 1950–69.

The decadal average of Palaearctic newcomers is 7·4, and of exclusively North American species 3·6. During the past 20 years over half of the species added have been Americans; four out of five new families, and 20 out of 25 new genera, belong to the Nearctic avifauna. The American invasion has been on a tremendous scale.

During these 20 years the relative positions of the Azores high pressure and Icelandic low pressure systems have changed. The mean centre of the Icelandic low is now some 10 degrees of latitude farther south than earlier in the century, and its position has moved to the west. The mean path of the Atlantic storm-track on the southern side of the depressions is probably more opportunely placed than previously to bring about wind-aided trans-Atlantic flights from the eastern coast of North America to the British Isles.

The number of full species admitted to the British and Irish lists of birds during the 17 decades which have elapsed since 1800 has varied from as few as four to as many as 27 per decade, the average being 13 (Table I).

There is a tendency for decades well above, or well below, the average to run together. During the first half of the nineteenth century an above-average rate was maintained, the decades 1830–39 and 1850–59 being especially good; but as a number of these birds later proved to be regular passage migrants and(or) winter visitors, the figures largely reflect the growth of our knowledge of their true status. Additions from 1870 to the end of the century were few and eclipsed

---

Systematics Association Special Volume No. 6, "The Changing Flora and Fauna of Britain", edited by D. L. Hawksworth, 1974, pp. 221–227. Academic Press, London and New York.

Table I. Numbers of new bird species admitted to the British and Irish lists since 1800 (decadal totals)

	Gaviiformes	Podicipediformes	Procellariiformes	Pelecaniformes	Ciconiiformes	Anseriformes	Falconiformes	Galliformes	Gruiformes	Charadriiformes	Columbiformes	Cuculiformes	Strigiformes	Caprimulgiformes	Apodiformes	Coraciiformes	Piciformes	Passeriformes	New species	New genera	New families	Palaearctic	Nearctic	Holarctic and Oceanic	Nearctic as % of total
1800–09	—	1	—	—	2	1	—	—	—	4	—	—	2	—	—	—	—	1	11	5	1	5	3	3	27
1810–19	—	—	—	—	2	3	—	—	—	4	—	—	1	—	—	—	—	5	15	4	1	12	0	3	0
1820–29	—	—	1	—	1	1	1	—	—	5	—	1	—	—	1	—	—	4	15	7	0	7	2	6	13
1830–39	—	—	4	—	—	5	1	—	—	6	—	—	—	—	1	—	—	4	21	8	1	7	6	8	30
1840–49	—	—	—	—	—	—	2	—	1	3	1	1	—	—	—	—	—	7	15	9	0	14	1	0	7
1850–59	1	—	3	—	—	2	—	—	1	8	—	—	—	1	—	—	—	5	21	6	1	9	7	5	33
1860–69	—	—	—	—	—	1	2	—	1	4	—	1	—	—	—	—	—	7	16	3	0	11	4	1	19
1870–79	—	—	—	—	1	—	—	—	—	1	1	1	1	1	—	—	—	2	4	0	0	1	2	1	50
1880–89	—	—	1	—	—	1	—	—	—	1	—	—	—	—	—	—	—	2	7	2	0	6	1	0	14
1890–99	—	—	—	—	—	—	—	—	—	1	—	—	—	—	—	—	—	6	8	4	1	6	1	1	13
1900–09	—	—	1	—	—	—	—	—	2	4	—	—	—	—	—	—	—	12	18	0	0	12	5	1	28
1910–19	—	—	—	—	—	—	—	—	—	1	—	—	—	—	—	—	—	5	6	1	0	5	0	0	0
1920–29	—	—	—	—	—	—	1	—	—	1	—	—	—	1	—	—	—	2	4	1	0	2	2	1	50
1930–39	—	—	—	—	—	—	1	—	—	1	1	—	—	—	—	—	—	2	4	0	1	1	1	0	25
1940–49	—	—	—	—	—	—	—	—	—	—	—	—	—	—	—	—	—	4	4	0	0	4	0	2	0
1950–59	—	—	—	1	1	2	—	—	1	6	1	—	—	1	—	1	—	15	27	10	2	9	16	2	59
1960–69	—	1	—	—	—	—	—	—	—	1	—	—	—	1	1	—	—	20	23	10	2	13	10	0	44
Totals	1	2	10	1	6	16	7	0	5	51	3	3	4	3	3	1	0	103	219	70	10	125	61	33	

only by the meagre totals during the 40 years from 1910 to 1949. The first decade of the twentieth century was better than average, and the two most recent ones have seen a remarkable recovery from the preceding lean period with a combined total of 50 new species.

The decadal average of Palaearctic newcomers is 7·4. Figures were fairly high down to 1869 for the reason given above. Except for 1900–09 they were then well below average until the two recent decades produced a combined total of 22 new species of Old World origin. An interesting fact is that seven out of nine of the Palaearctic species in the period 1850–59 have a Mediterranean or south-eastern source, whereas a majority in the following decade are of eastern or Siberian origin.

The decadal average of exclusively North American species is 3·6. Decades with well above-average recruitment from this source were 1830–39 (30% of the additions), 1850–59 (33%), and again the two most recent decades (59% and 44% respectively). During the past 20 years the sheer weight of the American invasion has been staggering: for example, there were over 130 individuals of 32 species during the three years 1960–62, Pectoral Sandpipers (*Calidris melanotos*) heading the list with about 20 records annually (Pyman *et al.*, 1961; Swaine *et al.*, 1962; Harber *et al.*, 1963). As a further instance, Wilson's Phalarope (*Phalaropus tricolor*), admitted in 1954, was irregular till 1960, and has occurred annually since, culminating in ten occurrences in 1971 (Smith *et al.*, 1972). Only one of the five new families, and five of the 25 new genera recorded in this century belong to the Old World avifauna. For analytical discussions of the seasonal and geographical patterns of American land-bird occurrences see Nisbet (1959, 1963) and Sharrock (1971).

In the last century the years 1850–55 seem to have been the most prolific with an average of about seven American birds annually in Europe (data from Dalgleish, 1880; Alexander and Fitter, 1955). The years in and around this period show the biggest concentrations of individuals of American species not admitted to the official lists, either on suspicion of being escapes from captivity or because ornithologists at that time did not believe that small land-birds could cross the Atlantic unaided. Though ship-assisted passage has always been a possibility (Durand, 1972) it is now generally accepted that wind-borne crossings frequently occur (Williamson, 1954; Williamson and Ferguson-Lees, 1960; Nisbet, 1963).

There is no doubt that bird-watching has enjoyed far more popularity, and has been carried out with far greater expertise, since World War II than at any other period in ornithological history. This intense vigour in seeking out and recording the rarer birds must have contributed much to the high figures for

newcomers, whether of Old or New World origin, in the two recent decades—
just as the intervention of the two World Wars must be held responsible in a
large measure for the paucity of records during 1910–49. The development of
the bird observatory network in Britain and Ireland has also played a significant
role; but it is interesting to note that its beginnings, the migration researches of
Dr William Eagle Clarke, the Misses Baxter and Rintoul, and others, on offshore
islands before and after World War I produced relatively few novelties. I
consider that all these factors, important biases as they are, do not adequately
explain the striking dominance of Nearctic species (and the large number of
individuals) since 1950. Climatic changes have undoubtedly played a major
role.

The period of climatic amelioration in north-west Europe which began in
the 1890s affected the status and distribution of the breeding birds of many
western European countries, particularly in Scandinavia, the British Isles and
Iceland (Salomonsen, 1948; Kalela, 1949). The westwards range-expansion of a
number of species into Finland, Sweden, Germany and Denmark resulted in a
greater incidence of vagrant appearances in Britain. Among these might be
mentioned the Arctic and Greenish Warblers (*Phylloscopus borealis* and *P.
trochiloides*; Williamson, 1967), Red-breasted Flycatcher (*Ficedula parva*;
Sharrock, 1973), Great Reed Warbler (*Acrocephalus arundinaceus*; Williamson,
1968), Red-flanked Bluetail (*Tarsiger cyanurus*; Mikkola, 1973) and Terek
Sandpiper (*Xenus cinerea*; Smith *et al.*, 1972). Equally there are a number of
southern species whose northwards range expansion has led to a greater
frequency of their appearance in Britain. Overshooting of adults in late spring
anticyclonic weather, and the dispersion of young birds in autumn are the
usual causes of this displacement.

With regard to the greater frequency of eastwards displacement from North
America, recent changes in the activity of Atlantic weather systems and the
mean path of the oceanic storm-track seem likely to have played an important
part. The warming-up of the first half of the twentieth century was occasioned
by the frequent transportation of tropical air to high latitudes in the warm
sectors of Atlantic depressions, and during much of this period the average
centre of minimum pressure was near to or even north of Iceland, with the mean
path of the depressions running from south-west to north-east. Assuming that
land-birds lost over the ocean are subject to a wind-assisted flight, the tendency
would be for American vagrants to reach Greenland rather than Britain and
Ireland, which would lie too far to the south of a wind-drifted track.

There is climatological evidence that a deterioration set in during the 1940s
and has continued to the present. Perry (1971) points out that the early part of

this century was a period of predominantly high zonal index, in contrast to the frequent low zonal index or less vigorous patterns of recent years; the Azores high has declined, and the Icelandic low increased, by an average of six millibars over this period. Lamb (1966) considers the earlier period to have been one of abnormal vigour in the general circulation, centred around 1925 (± 25 years). With the reduced intensity of the Azores high and the deepening of the Icelandic low their centres have been brought closer together, mainly by a southwards shift in the position of the low, which in the 1960s was commonly as much as ten degrees of latitude farther south than earlier in the century. Moreover, the Azores high has moved eastwards, and the Icelandic low westwards, on the annual mean chart. Irvine (1972) adduces evidence from the present climate of the Faeroe Islands which suggests that the average position of the low is now in that latitude.

It is worthwhile to look at the situation in Greenland in more detail. Since 1820 "first records" have been predominantly of species from North America, and down to 1890 the decadal average of new European taxa was less than two. During the next half-century the average was six, again dropping to two after 1940. Expressed as individual specimen records rather than new taxa, European vagrants had a much higher share of the decadal totals after 1890, reaching 60% during 1920–29 and 50% during 1930–39, the decline then continuing into the 1960s (data from Salomonsen, 1967). The European Fieldfare (*Turdus pilaris*) established a breeding population in south Greenland following a winter invasion in 1936 (Salomonsen, 1950) and three other European species began to breed regularly there in this period. The pattern fits closely the climatic one of an improvement culminating in the late 1930s with a deterioration since. With depressions moving north-eastwards into high latitudes there would be greater opportunities for drift-migrants in this part of the Atlantic to be carried to Greenland and Iceland in the backing winds of the northern sectors of the lows; but now that the depressions are operating farther south and west, opportunities will be fewer. The wind-borne path of Nearctic vagrants in the complementary wind systems north of the Azores high and south of the low centres frequently offers a more direct west-to-east displacement for birds whose place of origin is often the eastern U.S.A. (Nisbet, 1963) and whose terminus is frequently Ireland and the south-west peninsula of England (see maps in Sharrock, 1971).

A final point should also be made here. There is little doubt that the official list of the British Ornighologists' Union (1971), on which this analysis is based, is biased towards the exclusion of certain species on the grounds that they could have escaped from captivity. From the ornithologist's standpoint this present volume is timely; for within the foreseeable future it may become impossible to present a picture that attempts to correlate wild bird vagrancy with climatic change, so distorted will it become as a result of the increasing frequency of large-scale bird importations and the resulting escapes from dealers, zoos, waterfowl collections, falconers and the like.

REFERENCES

ALEXANDER, W. B. and FITTER, R. S. R. (1955). American land birds in western Europe. *Brit. Birds* **48**, 1–14.

BRITISH ORNITHOLOGISTS' UNION (1971). "The Status of Birds in Great Britain and Ireland." British Ornithologists' Union, Oxford.

DALGLEISH, J. J. (1880). List of occurrences of North American birds in Europe. *Bull. Nuttall Orn. Club* **5**, 65–74, 141–150, 210–221.

DURAND, A. L. (1972). Landbirds over the North Atlantic: unpublished records 1961–65 and thoughts a decade later. *Brit. Birds* **65**, 428–442.

HARBER, D. D. and SWAINE, C. M. (1963). Rarity Records Committee Report on rare birds in Great Britain in 1962. *Brit. Birds* **56**, 393–409.

IRVINE, S. G. (1972). An outline of the climate of the Faeroes. *Weather* **27**, 307–320.

KALELA, O. (1949). Change in geographical ranges in the avifauna of northern and central Europe in relation to recent changes in climate. *Bird Banding* **20**, 77–103.

LAMB, H. H. (1966). Climate in the 1960's. *Geogrl J.* **132**, 183–212.

MIKKOLA, H. (1973). The Red-flanked Bluetail and its spread to the west. *Brit. Birds* **66**, 3–12.

NISBET, I. C. T. (1959). Wader migration in North America and its relation to trans-Atlantic crossings. *Brit. Birds* **52**, 205–215.

NISBET, I. C. T. (1963). American passerines in western Europe. *Brit. Birds* **56**, 204–217.

PERRY, A. H. (1971). Changes in position and intensity of major Northern Hemisphere "centres of action". *Weather* **26**, 268–270.

PYMAN, G. A. (1961). Rarity Records Committee Report on rare birds in Britain in 1960. *Brit. Birds* **54**, 173–200.

SALOMONSEN, F. (1948). The distribution of birds and the recent climatic change in the North Atlantic area. *Dansk Orn. Foren. Tidsskr.* **42**, 85–99.

SALOMONSEN, F. (1950). The immigration and the breeding of the Fieldfare (*Turdus pilaris* L.) in Greenland. *Proc. Tenth Internat. Orn. Congr.*, 515–526.

SALOMONSEN, F. (1967). "Fuglene pa Grønland". Rhodos, Copenhagen.

SHARROCK, J. T. R. (1971). Scarce migrants in Britain and Ireland during 1958–67. Part 5—Pectoral Sandpiper, Sabine's Gull and American land birds. *Brit. Birds* **64**, 93–113.

SHARROCK, J. T. R. (1973). Scarce migrants in Britain and Ireland during 1958–67. Part 9—Aquatic Warbler, Barred Warbler and Red-breasted Flycatcher. *Brit. Birds* **66**, 46–64.

SMITH, F. R. (1972). Rarity Records Committee Report on rare birds in Great Britain in 1971. *Brit. Birds* **65**, 322–354.

SWAINE, C. M. (1962). Rarity Records Committee Report on rare birds in Great Britain in 1961. *Brit. Birds* **55**, 562–584.

WILLIAMSON, K. (1954). American birds in Scotland in autumn and winter 1953–54. *Scott. Nat.* **66**, 13–29, 200–204.

WILLIAMSON, K. (1967). "Identification for Ringers—The Genus *Phylloscopus*", revised edition. British Trust for Ornithology, Oxford.

WILLIAMSON, K. (1968). "Identification for Ringers—The Genera *Cettia*, *Locustella*, *Acrocephalus* and *Hippolais*", revised edition. British Trust for Ornithology, Oxford.

WILLIAMSON, K. and FERGUSON-LEES, I. J. (1960). Nearctic birds in Great Britain and Ireland in autumn 1958. *Brit. Birds* **53**, 369–378.

14 | British Amphibians and Reptiles

I. PRESTT

Central Unit on Environmental Pollution, Department of the Environment, London

A. S. COOKE

Monks Wood Experimental Station, The Nature Conservancy, Abbots Ripton, Huntingdon

and

K. F. CORBETT

British Herpetological Society, c/o Zoological Society of London, London NW1

Abstract: There are twelve indigenous British species: six amphibians and six reptiles. Their range of movement is limited and ability to recolonize is poor. There is no protective legislation, yet collecting pressure by dealers, pet-keepers and researchers is considerable for the rarer species. Past records of status are lacking. The commoner species have recently declined locally, primarily because of urbanization and intensive agriculture. The rarer species are declining seriously because of fragmentation of restricted habitat by afforestation, urbanization, military activity, mineral extraction, public pressure and agricultural reclamation.

National surveys indicate that the Common Frog and Common Toad decreased slightly in numbers in the 1950s but more severely in the 1960s. There have been no such surveys for newts, but it is believed they have decreased recently. Studies in south-east England confirm loss of Warty Newt colonies. All five species remain generally distributed. Inland the Natterjack has been reduced to one small colony from numerous southern heathland sites of 1950s and has also declined at some coastal sites.

The Viper, Grass Snake, Common Lizard and Slow Worm are still widespread although local declines have been documented. The local Smooth Snake and Sand Lizard have suffered from a loss of lowland heath. The Sand Lizard has disappeared from its coastal dune sites apart from studland and a threatened remnant population in Lancashire.

Systematics Association Special Volume No. 6, "The Changing Flora and Fauna of Britain", edited by D. L. Hawksworth, 1974, pp. 229–254. Academic Press, London and New York.

INTRODUCTION

The amphibians and reptiles are primitive vertebrates. Over most of the earth they have been superseded by more advanced forms, although in some regions (notably in the Tropics) large numbers of both species and individuals can be found. The British Isles (excluding the Channel Isles) have only twelve indigenous species: a frog, two toads, three newts, three lizards and three snakes. If we include the Channel Isles this adds another species of frog and two more lizards. We could also add to the list, as non-breeding visitors, the five species of marine turtle recorded in our coastal waters or stranded on our beaches.

Because they are easy to transport as spawn, young or adults, and are easy to keep in captivity, large numbers of introductions have been attempted in many parts of Britain using a variety of exotic species. Seven species have figured most prominently in these introductions. These are the Tree-frog (*Hyla arborea* (L.)), Midwife Toad (*Alytes obstetricans* (Laurenti)), Edible Frog (*Rana esculenta* (L.)), Marsh Frog (*R. ridibunda* Pallas), Green Lizard (*Lacerta viridis* (Laurenti)), Wall Lizard (*Lacerta muralis* (Laurenti)), and the Tesselated Snake (*Natrix tessellata* (Laurenti)). So far, all except the Marsh Frog and possibly the Tesselated Snake have only bred and survived for varying periods of time at the place of introduction. The Marsh Frog, originally introduced into Romney Marshes in 1935, has now become firmly established and has spread to many other parts of Kent and to East Sussex. Recent reports from Yorkshire suggest that the Tesselated Snake may also now be spreading.

Amphibians and reptiles have little aesthetic appeal, they are difficult to observe in the wild and they provide no sporting interest. Unlike birds, butterflies and flowering plants, therefore, they have received little attention (indeed, one of their principal attractions to man—their collection and sale as pets and laboratory specimens—has been unwelcome). As a result, detailed historical information about the two groups is either lacking or is scattered in rather obscure publications.

The first attempt to bring together information on their national distribution was made by Taylor (1948) in co-operation with the British Herpetological Society (B.H.S.). Updated versions of these distribution maps were published by Taylor in 1963. From 1965–70 the Biological Records Centre (B.R.C.), also in co-operation with the B.H.S., conducted a further national survey on a 10 km square basis, to assess the present distribution (B.R.C., 1973). Our principal knowledge of the distribution of the British herpetofauna is provided by these three projects. More detailed studies have been made recently to investigate reported declines in the Common Frog and Common Toad (Cooke, 1972a) and to record the loss of habitat and decrease in number of the Natterjack Toad, Sand Lizard and Smooth Snake in southern England (Corbett, in preparation). The Conservation Committee of the B.H.S. has recently received grants from Carnegie and the World Wildlife Fund to undertake detailed national surveys of the Warty Newt and the Natterjack Toad.

CHANGE IN STATUS

1. Amphibians

(a) Frogs and Toads

(i) The Common Frog (*Rana temporaria temporaria* (L.)). *Rana temporaria* is a trans-palaearctic species, *R. t. temporaria* being distributed over northern and central Europe including the British mainland, Ireland and the Isle of Man (Smith, 1969). The existing distribution maps tend to reflect the presence and absence of keen observers. The absence of the Frog from certain islands, such as the Outer Hebrides, is, however, clearly established.

Only in the last 20 years has there been any concern about what is generally assumed to be our most common anuran. In 1953, Smith remarked that he had received many accounts of local population decreases during the preceding "few years", and Taylor (1963), after completing his distribution survey, commented that the Frog was "not as abundant in many districts as it used to be". Later in the 1960s, the Frog was assumed to have suffered general population declines in Britain (see, e.g. Perring, 1966; Simms, 1969).

In 1970, two surveys were conducted in an attempt to determine changes in status of the Frog in the British Isles since 1950 (Cooke, 1972a). Details of local changes were accumulated from questionnaires returned by correspondents. Although the majority of these people appeared to have extensive knowledge of their local amphibians, most had not kept regular records so their replies were based largely on memory and opinion. The problems of inaccuracy and bias were discussed at length in Cooke's paper. In the absence of a better method of surveying past changes in status of the Frog, it was believed that the replies satisfactorily indicated population changes in certain regions of the British Isles during the 1950s and 1960s. In some parts, however, notably Eire, northern England and northern Scotland, coverage was negligible.

Two types of questionnaire were used. The Schools Survey (S.S.) Questionnaire was sent mainly to biology teachers via local education officers and requested simple information on local population changes. The second more detailed questionnaire, the Breeding Sites Survey (B.S.S.) Questionnaire, was sent to amateur and professional naturalists who were known to have studied their local amphibians or who had volunteered to help. This questionnaire asked for information on population trends in individual breeding sites.

Both types of questionnaire yielded information on overall changes in status in the correspondents' localities; 964 S.S. questionnaires and 99 B.S.S. questionnaires being received that contained sufficient information to permit the evaluation of a recent trend (increase, decrease, no change, or, in some instances, no longer found). For changes further back into the past there was less information. Locality data were summarized for convenience into five-year periods and the following number of useful replies were received: 1951–55, 248 replies; 1956–60, 360; 1961–65, 664; 1966–70, 1063. These replies indicated that any decrease in numbers experienced during the 1950s was slight, except in the northern Home Counties where decreases were more marked. During the early 1960s, Frog populations declined considerably over central, southern and eastern England, but decreases in central Scotland, Northern Ireland and south-west England were slight. During the last half of the decade, the decline spread to Scotland and Northern Ireland, while in England very severe decreases were noted in the south-east Midlands and northern Home Counties, although there was little change in the south-west and, rather surprisingly, in East Anglia. Elsewhere in England and Wales there were moderate or severe decreases. Information for 1971 and 1972 has been received from correspondents in a few areas. There seems to have been no appreciable change in populations around Bolton (Lancashire), Tamworth (Staffordshire), Dublin, or moorland in Aberdeenshire and Kincardineshire. In Huntingdonshire, frogs have recently been noted in several new sites, but in south-west Devon the species has suffered considerable declines in the last two years.

The extent of the population decline may have exceeded 99% in some areas. Savage (1961) regarded a density of five adult frogs per acre as a fair average for the British Isles, this estimate being based upon observations made in the 1930s and 1940s. Estimates of population density based on observations made in the last six years are much lower: Huntingdonshire, 0·01 frogs/acre; the Huddersfield area of Yorkshire, 0·02 frogs/acre; Isle of Wight, <0·08 frogs/acre; Norfolk, 0·1 frogs/acre (Cooke, 1972a). There are, however, regions such as the south-west and north of England, Ireland and Scotland where the Frog is still relatively abundant.

Useful information was received from B.S.S. correspondents on 292 frog-breeding sites (Cooke, 1972a). The Frog generally increased in the 1960s in the garden ponds studied, while in agricultural sites, it decreased severely in the early 1960s, but declines in the later part of the decade were less marked.

(ii) The Common Toad (*Bufo bufo bufo* (L.)). *Bufo bufo* is a palaearctic species, *B. b. bufo* being the European race (Smith, 1969). It is absent from Ireland, the Isle of Man, the Outer Hebrides and the Shetland Islands.

The Common Toad was included in the surveys described above for the Common Frog, and recent declines for this species in Britain were probably spatially and temporally similar to those of the Frog (Cooke, 1972a).

Virtually nothing is known about the population density of the Toad. Frazer (1966) estimated that the density of adult breeding males over five square miles of countryside in Kent was about 0·5/acre during the period 1955–61. He considered that this area was "not particularly suitable for toads", so the population was "probably not as heavy as compared with other areas". The present population density in the county of Huntingdon and Peterborough is, however, only about 0·02–0·04 adult toads/acre, but it is not yet possible to assess whether this is a typical density for the region, let alone the country.

(iii) The Natterjack Toad (*Bufo calamita* Laurenti). The Natterjack is widely distributed in western Europe, occurring at altitudes up to 4000 ft. In Britain it is now only found at sea level and since records were first obtained has been confined to the central and southern parts of the country, with its most northerly site being the Scottish coast of the Solway Firth. The Natterjack is primarily a coastal species, but also occurs inland usually on sandy heaths. Its numbers have decreased in most localities, but the declines have been most severe at inland sites.

The last major inland area for the Natterjack in Britain was on the London Basin and Wealden heaths to the south-west of London. Corbett's studies have shown that the decline of populations there was only recent, but was so severe that by 1972 of all previous known colonies, only one remained and that was reduced to four breeding pairs compared to the many hundreds of adults in past years. Prior to this decline the Natterjack was regarded as a locally abundant species in the area. The severity of the decline is illustrated by the changes recorded in the vicinity of Frensham. Here within a six mile radius to the south of the town there were twenty known colonies in the mid-1950s. These are all now extinct. Similarly, of the large population previously extending from Woking to Wisley, even the Send stronghold (Smith, 1969) which contained hundreds, is now extinct.

Elsewhere in southern England the disappearance of the Natterjack occurred at an earlier date. In Dorset, where apparently some of the colonies such as that at Bloxworth were very large, the last records were obtained about the turn of the century. It was recorded breeding up to 1946 in the New Forest, and at some other sites in south-west Hampshire until 1950, but has not been recorded here since.

The only other region of Britain which also previously contained several inland sites for the Natterjack Toad was eastern England. The most inland of these was on the heaths of Gamlingay in the extreme west of Cambridgeshire. Marr and Shipley (1904) reported the Natterjack to be abundant here in 1824, breeding in clay pits. It was still present in the early 1900s, but became extinct soon after. Other inland sites were at Tosback and the Breckland in Suffolk, and at Roydon Common in Norfolk. They are still present at Roydon, but may have been reintroduced to this site, and a small number may have survived at the Breckland site.

The most important of the coastal sites for the Natterjack Toad is the west Lancashire coast between Southport and Altcar. Much of this area is now under very considerable public pressure and as a result the habitat available to the Natterjack has been considerably reduced. The Natterjack is still present in large numbers however and in 1970 the population was estimated to be in the order of 10 000 adults (S. Smith, *in litt.*).

R. Wagstaffe who, with the late F. W. Holder, studied intensively the flora and fauna of this area from 1921 to 1970, confirms (*in litt.*) that until the early 1930s the Natterjack population could be estimated in hundreds of thousands. Until this time also the Natterjack occurred at a number of inland sites, the most inland of these being in the vicinity of Rufford about four miles from the coast.

Further to the south on the Cheshire coast there is one small surviving colony, but the Natterjack is now absent from North Wales. Forrest (1907) reported its status as "Common between Prestatyn and the River Conway" but it was apparently essentially confined to the coast. Small numbers are known to have persisted in North Wales until the late 1940s.

Further to the north the Natterjack Toad is still present in relatively large numbers on the Ravenglass sand-dunes and on both the English and Scottish shores of the Solway Firth. Information on its present status on the English side of the Solway is lacking, but it was evidently widely distributed along this coast in the late 1800s (Macpherson, 1892). A recent survey of the Scottish coast of the Solway (Boyd, 1971) confirmed numerous breeding colonies, with one estimated to contain about 300 adults.

The coasts, and adjoining inland areas, of Norfolk and Suffolk at one time possessed numerous (often very large) Natterjack colonies. E. A. Ellis (*in litt.*)

knows of twelve localities where Natterjack colonies were present, from Bawdsey in the south to Holkham in the north. All but three of these are now extinct. Ellis reports that although some local reductions had taken place by the turn of the century, the Natterjack was still present at most of these localities up to the 1930s. In some instances the numbers present were large, thus the colony some way inland at Reedham, by the river Yare, contained thousands of adults. The declines which resulted in the disappearance of most of the eastern coast Natterjack population were rapid and did not start until after the Second World War. While it is possible small numbers may yet remain at some of the localities, such as Reedham, the only important remaining sites are in north Norfolk at Winterton and Holkham. There is also a colony on the Lincolnshire coast.

(b) *Newts*

The three species of newt present in Britain are the Smooth Newt (*Triturus vulgaris vulgaris* (L.)), the Palmate Newt (*T. helveticus* (Razoumowski)) and the Warty Newt (*T. cristatus cristatus* (Laurenti)). All three are found throughout Europe and are relatively common and widespread in the British Isles. Only the Smooth Newt occurs in Ireland.

The general distribution of Warty and Smooth Newts in Britain is similar, with both species being present throughout most of England with fewer records from the south-west peninsula. Both have also only been recorded rather infrequently in Wales while the majority of the Scottish records are from central Scotland with few from elsewhere.

The Palmate Newt, the smallest European species, contrasts with the two previous species in being relatively uncommon in eastern England, but widely recorded throughout the south-west peninsula and Wales. There are also more records for this species from Scotland.

Little is known about changes in status of the newts. G. A. C. Bell (*in litt.*) believes that within the last 200 years the Smooth Newt must have increased because of the creation of small field ponds at the time of the Enclosures. He points out that had this species always been as common as it is now, then it would presumably have figured more prominently in folk lore. Beebee (1973) stated that although the Warty Newt disappeared recently from many places in East Anglia, Surrey, Hampshire, Sussex, London and Lancashire, there was "little evidence" for the other two species having declined. However, information received from correspondents indicates that obvious population declines for the Smooth and Palmate Newt have occurred in some places, e.g. around Bolton (Mrs E. Hazelwood, *in litt.*).

The most recent distribution maps compiled by the B.R.C. may give an indication of relative changes in status of the three species. Presence within the 10 km squares is divided into two classes: (1) present before 1960, but not recorded since and (2) recorded at some time during the period 1960–73 (and perhaps recorded before 1960 as well). For any area, the "intensity" of recording is unlikely to have remained the same for the two time periods, so it is unwise to assume, for instance, that many pre-1960 records and only a few post-1959 records indicate recent declines for a species. However, it is probably reasonable to assume that a person sending in records for one species of newt will also record the other species, if he knows they occur. Thus, within a given area, one can compare the factor,

$$\frac{\text{number of 10 km squares with post-1959 records}}{\text{number of 10 km squares with pre-1960 only records}}$$

for two different species and draw tentative conclusions on the change in the status of one relative to that of the other. Using the B.R.C.'s maps, the changes in status of the Palmate and Warty Newts have been compared with the change for the Smooth Newt for various areas (Table I). That dividing the factors for the Palmate or Warty Newt by the factor for the Smooth Newt gives an answer of less than unity for each area examined is taken to indicate that the Smooth Newt has done comparatively better than the other species throughout Britain. Relative to the Smooth Newt, the Palmate and Warty Newts have fared particularly badly in south-east England (see the comments of Beebee, 1973) and perhaps also in Wales, although Welsh coverage was poor. Thinking in absolute terms, if some population decline has occurred for the Smooth Newt, then the Palmate and Warty Newts appear to have decreased more severely in numbers. If one accepts that (1) the distribution maps can be used for this purpose and (2) coverage has been adequate, then conservationists should be more concerned about the Palmate Newt than about the Warty Newt. In England, the Palmate Newt seems to have fared worse than the Warty Newt and, since 1960, has been recorded in Britain in fewer 10 km squares. However, a comparison of pre- and post-1960 records could miss substantial local population decreases that have occurred within the last few years and such decreases have been reported for the Warty Newt in many parts of lowland England, e.g. of eleven colonies known within six miles of Brighton in the mid-1950s only one now survives (K. F. Corbett and T. Beebee, *in litt.*); and colonies are known to have been lost from near Portsmouth (J. Oakshatt, *in litt.*), from north-east Surrey, south and mid-Kent (G. Boyce, *in litt.*), and from several parts of Yorkshire (Howes, 1972).

TABLE I. Numbers of 10 km squares containing records for "pre-1960 only" and "1960–73", the information being extracted from the Biological Records Centre's maps for the distribution of newts in the British Isles. The factors may indicate relative changes in status for three species within any given area. See text for details

	Smooth Newt			Palmate Newt				Warty Newt			
	No. of records		Factor $(2)/(1)$	No. of records		Factor $(2)/(1)$	$\dfrac{\text{Factor}_\text{Palmate}}{\text{Factor}_\text{Smooth}}$	No. of records		Factor $(2)/(1)$	$\dfrac{\text{Factor}_\text{Warty}}{\text{Factor}_\text{Smooth}}$
	Pre-1960 only (1)	1960–73 (2)		Pre-1960 only (1)	1960–73 (2)			Pre-1960 only (1)	1960–73 (2)		
South-east England[a]	23	124	5·40	33	43	1·30	0·24	30	76	2·54	0·47
Rest of England	88	166	1·89	89	107	1·20	0·64	86	142	1·65	0·87
England	111	290	2·61	122	150	1·23	0·47	106	218	2·06	0·79
Wales	11	10	0·91	40	16	0·40	0·44	16	4	0·25	0·28
Scotland	40	20	0·50	62	24	0·39	0·78	28	10	0·36	0·72
Ireland	20	43	2·15	Absent				Absent			
Total (excluding Ireland)	162	320	1·98	224	190	0·85	0·43	150	232	1·55	0·78

[a] 100 km squares 41(SU), 51(TQ), 61(TR), 52(TL) and 62(TM).

2. Reptiles

(a) Lizards

(i) Common Lizard (*Lacerta vivipara* Jacquin). This is a widely distributed species extending across Europe and Palaearctic Asia. In Britain it is present throughout Scotland, England and Wales and is the only reptile found in Ireland.

The Common Lizard has been recorded in every mainland vice-county in Scotland, England and Wales and in at least twenty Irish vice-counties. It has also been recorded on Lundy, Anglesey and the Isle of Man and the inshore Scottish Islands, but records are lacking for the more distant Scottish islands.

No special studies of this species have been made, but it is unlikely to be endangered on a national scale. Its small size and viviparity enable it to exploit successfully a wide variety of habitats, including many greatly modified by man and often in close proximity to him.

(ii) The Slow Worm (*Anguis fragilis* L.). Like the Common Lizard, this species is widely distributed across Europe. In Britain it has been recorded from every English and Welsh vice-county and from all but two in Scotland. There are indications that it is more abundant in the south of Britain than in the north, and in the west more than the east (Smith, 1969).

No detailed survey has been made of the Slow Worm in Britain but it is not thought to be in danger on a national scale. There have been a number of reports of local decreases including two from as far apart as south-east England and Yorkshire. Howes (1972) states that, "Although the Slow Worm was reported as being 'common' to the south of the Doncaster district in Sherwood Forest and the Worksop areas (Victoria History of Notts., 1905) and to have occurred in the Askern district (Lankester, 1842), it is now exceedingly rare within the study area, only being sporadically recorded from a handful of localities". A study of a local population of this species on a six mile area of a scarp slope of the chalk near Portsmouth (J. Oakshatt, *in litt.*) has shown a severe decrease in numbers to have taken place during the last ten years in what was previously a dense population.

(iii) The Sand Lizard (*Lacerta agilis* (L.)). This species is found throughout most of Europe. The nominate race is common and generally distributed over central Europe, but in Britain is confined to the southern half of the country. In continental Europe it occupies a variety of habitats, presumably because the warmer summers impose few limitations on its egg laying, whereas in Britain it is restricted to mature lowland dry heath and sand-dunes.

Corbett has drawn the following conclusions from his detailed studies of the ecology and conservation of the Sand Lizard made during the last five years.

Despite being on the edge of its geographical range in Britain, it is well adapted to the two habitats it occupies. As a result it can be locally abundant with colonies having up to one hundred and forty adults per acre. Corbett's own observations of individual colonies covering five consecutive years, combined with those of other observers in the case of certain colonies to give total observed periods covering twenty-five years, show that the numbers in these large colonies remained remarkably stable with no evidence of major population fluctuations or crashes. The food supply remained sufficiently diverse and abundant to support these large numbers of lizards and a self-regulatory mechanism of adult predation of their young prevented higher densities.

In spite of this natural population stability and successful occupation of specialized niches, the British Sand Lizard populations have undergone alarming declines in terms of absolute numbers and distribution. To investigate the reasons and timing of these declines, in addition to the field studies of Sand Lizard colonies, a thorough search has been made of obscure local literature; contact has been made with amateur field herpetologists to obtain information on local sites, and historical studies have been made of the habitat. These investigations have shown that the main declines are recent and in all instances have primarily been caused by man's activities.

Since 1800 the major part of the British Sand Lizard population has been located in a small number of areas. In addition to these main areas, there were other (presumably much smaller) scattered sites. These included the coast of Cheshire, Flintshire, Sussex, and Kent, and inland sites in Wiltshire and Berkshire. The Sand Lizard is now extinct in all these. The main areas (for convenience termed "Zones") are: the south-west Lancashire coastal dunes (Zone I); the heaths of Surrey, north-east Hampshire, south-east Berkshire and north-west Sussex (Zone II); the south-west Hampshire heaths (Zone III); and the heaths and coastal dunes of south-east Dorset (Zone IV). Zones III and IV have only recently been separated from each other by the Poole/Christchurch urban development. Zones I and II have been separated from each other and from the originally combined Zones III and IV for at least 2000 years.

British summers became cooler about 2000 years ago and since then the Sand Lizard in Britain has probably been confined to a habitat with a sandy substrate in which to lay its eggs. This is not, as is often suggested, because sand is our warmest sub-soil (loam and chalk soils could provide an equally warm substrate) but because sand is more slowly colonized and overgrown by vegetation.

The present position is as follows. Zone I now carries only a remnant of less than a hundred adults with no major colonies surviving. R. Wagstaffe (*in litt.*)

confirms that until 1930 the Sand Lizard population could be estimated in thousands and was probably of the order of 8000–10 000 in total. As recently as the early 1960s there were fourteen major colonies, in addition to many minor ones, on both the dunes and nearby golf courses. The serious decline of this northern population is particularly to be regretted, as this population displays distinct differences in ecology and coloration to Sand Lizard populations in other Zones and so may represent a distinct race.

Populations in Zone II have also recently been seriously reduced and the remaining scattered Sand Lizards in this general area numbered under two hundred adults in 1971. In north Surrey and north-east Hampshire this decrease has been so severe that no breeding populations now remain. The nature of the decreases is well illustrated by the Frensham area. In the mid-1950s there were fifty-four colonies within three miles of Frensham itself, but by 1971 only two of these remained and these have now gone.

In Zone III, none of the twenty-two breeding sites known from the vast area of heath in the New Forest in the early 1950s and 1960s now remains. Searches have failed to reveal other existing breeding populations. Further west in this Zone and in Zone IV loss of heath has reduced and fragmented the Sand Lizard populations. In Zone IV the populations at Studland Peninsula (including even those in the National Nature Reserve) have undergone an estimated 80% decline during the last ten years.

As a result of the recent serious declines in Zones I and II, the vast majority of the surviving British Sand Lizards are now contained in the western edge of Zone III and in Zone IV.

(b) *Snakes*
(i) The Viper or Adder (*Vipera berus* (L.)). This is one of the most widely distributed reptiles in the world. In Britain it has been recorded throughout the mainland from the northern coasts of Scotland to the coasts of Kent and Cornwall in the south. With the exception of Anglesey, Arran, Jura, Islay, Mull and Skye, it appears to be absent from offshore islands. Within the British Isles the Viper occupies a wide variety of different habitats, including cliffs and marshes on the coast, mountains and moorlands in upland areas and chalk-downs, lowland heath, open woodland and boggy valleys elsewhere inland.

The information now available (Taylor, 1963 and B.R.C.) shows that by 1960 the Viper had been recorded in every vice-county from 1 to 109*, with the exception of 71 (Isle of Man). However, in nine of the vice-counties (three

---

* For vice-county names and boundaries see pp. 425–428.

in England and six in Scotland) there have been no records since 1948 and in seven others (three in England and two each in Wales and Scotland) there has been none after 1960. This means that recent confirmation of its presence is lacking for sixteen out of a total of one hundred and eight vice-counties in which it has been known to occur. This could indicate that a recent decrease has taken place or it could reflect a lack of recent surveys in these areas, particularly as the pre-1948 figures were obtained over a period of at least fifty years in contrast to the later results which were obtained in less than half this time. This difference in time scale could be at least partly offset, however, by the fact that in the post-1948 period the observers engaged in the work were better organized and had greater experience of this type of survey work.

A noticeable feature in the distribution of the Viper in England is the paucity of records from a wide diagonal belt of country extending from mid-Lancashire in the north-west, through south Lancashire, across the plain of the Midlands to Cambridgeshire in the south-east. The absence of the Viper in this stretch is not exceptional, as it contains large areas of some of the least suitable habitat for this species in England. The vice-counties of Cambridgeshire (29), Bedfordshire (30), Huntingdonshire (31), Warwickshire (38), Cheshire (58) and south Lancashire (59), from which no recent records have been obtained, all come within this belt. The lack of recent confirmation from these vice-counties, with the exception of Cheshire, is thus consistent with the fact that the Viper has apparently never been widespread in these areas during the 1900s and consequently will be more vulnerable to changing conditions and more difficult to record.

In mid-Cheshire and also in Hertfordshire, in the south of England, the Viper was recorded over relatively wide contiguous areas before 1947 so the almost complete absence of post-1947 records from these two is of note. It would be particularly valuable, therefore, to know whether or not this lack of recent records represents major decreases in these areas or simply a lack of recent surveys.

In Wales most of the records for the Viper have come from the north and down the west coast. Extremely few records have been obtained from the inland vice-counties including the eastern half of Merionethshire (48), the whole of Montgomeryshire (47), Radnorshire (43), Breconshire (42) and Carmarthenshire (44) and the western part of Glamorgan (41). Since 1960 there have been no records from the two vice-counties 44 and 50. The first (Carmarthen) had only one record prior to 1960, so a lack of recent records is not significant. In the second (Denbighshire and parts of Flintshire), however, there were numerous pre-1960 records, so this does represent a contrast with the earlier results and requires investigation as to whether there has been a decrease in the distribution or a lack of surveys.

In Scotland, while the Viper has been recorded in all the vice-counties except 84 (West Lothian), in the majority of cases the records are few in number and widely scattered. Only in vice-counties 73, 81, 83 and 87 (Kirkcudbright-shire, Berwickshire, Midlothian and South Perthshire with Clackmannan and parts of Stirling) have there been sufficient records to confirm its presence over relatively wide contiguous regions. Whilst therefore there are nine vice-counties in Scotland (76, 77, 83, 85, 93, 97, 98, 105) for which there are no recent records, it is only in the case of 83 (Midlothian) that this could indicate a significant decrease or a lack of recent surveys.

Our knowledge of the national status of the Viper can be summarized as follows. Over much of England and Wales it appears to remain generally widespread and locally common. In the interior of Wales and in a belt of country across the English Midlands, from mid-Lancashire in the West to Cambridgeshire in the East, it has probably never been common since about 1900 when recording first started. In these areas, particularly in the Midlands, its numbers may have been still further reduced but local confirmation is required. In three previously important areas (mid-Cheshire, Hertfordshire, Denbighshire and Flintshire) an almost complete lack of recent records should be checked as this could represent important decreases. It is widespread in Scotland with nothing to indicate any recent changes in its distribution, except possibly Midlothian where further surveys should be made.

(ii) The Grass Snake (*Natrix natrix* (L.)). Mertens (1947) recognized nine races of the Grass Snake. The race found in Britain (*Natrix natrix helvetica* (Lacepède)) also occurs in France, West Germany, Switzerland and Italy. In Britain it is confined to the southern half of the country (below Latitude 54° 40′), being absent from Scotland and only rarely recorded in Cumberland and Northumberland. The nominate race is present throughout northern and central Europe and in Sweden occurs as far north as Latitude 65°.

The Grass Snake occupies a more limited range of habitats than the Viper, as it remains in close proximity to water and lives at lower altitudes. While locally it is more mobile than the Viper and ranges over greater distances, it is essentially restricted to woods and hedgerows near water or to marshes.

The records (Taylor, 1963, and B.R.C.) show the Grass Snake has now been recorded from every English and Welsh vice-county with the probable exception of north Northumberland (68), north-west Yorkshire (65), Brecon-shire (42) and Radnorshire (43). Since 1960 it has not been recorded from seven vice-counties (three in Wales and four in England), i.e. seven out of a total of sixty-six in which it has been known to occur. Subject to the reservations discussed when this same evaluation was made for the Viper, this could indicate some recent decreases.

In two of the Welsh vice-counties, Cardiganshire (46) and Anglesey (52), there had only been few records prior to 1960 and so the absence of recent records would not indicate any major change. In the third, Denbighshire, parts of Flintshire and Caernarvonshire (5), however there were numerous records prior to 1960 and the lack of any recent observations would mean that a decrease has occurred. In two of the four English vice-counties, south Lancashire (59) and mid-Lancashire (60), the Grass Snake has always been scarce (indeed Taylor considered it possible that the records for south Lancashire could be attributed to introduced specimens). It was, however, relatively widely distributed in the other two vice-counties, Durham (66) and Derbyshire (57), and their lack of recent records would indicate a change in its status. There is also an absence of recent records for the Grass Snake from the western part of Nottinghamshire (56), Cheshire (58), the northern part of Staffordshire (39), East Suffolk (25), Buckinghamshire (24) and Shropshire (40), all of which are areas in which it was widely recorded prior to 1960.

The general situation for the Grass Snake in Britain appears to be that it is not in any immediate danger and remains widely distributed and locally common. There are, however, recent indications from a number of widely scattered localities that it may now be less common or possibly absent in certain areas in which it was previously widespread. Surveys are needed in these areas to confirm the situation.

(iii) The Smooth Snake (*Coronella austriaca* (Laurenti)). This species is widely distributed in Europe, its range extending from Scandinavia in the north to the northern parts of Greece in the south (Smith, 1969). Although the Smooth Snake, like the Viper, is viviparous it has never been recorded outside southern England and here its best known and largest breeding populations have always been located on mature dry lowland heath. Its association with this habitat is thought to be primarily linked to its preference for reptile prey, the lowland heath representing our optimum reptile habitat. Unlike the Sand Lizard it can survive in fringe and sub-optimum habitats, but only in small numbers which are unlikely to breed. It has a much longer life span in the wild (over twelve years) than the Sand Lizard.

Since the species was first recorded, its only two important strongholds in England have been (i) the London Basin and Wealden heath area and (ii) the

I

TABLE II. Dates of sightings of turtles around the coasts of the British Isles. (Data abstracted from original material accumulated by Taylor and summarized in 1963).

	Hawksbill	Atlantic Loggerhead	Kemp's Loggerhead	Green Turtle	Leathery Turtle	Unidentified Turtles	Total
Pre-1920	2	5	—	—	2	3	12
1920–29	—	3	2	—	—	2	7
1930–39	—	7	6	—	—	2	15
1940–49	—	7	4	—	1	3	15
1950–59	2	8	1	1	11	11	34
1960	—	1	—	—	1	—	2
Post-1960	No data	No data	No data	No data	No data	No data	No data
Total	4	31	13	1	15	21	85

heaths around the Christchurch and Poole conurbations. Studies by Corbett from 1968 to 1972 have shown that in both these areas its numbers and distribution have recently undergone a serious decrease. It is the most secretive of our reptiles which makes information on the numbers surviving in local populations difficult to obtain; however contemporary estimates by K. Corbett, R. Doughthwaite, J. Horsey, A. Phelps, M. Preston, D. Street and J. Webster (*in litt.*), at present studying this species, suggest a surviving (and still decreasing) English population of between 1000 and 3000 adults.

(c) Marine Turtles

Taylor (1963) listed five species of marine turtles that have been found stranded on the shore or caught in home waters: the Hawksbill (*Eretmochelys imbricata* (L.)), the Atlantic Loggerhead (*Caretta caretta* (L.), Kemp's Loggerhead (*Lepisdochelys kempi* (Garman)), the Green Turtle (*Chelonia mydas* (L.)) and the Leathery Turtle (*Dermochelys coriacea* (L.)). The dates of the reports have been extracted from Taylor's original data held by the Nature Conservancy and are summarized in Table II. Reports became more frequent during the twentieth century, particularly during the 1950s, perhaps because (1) more turtles strayed into British waters or (2) the likelihood increased of a wandering turtle being seen and reported. The figures suggest that the Atlantic Loggerhead and Kemp's Loggerhead were the most commonly encountered species in the 1920s, 1930s and 1940s, while the former and the Leathery Turtle were most often sighted in the 1950s. There were eight sightings of Leathery Turtles in 1959 alone.

REASONS FOR CHANGES IN STATUS
1. Frogs and Toads

The recent surveys of the Common Frog and Common Toad, the B.S.S. and the S.S., seem to have had a rural and a suburban bias respectively (Cooke, 1972a). Less than 14% of the breeding sites referred to in the B.S.S. were in towns with a human population density exceeding 10 000. Correspondents to the B.S.S. believed that local decreases were mainly caused by destruction of wetland habitat by man and by sites becoming unsuitable either because of natural processes or because of human interference. S.S. correspondents were in little doubt that urbanization was to blame. For instance, loss of breeding sites specifically because of housing development was apportioned 32% of the blame for local Frog declines by S.S. correspondents, but only 9% by B.S.S. correspondents. Such differences of opinion are consistent with the suggestion that suburban and rural views respectively were being sampled.

For the Frog, an attempt has been made in Table III to list in approximate order of decreasing importance factors controlling (a) the production and survival of spawn, (b) tadpole numbers, (c) adult numbers. These lists, which probably contain the critical factors but are not exhaustive, have been compiled taking into account evidence from correspondents and from published literature. For most factors, references to local observations or to general discussions are quoted in the Table. Relevant American, Dutch and Swiss references and references to other anurans have been included. The decline of amphibian species is a world-wide problem. Heusser (1968) blamed loss of breeding sites and death on the road for amphibian losses in Switzerland, and the former is likely to be the major factor responsible for reducing British populations over the last twenty-five years. Although three separate lists have been drawn up in Table III, the situation is of course circular. For example, complete loss of one season's tadpoles will mean no adults will result from the spawn of that year. Other important factors on a national scale are probably (a)ii, (b)i, ii and iii, and (c)i, ii and iii. Collection of adults by schools for their own use and by others to keep as pets ((c)iv) could be serious in urban areas where many breeding sites are readily accessible. B.S.S. correspondents reported that sites

TABLE III. An attempt, for Common Frog populations in Britain, to list in order of decreasing importance factors limiting (a) the production and survival of spawn, (b) tadpole numbers, (c) adult numbers

Factors	Literature concerning the Common Frog and other anurans in which these factors are referred to or discussed
(a) Factors limiting the production and survival of spawn	
i. [a]Loss of breeding site	Yalden (1965), Heusser (1968), Culley and Gravois (1970), Cooke (1972a), Beebee (1973)
ii. [a]Collection by children	Yalden (1965), Cooke (1972a)
iii. Site dries out before spawn hatches	
iv. Spawn infertile or diseased	Van Gelder and Oomen (1970), Van Gelder and Kalkhoven (1971)
v. Spawn eaten by predators	Martof (1956), Savage (1961), Yalden (1965), Heusser (1970a, 1971), Grubb (1972)
vi. [a]?Professional collection	

Factors	Literature concerning the Common Frog and other anurans in which these factors are referred to or discussed
(b) Factors limiting tadpole numbers	
i. Predation	Martof (1956), Savage (1961)
ii. Food shortage	Savage (1961)
iii. Site becomes unsuitable in summer—dries out, oxygen depleted, etc.	Smith (1953), Martof (1956), Hazelwood (1971*b*)
iv. *a*Collection by children	Savage (1961), Cooke (1972*a*)
v. Disease, parasites etc.	Savage (1961), Bell (1970), Culley and Gravois (1970), Gibbs, Nace and Emmons (1971)
vi. *a*Pollution	Hazelwood (1970), Cooke (1972*a, b*)
vii. Professional collection	
(c) Factors limiting adult numbers	
i. Predation	Martof (1956), Savage (1961), Frazer (1966), Smith (1969), Bell (1970)
ii. *a*Road mortality	Pickles (1942), Moore (1954), Hodson (1966), Heusser (1968)
iii. *a*?Death during or just after hibernation due to adverse weather or human activities	Ashby (1969), Bell (1970), Heusser (1970*b*)
iv. *a*Amateur collection—by schools, children and pet keepers	Hazelwood (1971*a*), Kelly and Wray (1971), Beebee (1973)
v. *a*Professional collection	Smith (1953), Wheeler, Malenoir and Davidson (1959), Savage (1961), Gibbs *et al.* (1971), Cooke (1972*a*)
vi. *a*Pollution	Pickess and Snow (1966), Ferguson and Gilbert (1967), Hazelwood (1970), Shea (1970), Cooke (1972*a*)
vii. Disease, parasites, etc.	Savage (1961), Smith (1969), Gibbs *et al.* (1971)
viii. Food shortage	Savage (1961), Heusser (1969)

a Factors which have increased in importance recently.

on wasteland, common land, public parkland and golf courses comprised only 13–16% of the total number of sites during the 1960s, while such sites reported to the London Natural History Society made up 32% of the breeding sites within a 20 mile radius of St Paul's Cathedral in the 1960s.

One should not be over-pessimistic about the fate of our commoner amphibians. More than 11% of the B.S.S. and S.S. correspondents reported local increases in Common Frog populations in the late 1960s. Creation of wetland habitat, particularly garden ponds, was believed to be largely responsible. Information on the extent of the use of garden ponds has been extracted from completed B.S.S. questionnaires and is presented in Table IV. Taking into

TABLE IV. The increasing use of garden ponds for breeding by the Common Frog and Common Toad, 1950–70. These sites were recorded by correspondents to the Breeding Sites Survey described by Cooke (1972a)

	Common Frog			Common Toad		
	Total no. of breeding sites recorded	Garden ponds		Total no. of breeding sites recorded	Garden ponds	
		No.	% of total		No.	% of total
1950	140	12	9	52	2	4
1960	182	26	14	74	5	7
1965	243	43	18	86	11	13
1970	268	53	20	94	15	16

account every site in which breeding was noted, the percentage of garden sites increased for both the Common Frog and the Common Toad from 1950 to 1970. Most garden ponds are rather small and so cannot support large breeding populations, but since frog colonies in any type of site in Britain tend to be relatively small, pond owners can make an appreciable contribution to the conservation of this species by encouraging breeding in their gardens. Toads, on the other hand, frequently form colonies in field ponds of more than 1000 individuals, and it is the conservation of these big breeding gatherings that will most benefit this species.

The Natterjack, like some other amphibians and reptiles, has suffered from destruction of habitat, collecting, gassing of rabbit burrows, fires and other human pressures. At its inland sites, as in the case of the Sand Lizard and Smooth Snake, it has particularly suffered from the destruction of heathland. At its last inland stronghold, on the heaths of the south-west of London, its decline has been recent and rapid and coincided with the declines of the other two species. Its coastal sites have suffered from the increasing use of sand-dunes by tourists.

In contrast to the Sand Lizard and Smooth Snake however man's recent obvious activities may not explain all the decreases, particularly in eastern England. It is of course difficult to speculate what produced the losses of colonies in places as far apart as Dorset and Cambridgeshire in the early 1900s. Certainly in Dorset large areas of heath still remain. Also in eastern England, as Ellis (*in litt.*) points out, while some of their habitat has been destroyed much of it still appears relatively undisturbed. Ellis mentions several factors which could have played a contributory part, e.g. the severe winters of 1947 and 1963, the sea flooding of 1953 and the use of heavy machinery. Neither Ellis nor M. Wilson (*in litt.*) have found Natterjacks destroyed by toad-fly larvae (*Lucilia bufonivora*), but Ellis considers that in a favourable climatic period the toad fly might have taken toll of the Natterjack on a grand scale.

It is well established that the Natterjack may suffer bad breeding seasons because of drying-out of their breeding pools, particularly those in sand-dunes. These breeding failures can certainly result in sudden local population decreases, but little is known of how temporary or long lasting their effect may be on the populations. The Southport populations have failed to breed successfully during the last two seasons (S. Smith, *in litt.*) and their present population is much reduced. The Winterton population has also failed in the last two seasons (J. Pickett, N. Richards, and J. Woolston, all *in litt.*). This breeding failure could be a more critical factor now the remaining populations are becoming smaller and more confined.

2. Newts

It is generally assumed that most newt population declines have resulted from loss of breeding ponds. The Warty Newt is thought to be more particular in its choice of breeding site, tending to avoid the small, shallow pools often used by the other species (Bell, 1970; Beebee, 1973). Warty Newts, however, seldom breed in ponds in which Smooth or Palmate Newts cannot also be found. One of us (A.S.C.) has compiled from the literature and from personal and Biological Records Centre records, a list of 112 Warty Newt breeding sites, and, of these, 89 (80%) contained Smooth Newts, 25 (22%) contained palmate newts and 94 (84%) had at least one other species breeding in them. Apparent absence of the other species from many ponds may have been due to failure to catch or see the animals rather than to real absence. Thus loss of these sites will affect all three species, but since the Smooth and Palmate Newts are more catholic in their choice of breeding ponds, survivors of these species are more likely to be able to find other suitable ponds to colonize. Collectors were also blamed by Beebee (1973) for the decline of the Warty Newt in south-east England.

3. Reptiles

The principal cause of the decline of the Smooth Snake and Sand Lizard is the destruction of their primary habitat—mature lowland heath and coastal sand-dunes. The lowland heath habitat most suitable to reptiles is composed of deep stands of Heather (*Calluna vulgaris*) with the individual plants being 10–30 years in age. In the absence of disturbance by man, or by invasion from alien pine (*Pinus*), this habitat appears to reach a permanent natural equilibrium.

Man-induced changes come about as a result of public pressure exposing the top soil and, most significantly, by fires. The latter allows encroachment of *Betula*, *Ulex*, *Pteridium aquilinum* and grasses (*Molinia* and *Agrostis* species).

The destruction and fragmentation of heathland is caused by afforestation, urbanization, agricultural reclamation, military activity and mineral extraction (in that order of importance). The recolonizing ability of reptiles being poor, the fragmentation of heaths (*cf.* Moore, 1962) renders local populations more vulnerable to extinction from fire, collecting, and the gassing of rabbit burrows. When fragmentation has resulted from urbanization there is the additional danger of the destruction of isolated reptile populations by children and domestic cats.

Rabbits perform the valuable function of exposing areas of loose sand suitable as lizard breeding sites and they also keep back gorse and grasses, so the loss of rabbits from myxomatosis has had a detrimental effect. The gassing of rabbit burrows now carried out to prevent recolonization by rabbits can result in the death of reptiles using rabbit burrows for hibernation, especially as gassing is usually carried out in winter.

In the New Forest the practice of burning the heath to improve grazing is almost certainly responsible for the decline of reptiles, as it has both reduced and fragmented the areas of mature dry heath. Unfortunately the alternative practice of cutting the heather is little better. While it does eliminate direct mortality by burning, it still removes the necessary cover and food. After cutting or burning, the habitat takes ten years to recover, yet a Sand Lizard has an average life span of only five years in the wild. Areas of mature heath will have to be safeguarded if this species is to survive here.

Contrary to the hope expressed by Moore (1962), many isolated but relatively undisturbed remnants of heath have not remained suitable for the Sand Lizard and Smooth Snake. This is principally because of the introduction of mechanical cutting on a wide scale, on verges, on golf courses, in parks, and in gardens, destroying the old deep heather. Also the deep heather in many of the rides and under the trees in conifer plantations now established on old heathland e.g. the Hurne, Ringwood, Purbeck and Wareham Forests, is being rendered unsuitable for reptiles as the trees mature and suppress the ground vegetation.

As reptile populations become more isolated and reduced in size, the effect of collecting becomes more severe. Collecting is carried out by research workers,

TABLE V. Changes in status of several species found or once found in Guernsey and Jersey

Island	Species	Change in status	Sourse of information
Guernsey	Common Frog and Agile Frog	Gradually declined over last 50 years because of habitat loss due to drainage and building	N. Jee (*in litt.*)
	Common Toad	Introduced in 1953, perhaps now extinct	N. Jee (1967, and *in litt.*)
	Smooth Newt	Introduced in 1930, but now suffering from flooded quarries being filled in	N. Jee (1967)
	Green Lizard	May have increased since 1902	Frazer (1949)
Jersey	Common Frog	Possibly recently introduced	Mrs F. le Sueur (*in litt.*)
	Agile Frog	Once quite common, but now very scarce	Mrs F. le Sueur (*in litt.*)
	Smooth Newt	Became extinct in the twentieth century	Frazer (1949)

dealers and pet-keepers, the last of these on occasions even collecting British specimens for use as food for more exotic species being kept as pets. As an example of the detrimental effect collecting can have, the recent 80% decline in the Sand Lizard at Studland is considered to have been caused by professional collectors.

The loss of coastal habitat is mainly as a result of urbanization and recreational public pressure.

THE CHANNEL ISLANDS

The Channel Islands are best considered separately since their herpetofauna includes the Agile Frog (*Rana dalmatina* Bonaparte), the Green Lizard (*Lacerta viridis* (Laurenti)) and the Wall Lizard (*L. muralis* (Laurenti)), species that are not indigenous to the mainland of Britain. Changes in status of several species on the two main Islands, Guernsey and Jersey, are summarized in Table V.

CONCLUSIONS

The amphibians and reptiles provide examples of the two extremes of the conservation dilemma. Of the twelve indigenous British species, nine are widespread and certainly still common in many places although recent local declines have occurred. With these species the problems are (a) to assess the severity of these local declines in relation to future survival in Britain; and (b) to ensure that conservation action is taken before the declines accelerate beyond the point when they may be difficult or even impossible to halt.

The other three species are in a critical situation and for two (the Sand Lizard and Smooth Snake) the stage has already been reached when without almost continual conservation action being taken, they will become extinct in Britain in the near future.

The recently formed Conservation Committee of the B.H.S., working in conjunction with the Working Party set up by the Nature Conservancy, has already initiated work on site protection and management in support of national and regional surveys and recording. Some seventy key reptile sites have now been selected. Efforts will be made to get these protected as National, Local or County Trust Reserves or as Sites of Special Scientific Interest. A programme of habitat clearance and improvement, carried out by the Conservation Corps and B.H.S. members, is well advanced. In the seriously depleted Zone II there have been some experimental reintroductions of Sand Lizards to renovated heath sites. Experimental reintroductions are also being tried in Zones II, III and IV for the Natterjack, where it has recently become extinct.

REFERENCES

ASHBY, K. R. (1969). The population ecology of a self-maintaining colony of the common frog (*Rana temporaria*). *J. Zool.* **158**, 453–474.

BEEBEE, T. J. C. (1973). Observations concerning the decline of the British amphibia. *Biol. Conservation* **5**, 20–24.

BELL, G. A. C. (1970). The distribution of amphibians in Leicestershire. *Trans. Leicester lit. phil. Soc.* **64**, 122–142.

BIOLOGICAL RECORDS CENTRE (1973). "Provisional Atlas of the Amphibians and Reptiles of the British Isles." Natural Environment Research Council, London.

BOYD, M. (1971). "Survey of the Distribution of the Natterjack Toad on the Dumfriesshire Coast." [Unpublished report.] Nature Conservancy, Edinburgh.

Cooke, A. S. (1972a). Indications of recent changes in status in the British Isles of the frog (*Rana temporaria*) and the toad (*Bufo bufo*) *J. Zool.* **167**, 161–178.

Cooke, A. S. (1972b). The effects of DDT, dieldrin and 2,4-D on amphibian spawn and tadpoles. *Environ. Pollut.* **3**, 51–68.

Corbett, K. F. [In preparation]. "The Ecology and Conservation of the British Sand Lizard". Ph.D. thesis, University of London.

Culley, D. D. and Gravois, C. T. (1970). Frog culture. *Am. Fish Farm.* **1**, 5–10.

Ferguson, D. E. and Gilbert, C. C. (1967). Tolerances of three species of anuran amphibians to five chlorinated hydrocarbon insecticides. *J. Miss. Acad. Sci.* **13**, 135–138.

Forrest, H. E. (1907). "The Vertebrate Fauna of North Wales." Witherby, London.

Frazer, J. F. D. (1949). The reptiles and amphibia of the Channel Isles, and their distribution. *Br. J. Herpet.* **2**, 51–53.

Frazer, J. F. D. (1966). A breeding colony of toads (*Bufo bufo* (L.) in Kent. *Br. J. Herpet.* **3**, 236–252.

Gibbs, E. L., Nace, G. W. and Emmons, M. B. (1971). The live frog is almost dead. *BioScience* **21**, 1027–1034.

Grubb, J. C. (1972). Differential predation by *Gambusia affinis* on the eggs of seven species of anuran amphibians. *Am. Midl. Nat.* **88**, 102–108.

Hazelwood, E. (1970). Frog pond contaminated. *Br. J. Herpet.* **4**, 177–185.

Hazelwood, E. (1971a). Frog pond desecrated. *Br. J. Herpet.* **4**, 237–239.

Hazelwood, E. (1971b). Amphibians on the dunes at Ainsdale, Lancs. "Obituary". *Br. J. Herpet.* **4**, 239.

Heusser, H. (1968). Wie Amphibien schützen? *Nat. forsch. Ges. Schafthausen, ser.* II, **3**, 1–12.

Heusser, H. (1969). "The Ecology and Life History of the European Common Toad *Bufo bufo* (L). An abstract of a five year study." Zentralstelle der Studentenschaft, Zürich.

Heusser, H. (1970a). Laich-Fressen durch Kaulquappen als mögliche Ursache spezifischer Biotoppräferenzen und kurzer Laichzeiten bei europäischen Froschlurchen (Amphibia Anura). *Oecologia* **4**, 83–88.

Heusser, H. (1970b). Ansiedlung, Ortstreue und Populationsdynamik des Grasfrosches (*Rana temporaria*) an einem Gartenweiher. *Salamandra* **6**, 80–87.

Heusser, H. (1971). Laich-Räubern und—Kannibalismns bei sympatrischen Anuren—Kaulquappen. *Experientia* **27**, 474.

Hodson, N. L. (1966). A survey of road mortality in mammals (and including data for the Grass snake and Common frog). *J. Zool.* **148**, 576–579.

Howes, C. A. (1972). "The History and Distribution of Reptiles and Amphibians in South-east Yorkshire and the Doncaster District." [Unpublished report]. Museum and Art Gallery, Doncaster.

Jee, N. (1967). "Guernsey's Natural History." Guernsey Press, Guernsey.

Kelly, P. J. and Wray, J. D. (1971). The educational uses of living organisms. *J. biol. Educ.* **5**, 213–218.

MacPherson, H. A. (1892). "A Vertebrate Fauna of the Lakeland." Douglas, Edinburgh.

Marr, J. E. and Shipley, A. E. (1904). "Handbook of the Natural History of Cambridgeshire." Cambridge University Press.

MARTOF, B. (1956). Factors influencing size and composition of populations of *Rana clamitans*. *Am. Midl. Nat.* **56**, 225–245.

MERTENS, R. (1947). Studien zur Eidonomie und Taxonomie der Ringelnatter (*Natrix natrix*). *Abh. Senckenb. Naturf. Ges.* **476**, 138.

MOORE, H. J. (1954). Some observations of the migration of the toad (*Bufo bufo bufo*). *Br. J. Herpet.* **1**, 194–224.

MOORE, N. W. (1962). The heaths of Dorset and their conservation. *J. Ecol.* **50**, 369–391.

PERRING, F. H. (1966). Where have all the frogs gone? *Wild Life Obsr* **19**, 10–11.

PICKESS, B. P. and SNOW, W. F. (1966). Observations on the effects of pollution of the East Stream of the Reserve. *Ruislip Dist. nat. Hist. Soc.* **15**, 10–12.

PICKLES, W. (1942). Animal mortality on three miles of Yorkshire roads. *J. Anim. Ecol.* **11**, 37–43.

SAVAGE, R. M. (1961). "The Ecology and Life History of the Common Frog." Pitman, London.

SHEA, K. P. (1970). Dead stream. *Environment, St. Louis* **12**, 12–15.

SIMMS, C. (1969). Indications of the decline of breeding amphibians at an isolated pond in marginal land, 1954–1967. *Br. J. Herpet.* **4**, 93–96.

SMITH, M. A. (1953). The shortage of toads and frogs. *Country Life* **114**, 770–771.

SMITH, M. A. (1969). "The British Amphibians and Reptiles", 4th edition. Collins, London.

TAYLOR, R. H. R. (1948). The distribution of Reptiles and Amphibia in the British Isles, with notes on species recently introduced. *Br. J. Herpet.* **1**, 1–38.

TAYLOR, R. H. R. (1963). The distribution of amphibians and reptiles in England and Wales, Scotland and Ireland and the Channel Isles: a revised survey. *Br. J. Herpet.* **3**, 95–115.

VAN GELDER, J. J. and KALKHOVEN, J. T. R. (1971). Eieren van der knofwokpad (*Pelobates fuscus* Laur.) in de Hatertse en Overasseltse Vennen. *Natuurh. Maandblad.* **60**, 39–44.

VAN GELDER, J. J. and OOMEN, H. C. J. (1970). *Rana arvalis* Nilsson: Reproduction, growth, migration and population studies. *Neth. J. Zool.* **20**, 238–252.

WHEELER, A. C., MALENOIR, G. and DAVIDSON, J. (1959). A first report on the report on the reptiles and amphibians in Epping Forest. *Essex Nat.* **30**, 179–188.

YALDEN, D. W. (1965). Distribution of reptiles and amphibians in the London area. *Lond. Nat.* **44**, 57–69.

15 | Changes in Composition of the Terrestrial Mollusc Fauna

A. SOUTH

Department of Biological Sciences, City of London Polytechnic, London

Abstract: A brief survey of earlier nineteenth-century records of British terrestrial molluscs, including the first critical check-list (1840), is given. The seven editions of the census of non-marine Mollusca, published by the Conchological Society between 1885 and 1951, are described and their limitations discussed. A brief description is included of the current 10 km square mapping scheme organized by the Society.

A summary is given of additions to, and extinctions from, the British fauna since 1840. The probable causes of these additions, including the possible effect of recent climatic changes, are discussed. Some examples of changes in the distribution of native species, and possible reasons for these changes, are given. The significance of those man-made factors known to affect numbers and distribution of slugs and snails is discussed with special reference to the development of slugs as agricultural pests.

INTRODUCTION

Snails are among the few terrestrial invertebrates that leave readily identifiable traces in Quaternary deposits. For this reason there have been a number of studies made on changes in land mollusc faunas in the Post-glacial or Holocene period, and these have been reviewed by Evans (1972). Snails have been used as indicators of Post-glacial environmental changes. Kerney (1966a) reviewed the effect of man on the snail fauna over this period. The present paper attempts to determine what changes have taken place since the first critical check-list was published in 1840 and to try to assess how much of this change is due to the continuing deterioration in climate from the thermal optimum and how much is due to the ever increasing pressure of man on his environment as the result of increasing population, urbanization, industrial growth and modern agriculture.

Systematics Association Special Volume No. 6, "The Changing Flora and Fauna of Britain", edited by D. L. Hawksworth, 1974, pp. 255–274. Academic Press, London and New York.

The generic and specific nomenclature follows that of Ellis (1951), except that
Acicula fusca (Montagu) is used instead of *Acme fusca* and *Cochlicopa lubricella*
(Porro) instead of *C. minima*. The generic name *Cepaea* Held is used in place of
Helix for *C. hortensis* and *C. nemoralis* and *Hygromia liberta* (Westerlund) is
taken as a synonym of *H. hispida* (Linné) (Ellis, 1969: 279). The nomenclature
for more recent introductions than Ellis (1951) follows that of Evans (1972).
While it is clear that three species are included in the *Arion fasciatus* agg. (see
Kerney, 1966b), no attempt has been made to separate the species as there are
few reliable individual records until comparatively recently. The extensive
synonymy that exists in the earlier records has been elucidated with the aid of
later censuses and Ellis (1969).

THE HISTORY OF RECORDING

The first census of British species was compiled by Gray (1840) in his revised
edition of Turton's book. A check-list had been published three years previously
by Alder (1837) but this gave no information about the geographical distribution of each species. Gray listed the British species and summarized their geographical distribution in the form of a table. He divided the British Isles into
eight regions of rather differing size, one region representing the whole of
Scotland while another represented only the Isle of Man. He was, however,
hindered by the paucity of records available to him. Gray also listed the various
additions to the British list of slugs and snails over the previous two hundred
years.

Jeffreys (1862) included a list of British species together with some information
on their foreign distribution but little information on their British distribution.
Taylor (1894–1921) included the known geographical distribution, including
maps, for many slugs and snails in his monograph which began to appear in
parts in 1894 but was unfortunately never completed.

The first systematic records are found in the census of non-marine mollusca
published by the Conchological Society in 1885 (Taylor and Roebuck, 1885)
and based on a check-list published two years previously (Anon., 1883). The
history of this census has been described by Kerney (1967). Six further editions
of the census were published (Williams, 1889; Adams, 1896; Williams, 1901;
Adams, 1902; Roebuck, 1921; Ellis, 1951) while a supplement to the seventh
edition was published in 1966 (Kerney, 1966b).

The Conchological Society's censuses were based on the Watsonian vice-
counties for England, Wales and Scotland and on the scheme devised by
Babington (1859), later modified by Praeger (1896), for Ireland. Distributions
were shown in tabular form only in earlier editions but in the sixth and seventh

editions small distribution maps were also given for most species. Each record for these censuses was only accepted if the actual specimen was seen and verified by the referees of the Society.

The seven editions of the census have provided a great deal of information about the distribution of British non-marine mollusca and Ellis (1951) considered that, with the exception of a few critical or recently discovered species, the distribution had been fully worked out on a vice-comital basis by 1951, and, fourteen years later Kerney (1966b) was able to support this conclusion.

This census is of rather limited value, however, when studying the changes that have taken place in the composition of the terrestial slug and snail fauna for a number of reasons. Vice-counties are of limited value as recording units because of their large size and varying shape, one county often including areas that differ widely in geology, climate and other factors. The true range of some rather local species may be exaggerated when vice-counties are used, and several examples are cited by Kerney (1967). *Azeca goodalli* is shown as having a range similar to that of *Marpessa laminata*, although the latter is much the commoner species. He also shows how in the case of *Abida secale*, the use of vice-counties grossly inflates the true range of very local species. Some allowance was made for this in the 1951 census by the use of dot-maps for the distribution of a few species.

Another disadvantage of these censuses is that because records are essentially cumulative, they tend to obscure extinctions and declines in the range of species during the period covered by the seven editions of the census (1884–1951). Furthermore, increases in the range of a species may indicate a genuine extension of its range. Some records are based on data at least 80 or 90 years old and so the 1951 census could hardly give an accurate description of the true distribution in 1951. It is still possible, however, to draw some conclusions about additions and deletions from the British fauna using the seven editions of the census (Table I).

In 1961, the Conchological Society decided to adopt the 10 km square system for their records in conjunction with the Nature Conservancy's Biological Records Centre at Monks Wood Experimental Station. Several species maps have already been published, for example those for *Monacha cantiana* and *M. cartusiana* (Kerney, 1970). Two symbols are used on these maps to distinguish pre- and post-1950 records and this will provide considerably more information on changes in distribution as further maps are produced. It is understood that a provisional set of species maps will probably be produced in 1974 (Kerney, 1971).

The original authors of the Conchological Society's census emphasized how little of Britain had been examined for records, especially in the north and west, and how supposedly ubiquitous slugs like *Agriolimax reticulatus* had been

TABLE I. Summary of additions and deletions from successive censuses between 1840 and 1951

Census date	1862	1885	1901
Total number of species recorded	73[a]	80	83
Species added since previous census	*Geomalacus maculosus* (A)[b] *Milax gagates* (A) *Lenhmannia marginata* (A) *Vertigo moulinsiana* (E)	*Testacella maugei* (A) *Limax cinereoniger* (A) *Agriolimax laevis* (A, B) *Cepaea hortensis* (B) *Oxychilus helveticus* (D) *Vertigo lilljeborgi* (E)	*Testacella scutulum* (A) *Arion intermedius* (A) *A. fasciatus* (A) *A. subfuscus* (A) *Oxychilus draparnaudi* (G)
Species recorded in previous census but not in present census	*Agriolimax laevis* (A, B) *Clausilia dubia* (B) *Cepaea hortensis* (B) *Helix aperta* (H) *Hygromia limbata* (H)		*Vertigo lilljeborgi* (E)

[a] The total number of species recorded in the 1840 census was 72.
[b] For explanation of (A)–(H), see text.

recorded from only twenty-three vice-counties. This problem has persisted until the present day, and Kerney (1972) has shown how Wales and Scotland remain poorly covered despite the considerable efforts that have been made to remedy this situation.

ADDITIONS AND EXTINCTIONS

Additions to successive censuses from 1840 until the present time are summarized in Table I. It can be seen that over the past 132 years the total number of terrestrial species recorded has increased by about 50%. This summary does not

1902	1921	1951	post-1951
84	95	108	111

Helicella elegans (H)	*Limax tenellus* (A)	*Arion lusitanicus* (A)	*Milax insularis* (A)
		A. rufus (A)	
	Clausilia dubia (B)	*Milax budapestensis* (A)	*Columella aspera* (C)
		Agriolimax agrestis (A)	
	Vallonia costata (C)	*A. caruanae* (A)	*Vitrea diaphana* (F)
	V. excentrica (C)		
	Helicella gigaxi (C)	*Catinella arenaria* (C)	
		Succinea elegans (C)	
	Vertigo lilljeborgi (E)	*Cochlicopa lubricella* (C)	
		Carychium tridentatum (C)	
	Vitrina pyrenaica (F)	*Truncatellina britannica* (C)	
		Vitrea contracta (C)	
	Fruticicola fruticum (H)		
	Hygromia limbata (H)	*Hygromia cinctella* (H)	
	Helicella neglecta (H)		
		Vertigo genesii (F)	
		Vitrina major (F)	
	Hygromia umbrosa (H)		*Helicella neglecta* (H)

include species originally recorded separately and now known to be synonyms of other species. A list of species omitted at each census but recorded in the previous census is also given as a guide to possible extinctions. The significance of the additions and omissions can be best considered under the following groupings:

1. Group A: Slugs

According to Quick (1960), the first British records for 12 of the 24 species of slugs recorded were pre-1840. Gray (1840) omitted four of these twelve (*Testacella maugei*, *Testacella scutulum*, *Arion fasciatus* and *Arion subfuscus*) from his distribution list, and although these species were probably well established by 1840 they were not included until later censuses. These species were included by Gray, however, as either synonyms or varieties of existing species, or, in the case of *T. maugei*, in a supplementary list of introduced species restricted to artificial habitats like greenhouses, etc. Gray also included another species, *Limax cinereoniger*, as a synonym of *Limax maximus*. although it was not recorded separately until 1885.

Four additional slugs were first recorded in Britain between 1840 and 1848, but two of these, *Arion intermedius* and *Limax tenellus*, which were first recorded by Alder (1848) were not included in a census until much later. Taylor included *L. tenellus* in his monograph and it seems likely that this rare and local species was simply overlooked. There is little doubt that all four species were well established in the British Isles before 1840. Another species that was overlooked from the first British record in 1893 until the 1941 census was *Arion lusitanicus*. Since the 1951 census, however, this species has been reported from at least eleven new and widely dispersed localities, and will probably prove to be more widespread, having been confused with *Arion ater* in the past. *Arion rufus* has also been distinguished only recently (in the 1951 census) from the latter species, and some authors, e.g. Quick (1960), consider it to be a subspecies of *A. ater*. *A. rufus* has also been reported from at least eleven new localities since 1951 and is likely to be widely distributed, especially in southern England. Quick (1960) considered that *A. lusitanicus* and *A. rufus* were probably recent intro-ductions in Britain, although he considered that the former species might be indigenous to Ireland.

Agriolimax agrestis sensu stricto was first recorded from the Norfolk Broads by Ellis (1941) and has since been reported from several localities in Scotland and England. The name *A. agrestis* was used widely to include *A. reticulatus* and possibly *Agriolimax caruanae* as well until about 1940, and it is, therefore, difficult to assess which records belong to *A. agrestis* seg. and whether this species is a recent introduction or not. Since it is now known from localities that are widely separated from one another, e.g. the Norfolk Broads, West Riding of Yorkshire, Outer Hebrides and Moray, it seems likely that *A. agrestis* is a native species rather than a recent introduction. *A. caruanae* has almost certainly been confused with other species of *Agriolimax*. It was discovered in Cornwall by Oldham in 1930 (Quick, 1960) and had been recorded from over twenty localities by 1966. It is now clear that *A. caruanae* is generally distributed and abundant throughout the British Isles. It was thought to be a largely synanthropic species, but Ellis (1964) found it to be ubiquitous in west Cornwall. There is also some evidence of a fossil record for this slug (Hayward, 1954), and Ellis (1951) suggested that *A. caruanae* is probably native in the south-west and introduced elsewhere in Britain, where it is especially common in gardens and on arable land. Kerney (1968) suggested that there is evidence for a rapid spread of this species as it is unlikely to have been overlooked for so long, and McMillan (1969) considered that it was probably a recent introduction into Ireland.

The first British record for *Milax budapestensis* was in 1930 and it had been reported from 39 vice-counties by the 1951 census, in England and Wales, mainly from gardens and cultivated land. The number of localities has now doubled since the last census and it has been found in Ireland and Scotland. It is difficult to establish whether *M. budapestensis* is indigenous and there is little evidence to suggest that earlier records were confused with other *Milax* spp. although some of the examples of *Milax gagates* figured by Taylor in his monograph resemble this species more closely than *M. gagates*. There is little doubt, however, that *M. budapestensis* has been spread largely through the agency of man, and Dobson (1963) gives evidence that its distribution has expanded in the Glasgow area over the past thirty years.

2. Group B

Three of the species listed by Gray (1840) were overlooked in some later censuses and so appear in the lower part of Table I. *Cepaea hortensis* and *Agriolimax laevis* were not recorded again until the 1885 census. *Clausilia dubia* did not reappear until 1921, probably because of its scarcity and probable confusion with *Clausilia bidentata*.

3. Group C

Vitrea contracta, *Cochlicopa lubricella* and *Carychium tridentatum* are now known to be widespread, although not recorded until the 1951 census. Until this date they had probably been included in composite species. The specific status of a fourth species, *Truncatellina britannica*, is still uncertain (Ellis, 1969) but it has probably been included in the *Truncatellina cylindrica* agg. in censuses prior to 1951. It has become increasingly apparent that a new British species, *Columella aspera*, is widespread in Britain (Kerney, 1972) although probably included previously in the *Columella edentula* agg. *Vallonia costata* and *Vallonia excentrica* were similarly included in the *Vallonia pulchella* agg. until 1921, and are now known to be common. *Helicella gigaxi* was also confused with *Helicella caperata* until the 1921 census. Ellis (1949) considered that the two succineid snails, *Succinea elegans* and *Catinella arenaria*, had been confused with other *Succinea* spp. previously, and Quick (1943) suggested that the specimens collected by Jeffreys in 1862 from South Wales might be *C. arenaria* and not *Succinea oblonga* as they were labelled.

4. Group D

Taylor (1885) stated that *Oxychilus helveticus* was included by Gray in 1840 as var. *glaber* Studer of *Oxychilus alliarius* and so was known in 1840.

5. Group E

According to Ellis (1949), *Vertigo lilljeborgi* was discovered by Jeffreys in 1845 and it is not clear why it was omitted from the 1901 census. There is no reason to suppose that this species or *Vertigo moulinsiana*, recorded by Jeffreys in 1862, were not established by 1840.

6. Group F

Vertigo genesii and *Vitrea diaphana* are rare boreal-alpine species showing a relict distribution in Britain and are both probably survivors from early post-glacial times. *Vitrina pyrenaica* a Pyrenean species, was discovered in Ireland in 1908, and had probably been overlooked because of its very restricted distribution. Fogan (1969) considered it likely that this snail might be indigenous and not an introduction, as previously believed by Ellis (1951).

Another rare species that has probably also been overlooked is *Vitrina major*, which was discovered by Boycott (1922). He considered that this species had in fact been previously recorded in Britain by Jeffreys in 1830.

7. Group G

Oxychilus draparnaudi belongs to a group of slugs and snails called anthropophiles by Boycott (1934) and synanthropic species by Kerney (1966a). These are species which show evidence of a recent extension of their range as the result of human activity and which, if native, occur in the wild only in south-western Britain. The significance of this recent spread is discussed below. *O. draparnaudi* was first recorded in the census of 1901 from fifteen vice-counties although in an earlier edition of William's book (1892) it is recorded only from five localities and was not mentioned in the 1885 census. It was almost the last of the zonitid snails to be discovered and yet it is the largest. It is unlikely that it was overlooked and is probably a relatively recent introduction.

8. Group H

Several slugs and snails have been introduced into this country either as greenhouse aliens that have not become naturalized in the open or as adventives or casual introductions that have not become established. Ellis (1951) records about twenty species in each category, some being known from several localities.

Many of the greenhouse aliens have been taken from botanic gardens at Kew and Edinburgh, and Verdcourt (1969) described what was probably a new species of *Gulella* (Streptaxidae) and probably still unknown in its native habitat. There are a number of other lists of accidental introductions, including that of Paul (1963) who cites importations of snails with bananas. He stresses the ability of snails to survive the often severe conditions of transport, although in most instances introduced species do not survive in this country once they have arrived. Gray (1840) included a list of eight greenhouse aliens or adventives together with a much longer list of doubtful records, some of which he considered to have been introduced by unscrupulous collectors.

Some introduced species have become sufficiently well established to warrant their inclusion in the census. The two most successful introductions in recent times have been *Hygromia limbata* and *Hygromia cinctella* into South Devon. *H. limbata* was first noticed in 1917 by Kennard (Kennard and Woodward, 1918). *H. cinctella* was later found to be common in the same area by Comfort (1951). Both snails appear to be well established in South Devon but have not been recorded elsewhere in recent times, although Gray (1840) refers to a colony of *H. limbata* near Hampstead, London, which had probably been introduced and which had died out by the time the 1862 list was published.

An apparently isolated colony of *Helicella neglecta* discovered at Luddesdown, Kent, in 1915 (Kennard and Woodward, 1917a) was probably an introduction and was subsequently eliminated by a collector in 1922 (Ellis, 1969). *Helicella elegans* is also an introduced species (Ellis, 1951) and was first recorded near Dover, Kent, and later near Chaldon, Surrey. There are several unconfirmed earlier records for this snail, including one from Cumberland (Gray, 1840). *Fruticicola fruticum* was also discovered in Kent near Dover in 1908 and later, in 1911, near Penshurst, also in Kent. Kennard and Woodward (1917b) considered it to be a recent introduction and Kerney (1966b) stated that it was probably extinct by that time. Gray (1840) added *Helix aperta* a Mediterranean species, to the British list on the basis of a single crushed individual found in mud under the side of a hedge in Guernsey. Since there have been no other British records for this species it has little claim to a place in the British list. *Milax insularis*, another Mediterranean species, was recorded from Bexhill, East Sussex, by Quick (1960) but there have been no further records for this species. Another example of a successful adventive species was the established colony of the Central European snail, *Hygromia umbrosa* at Margate, Kent, recorded by Taylor in 1914 and which had apparently died out by the 1921 census.

It can be said in conclusion that although Table I shows a great deal of change, in the majority of instances these represent only apparent changes in the fauna. There is no evidence that any indigenous slug or snail has become extinct since 1840, and it seems likely that there have been no extinctions over a much longer period than this. Several introduced species have become established during this period although several of these subsequently became extinct.

REASONS FOR THE SUCCESSFUL INTRODUCTIONS

It is clear from the list given by Ellis (1951) that exotic species are often introduced into Britain either accidentally or intentionally. It is therefore not surprising that some of these, especially European species, have become established for varying periods of time. In addition to introductions by unscrupulous collectors, mainly in the nineteenth century, some species have undoubtedly been introduced as food. Taylor (1894–1921), for example, cited in his monograph an instance of a restaurateur in Soho importing a hundred thousand snails, mainly from France, and stated that the fashion was spreading. Although still the subject of some controversy, it seems likely that both *Helix pomatia* and *Helix aspersa* were introduced by the Romans or earlier settlers into Britain as food (Ellis, 1969). Stelfox and McMillan (1966) put forward the idea that the snail *Theba pisana* might have been introduced as food by settlers on the Dublin coast. The introduction of *H. aperta* recorded above might be related to the popularity of this snail as a food. Living *Achatina fulica* (Giant African snail) are sold as food in London, although it would be impossible for this snail to become acclimatized. The species has provided the most spectacular example of the way that a snail can become widely dispersed through trade (Mead, 1961).

Snails can also resist the problems presented by transport, and Mead also gives examples of the ability of many species to survive without food or water for very long periods of time. One of the best known examples in this country is that of a *Helix desertorum* in the British Museum which, after four years' attachment to a tablet, emerged and moved about when taken into a moist atmosphere. It is also clear, from the paper by Stelfox (1969) on "*Marpessa laminata* bred in a small box for fifty-four years", that some species can survive and breed under the most adverse conditions.

Slugs can also withstand the adverse conditions associated with dispersal through trade routes, although they have no shell. Since the first report of an introduced slug prior to 1822, at least fifteen European species have been found in North America, mainly in urban areas and on arable land. Three species, *Agriolimax reticulatus*, *Arion fasciatus* and *A. subfuscus*, have become more widely distributed in natural habitats (Chichester and Getz, 1969). These authors considered that most of these introductions were made on plant materials imported from Europe. It is difficult to accept their suggestion, however, without any evidence, that some of these slugs have dispersed actively along roadside verges for any distance without the participation of man. It seems more likely that this dispersal has been accomplished passively by vehicles. It is perhaps interesting that there is no evidence for the establishment of indigenous North American slugs and snails, like the arionid pest species *Prophysaon*, into this country, although the opportunities for introduction exist. In 1967 I was given two individuals of an unidentified arionid slug taken from pre-packaged carrots imported from Texas (U.S.A.) and cooled in transit. These slugs survived for some time in the laboratory.

The reason why more introduced species do not establish themselves is probably that environmental conditions are unsuitable. Boycott (1934) observed that "Mollusca are obviously ill adapted to spread far and quickly by their own proper motions", and it is also clear from Gray (1840) that attempts at the artificial introduction or dispersal of British and foreign species are rarely successful. Taylor says that *Helix pomatia* is difficult to establish in localities where it does not occur. Cockerell (1884) failed to establish *Monacha cartusiana* at Chislehurst, Kent, using individuals taken from Sandwich. However, Stratton (1949) records the success of Boycott in his, perhaps unwise, attempts to establish several species, including *H. pomatia*, in his Hertfordshire garden in the 1920s.

Successful establishment of *Hygromia limbata*, *H. cinctella*, *Helicella neglecta* and *H. elegans* in southern England during the latter part of the last century or the first part of this century, may result from changing environmental conditions. These species are all found in south or south-west France below the Loire. Biggs (1957) stated that *H. cinctella* had been extending its range across France over the past century as far as Brittany, and Milman (1951) stated that *H. limbata* had become naturalized in north-west France. Milman also suggests that these two species were probably introduced into South Devon, either with cargoes of cider apples from France or in imported plants from the continent. It is also interesting to note, since *Arion lusitanicus* might be a recent introduction, that this species has a similar distribution in France to that of the above snails, i.e. south of the Loire (Chevallier, 1969). A possible reason for the successful establishment of these species, especially *H. limbata* and *H. cinctella*, over the past century may be due to the warming of the climate that took place between about 1890 and the 1930s (Lamb, 1965). Even small changes in climate can be important (Lamb, 1970); an increase of 1°C in the summer climate of Britain, for example, will give southern Britain the type of climate now experienced in the vicinity of the Loire in France. The successful introductions have been in southern England, and it is suggested that the establishment of these species has been due to the slight warming up of the climate that occurred. Since this trend has now been reversed it will be interesting to see whether the colonies of introduced species will now show a decline. It would also be interesting to know of evidence for similar introductions for the previous climatic optimum in the twelfth and thirteenth centuries. The brief establishment of the Central European *H. umbrosa* at Margate might also be the result of this temporary amelioration in the climate.

CHANGES IN THE DISTRIBUTION OF SOME NATIVE SPECIES

It is only over the past 20–30 years that attempts have been made to estimate numbers of slugs and snails accurately. There is, therefore, very little quantitative information available relating to changes in numbers or, indeed, any other aspects of biology, such as fecundity, longevity or the influence of physical factors. In order to try to find evidence for changes in distribution over the past century, the percentage increase in vice-county records between successive censuses was calculated for 65 species of common slugs and snails which were first recorded before the 1885 census. In two-thirds of the species examined, the percentage increases did not differ significantly from the mean increase between each successive census. Almost one-third of species showed a significantly lower rate of increase, but these were either ubiquitous species like *Agriolimax reticulatus* or locally common species with a more restricted range like *Pomatias elegans*. Three species showed a significantly greater percentage rate of increase and they were *Limax cinereoniger*, *Acicula fusca*, and *Vertigo substriata*. There is, however, little evidence that these slugs and snails have increased their range recently and this result has probably arisen by chance, as might be expected from a sample of 65 species. It would appear that there is little information to be gained about changes in distribution over the past century, from the seven editions of the census.

Examination of the distribution maps in Ellis (1951) indicates that about ten species appear to reach the northern limit of their distribution in Britain, although some, for example *Monacho cantiana* and *Pomatias elegans*, are fairly common within their range. Several of these species are now on the wane in Britain by comparison with their distribution in the mid-Post-glacial and this has been put forward as evidence of a decline in temperatures since that time (Evans, 1972). Kerney (1968) suggested that changes in the distribution of *P. elegans* are due to its intolerance of winter cold. He showed that the northern and eastern colonies of this species are being slowly eroded and a comparison of pre- and post-1850 records (Evans, 1972) shows that this contraction in range has been continued over the past century. This is not true for all of these species and the distribution of *M. cantiana*, a vigorous colonizer of roadside verges, has not altered appreciably during the past century (Kerney, 1970) although the closely related *M. cartusiana* is still declining.

Two other species in this group, *Ena montana* and *Helicodonta obvoluta*, were listed as anthropophobes by Boycott (1934). These are species which appear to be intolerant of human interference and are rapidly eliminated by disturbance of their habitats. They are rare in Britain at the present time and occur almost exclusively in old-woodland habitats, although in France both species are common hedgerow snails. Kerney (1968) found that shells of *E. montana* were common in Neolithic and Bronze Age layers at archaeological sites in places where human interference had already been present. They are waning species, and a distribution map in Evans (1972) shows that *H. obvoluta* has disappeared from the most westerly parts of its range during the past century. Davis (1953) suggested that its former strongholds were being diminished as the remnants of beech forests were destroyed. It is difficult to assess the truth of this statement, as there is little evidence to suggest that other anthropophobic old-woodland species like the slugs *Limax cinereoniger* and *L. tenellus* or the snail *Acanthinula lamellata* have declined over the past century. These slugs, although local and rare, are generally distributed throughout most of Britain and are essentially forest species on the continent. *A. lamellata*, a local snail, also appears to be generally distributed over much of northern Britain. Most of these anthropophobic species are unable to colonize new woodland and are still mainly restricted to the relics of ancient woodlands. The decline in *E. montana* and *H. obvoluta* can probably be largely attributed to climatic change as in *P. elegans* although it is surprising that the temporary warming of the climate towards the end of the nineteenth century did not appear to affect this decline. This emphasizes how little is known about the effects of temperature on slugs and snails.

Taylor, in his monograph, took a rather pessimistic view of the fauna and cited a number of slugs and snails as waning species. He considered that *A. subfuscus* was declining in eastern Britain, but this now seems unlikely (Ellis, 1969). He also referred to the rare slug *Geomalacus maculosus*, a species restricted to Cork and Kerry in south-west Ireland, as a waning species, and yet in 1967 it was still found to be common in suitable habitats in that area (Mordan, personal communication). *T. maugei* was considered by Taylor as a retreating species, and yet this synanthropic slug has since extended its range. It is more likely that it was a recent introduction.

One group of slugs and snails that has shown an increase in distribution over the past hundred years are the anthropophiles (Boycott, 1934) or synanthropic species (Kerney, 1966a). These are species showing a close relationship with man and living in close proximity to houses, gardens or cultivated land, and they appear to be spread directly by man. Species included in this category by both Boycott and Kerney are *Testacella* spp., *Limax flavus*, *Helix aspersa*, *O. draparnaudi*, *Hygromia striolata* and *Milax* spp. Boycott included other species like *A. reticulatus* which have a much wider range and are not so closely linked with the activities of man. While some synanthropic species may be indigenous, being more widespread away from human habitation in south-west England, several, such as *O. draparnaudi* and *H. aspersa* are aliens (Evans, 1972) and did not occur with any certainty in pre-Roman times. Others like *H. striolata* have a fossil record but appear to have become extremely rare and then recovered as synanthropic species in Roman times or even later. It has been suggested above that *T. maugei* was a recent introduction and Taylor also described *T. scutulum* as "especially prevalent in and characteristic of the Metropolitan district" while Castell (1962) agreed with this conclusion.

Limax flavus, a species known in Britain since at least the seventeenth century, is still restricted and "domesticated to about the same degree as the house mouse or cricket" (Boycott, 1934). *Hygromia striolata*, on the other hand, has become sufficiently widespread in recent times to become a pest species (Anon., 1963); and yet, as Kerney (1968) pointed out, "it may even be that *Hygromia striolata* could not continue to exist in these northern parts of Britain if human inter-ference ceased." It is interesting in this context to note that Smith (1970) showed from a study of the slugs and snails in an area of the Scottish Highlands that total numbers and species diversity decreased away from human-influenced habitats. The two recently introduced snails, *H. limbata* and *H. cinctella*, show a synanthropic distribution in South Devon, and it is possible that these species will show a similar extension of range to that of other synathropic species.

De Leersnyder and Hoestlandt (1958) discussed the recent extension in range of the snail *Cochlicella acuta* into Sussex, east of Brighton, and also the discovery of two new stations in Kent, isolated from the Sussex populations, at Greenhithe and near Sandwich. Although the Greenhithe colony has since become extinct, that at Sandwich appears to be stable, and the authors con-sidered that its introduction might date back to the 1914–18 war and be the result of increased rail and road traffic at that time. Since Ellis (1969) considered that snails in this colony represented a larger race than the indigenous *C. acuta*, it is possible that the colony was introduced by troops returning from the continent at that time rather than an eastern extension of the range of the Sussex population. De Leersnyder and Hoestlandt considered that a slow rise in temperature of the North Sea facilitated the establishment of the Sandwich colony and suggested that the much lower absolute winter temperatures experienced at Greenhithe explained the extinction of the snail from that area.

M. cantiana provides another example of the accidental spread of a snail by man. Gray (1840) described how this snail was introduced with ballast on the banks of the Tyne by colliers and how it spread rapidly in hedgerows. The distribution map given by Kerney (1970) shows that *M. cantiana* is still found in the area, although it has declined since 1840.

THE INFLUENCE OF MAN

Kerney (1966a) and Evans (1972) have reviewed the ways in which man has influenced the native land mollusc fauna since the development of Neolithic farming communities. The major changes have resulted from a disturbance of habitats and an increased dryness due to forest clearance, agriculture and settlement. Xerophile species, especially those characteristic of grassland such as several species of *Helicella*, have been favoured at the expense of woodland and hygrophile species. The landscape has also become more diverse with a mosaic of different habitats and this has produced a more varied mollusc fauna.

The pressures of a rapidly increasing population over the past two hundred years have had a marked effect on the landscape, and, more recently, the intensification of agricultural techniques has modified this even further. There is, unfortunately, little information on any changes that have taken place in numbers and distribution of slugs and snails over this period, since vice-county units are too large to indicate local changes. It is possible, however, to draw some general conclusions about changes from the literature that is available.

Boycott (1934) concluded that shelter was one of the most important factors determining the distribution of land molluscs. In arable regions, hedges, ditches, roadside verges and railway embankments provide refuges for many slugs and snails in an environment that would otherwise be unsuitable. It is clear from Ellis (1969) that hedges and roadside banks form a major habitat for many species, particularly helicids and zonitids. It may even be true that these habitats made possible the relatively recent spread of helicellids referred to by Kerney (1966a). Ellis (1969) considered that the xerophile snail *Helicella virgata* had greatly extended its range, partly because of the facilities for dispersal offered by railway embankments. Some recently introduced snails such as *H. elegans*, *H. neglecta* and *Fructicicola fruticum* have come from hedgerows and roadside banks. The recent loss of hedges (Hooper, 1970) will undoubtedly have affected numbers of even the more common species especially in those drier parts of eastern England where the rate of hedgerow removal has been greatest. Recent chemical and mechanical developments in roadside verge management also tend to reduce the available shelter for snails. Unfortunately little is known about the effect of this management, or of lead and other pollutants emitted by vehicles on the mollusc fauna of roadside verges.

Examination of Post-glacial deposits has shown that grazing tends to favour xerophile species (Evans, 1972) and this has been demonstrated experimentally by Morris (1968). Chappell *et al.* (1971) showed that trampling on chalk grassland has a similar effect and that increased public pressure, including the use of cars, reduces the species diversity of snail faunas. The effect of grazing may be reversed, and Stratton (1963) recorded a decline in numbers of the xerophile *Helicella virgata*, a species associated with short downland turf, as the result of cessation of grazing after the decline of rabbits due to myxomatosis. Evans (1972) also listed evidence for a decline in numbers of the related species *H. itala* resulting from the cessation of grazing by rabbits and sheep.

Arable farming tends to reduce species diversity and few snails are found on cultivated land. Several slugs, especially *Agriolimax reticulatus*, *A. hortensis* and *Milax budapestensis*, have become adapted to these conditions and although numbers may be reduced by intensive cultivations (Runham and Hunter, 1970), under normal conditions these slugs have become established as pests in areas where conditions are suitable (South, 1973). *A. reticulatus* is especially abundant on arable land, and Evans (1972) cited archaeological evidence that agriculture has had a beneficial effect on this species. The reasons for this are not clear since although this slug is widely distributed, numbers in natural habitats appear to be relatively low (Mordan, 1973). *A. reticulatus* is less exacting in water requirements than most other slugs (South, 1965; Lyth, 1972) and this, together with the absence of competition, may be responsible for its success on arable land. This species has a long history as a pest and was cited by Gilbert White in his "Natural History of Selborne."

Examination of advisory leaflets shows that while *Agriolimax reticulatus* and *Arion hortensis* were established as pests by 1905, *Milax* spp. were not included until 20–30 years later. This is further evidence for a recent extension in distribution of the synanthropic *Milax* spp. It is difficult to assess, however, whether the increased awareness of the extent of slug damage over the past 20–30 years represents an increased efficiency in detecting the damage or a genuine increase in numbers of slugs possibly due to changes in the pattern of agriculture. There is evidence, for example, that the recent development of slit seeding results in increased numbers of *A. reticulatus*.

Little is known about the effect of pesticides on land molluscs. Slugs are difficult to kill by contact poisons as they tend to slough off noxious substances with mucus (Runham and Hunter, 1970). Davis and French (1969) showed that slugs and worms acquired much higher levels of DDT residues than ground-beetles and that these were more persistent. However the significance of these residues is not yet understood.

Destruction of habitat through urbanization has undoubtedly affected the distribution of many snails. Castell (1962) described how several localities in London for the very rare snail *Laciniaria biplicta* have been destroyed over the past two hundred years. Kennard (1923) gave other examples of the way in which building operations have destroyed the recorded localities for a number of rare species. It is difficult to assess the local effect on numbers of common species. The distribution of *H. pomatia* does not appear to have changed significantly. However, many localities for this snail coincide with areas that are popular with commuters and it seems likely that there have been local reductions in numbers of this snail due to building operations. One such instance has been recorded at Lullingstone, Kent (Anon., 1954).

CONCLUSIONS

There is no evidence that any indigenous land mollusc has become extinct over the past 130 years, and Kerney (1968) concluded that none had become extinct in Britain since the climatic optimum about 5000 B.C., when the native fauna of land and freshwater Mollusca was probably complete. The range of some species has, however, continued to decline over the past century, a process that began after the climatic optimum with the change in temperature. Changes in microclimate due to the activities of man have increased over the past century, and there seems little doubt that there have been local declines in some species and increases in others, although this is not evident from the censuses since the vice-county units are too large to detect local changes in distribution. These changes may be shown more clearly when species maps based on the 10 km square system are published.

Few of the species that are introduced either accidentally or deliberately manage to establish themselves, but at least two snails have been successful over the past century, possibly assisted by an amelioration in climate at the end of the nineteenth century. Several species, possibly introductions in historic time, have recently shown a considerable increase in range as the result of their association with man. At least one species, *Agriolimax reticulatus*, can tolerate the drier conditions and disturbance caused by modern farming methods and has become a serious pest in some areas.

Kerney (1966a) considered that the total snail fauna of Britain at the time of the climatic optimum in the Atlantic Period, although locally diverse, was smaller than today and more uniform over great areas, woodland species being favoured as a result of the extensive forest cover at that time. Since that time the effect of agriculture and felling of forests has given rise to a mosaic of habitats each with a distinct population of slugs and snails with a resulting increase in species diversity.

It now appears that over the past 100–200 years the considerable increase in the pressure exerted by man on the countryside has tended to reverse the process and the fauna is again becoming less diverse over much of the country. The original woodland species are becoming more restricted, and a new group, the synanthropic slugs and snails, many of which are recent introductions, are replacing them in those areas where man has had the greatest influence.

REFERENCES

ADAMS, L. E. (1896). "The Collector's Manual of British Land and Freshwater Shells", 2nd edition. Taylor Bros., Leeds.

ADAMS, L. E. (1902). The census of the British land and freshwater Mollusca. *J. Conch.* **10**, 217–224.

ALDER, J. (1837). Notes on the land and fresh-water Mollusca of Great Britain, with a revised list of species. *Mag. Zool. Bot.* **2**, 101–119.

ALDER, J. (1948). A catalogue of the mollusca of Northumberland and Durham. *Trans. Tyneside Nat. Fld Club*, **1**, 97–209.

ANON. (1883). The Conchological Society's list of British land and freshwater Mollusca. *J. Conch.* **4**, 45–52.

ANON. (1954). Field Meeting. *J. Conch.* **24**, 25.

ANON. (1963). "Slugs and Snails". *U.K. Min. of Agriculture Advisory Leaflet* no. 115.

BABINGTON, C. C. (1859). Hints towards a *Cybele Hibernica*. *Nat. Hist. Rev., Dublin* **6**, 533–537.

BIGGS, H. E. J. (1957). Notes on *Hygromia cintcella* (Draparnaud). *J. Conch.* **24**, 177–178.

BOYCOTT, A. E. (1922). *Vitrina major* in Britain. *Proc. malac. Soc. Lond.* **15**, 123–130.

BOYCOTT, A. E. (1934). The habitats of land Mollusca in Britain. *J. Ecol.* **22**, 1–38.

CASTELL, C. P. (1962). Some notes on London's molluscs. *J. Conch.* **25**, 97–117.

CHAPPELL, H. G., AINSWORTH, J. F., CAMERON, R. A. D. and REDFERN, M. (1971). The effect of trampling on a chalk grassland ecosystem. *J. appl. Ecol.* **8**, 869–882.

CHEVALLIER, H. (1969). Taxonomie et biologie des grands *Arion* de France (Pulmonata: Arionidae). *Malacologia* **9**, 73–78.

CHICHESTER, L. F. and GETZ, L. L. (1969). The zoogeography and ecology of arionid and limacid slugs introduced into northeastern North America. *Malacologia* **7**, 313–346.

COCKERELL, T. D. A. (1884). Colonization of land snails at Chislehurst. *J. Conch.* **4**, 238.

COMFORT, A. (1951). Distribution of *Hygromia cinctella* (Draparnaud) at Paignton. *J. Conch.* **23**, 136.

DAVIS, A. G. (1953). On the geological history of some of our snails, illustrated by some Pleistocene and Holocene deposits in Kent and Surrey. *J. Conch.* **23**, 355–364.

DAVIS, B. N. K. and FRENCH, M. C. (1969). The accumulation and loss of organochlorine insecticide residues by beetles, worms and slugs in sprayed fields. *Soil Biol. Biochem.* **1**, 45–55.

DE LEERSNYDER, M. and HOESTLANDT, H. (1958). Extension du gastropode méditerranéen, *Cochlicella acuta* (Müller), dans le sud-est de l'Angleterre. *J. Conch.* **24**, 253–264.

DOBSON, R. M. (1963). Observations on the occurence of keeled slugs (*Milax*, Gray) in the Glasgow area. *Glasgow Nat.* 18, 246–248.

ELLIS, A. E. (1941). The mollusca of a Norfolk Broad. *J. Conch.* **21**, 224–243.

ELLIS, A. E. (1949). Recorder's Report: non-marine Mollusca. *J. Conch.* **23**, 60.

ELLIS, A. E. (1951). Census of the distribution of British non-marine Mollusca. *J. Conch.* **23**, 171–244.

ELLIS, A. E. (1964). *Arion lusitanicus* Mabille in Cornwall. *J. Conch.* **25**, 285–287.

ELLIS, A. E. (1969). "British Snails", 2nd edition. Clarendon, Oxford.

EVANS, J. G. (1972). "Land Snails in Archaeology." Seminar Press, London and New York.

FOGAN, M. (1969). *Vitrina (Semilimax) pyrenaica* (Ferussac) in Kerry North. *Ir. Nat. J.* **16**, 175.

GRAY, J. E. (1840). "A Manual of the Land and Fresh-water Shells of the British Islands . . . by William Turton, M.D.", new edition. London.

HAYWARD, J. F. (1954). *Agriolimax caruanae* Pollonera as a Holocene fossil. *J. Conch.* **23**, 403–404.

HOOPER, M. D. (1970). The botanical importance of our hedgerows. *In* "The Flora of a Changing Britain" (F. Perring, ed.), pp. 58–62. Classey, Middlesex.

JEFFREYS, J. G. (1862). "British Conchology", vol. 1. London.

KENNARD, A. S. (1923). The Holocene non-marine Mollusca of England. *Proc. malac. Soc. Lond.* **15**, 241–259.

KENNARD, A. S. and WOODWARD, B. B. (1917a). On the occurrence in England of *Helicella neglecta* (Drap.). *Proc. malac. Soc. Lond.* **12**, 133.

KENNARD, A. S. and WOODWARD, B. B. (1917b). On the occurrence of *Eulota fruticum* (Mull.) in Kent. *Proc. malac. Soc. Lond.* **12**, 124.

KERNEY, M. P. (1966a). Snails and man in Britain. *J. Conch.* **26**, 3–14.

KERNEY, M. P. (1966b). Census of the distribution of British non-marine Mollusca: supplement to the 7th Edition. *J. Conch.* **25**, 1–8.

KERNEY, M. P. (1967). Distribution mapping of land and freshwater Mollusca in the British Isles. *J. Conch.* **26**, 152–160.

KERNEY, M. P. (1968). Britain's fauna of land Mollusca and its relation to the Post-glacial thermal optimum. *Symp. zool. Soc. Lond.* **22**, 273–291.

KERNEY, M. P. (1970). The British distribution of *Monacha cantiana* (Montagu) and *Monacha cartusiana* (Müller). *J. Conch.* **27**, 145–148.

KERNEY, M. P. (1971). 10-kilometre square mapping scheme for non-marine Mollusca. *The Conchologist's Newsletter*, March 1971.

KERNEY, M. P. (1972). 10-kilometre square mapping scheme for non-marine Mollusca. *The Conchologist's Newsletter*, March, 1972.

LAMB, H. H. (1965). Britain's changing climate. *In* "The Biological Significance of Climatic Changes in Britain" (C. G. Johnson and L. P. Smith, eds), pp. 3–31. Academic Press, London and New York.

LAMB, H. H. (1970). Our changing climate. *In* "The Flora of a Changing Britain" (F. Perring, ed.), pp. 11–24. Classey, Middlesex.

LYTH, M. (1972). "Aspects of the Spatial Distribution of Slugs (Gastropoda: Pulmonata) with some Reference to their Water Relations." Ph.D. thesis, University of London.

McMILLAN, N. F. (1969). The slug *Agriolimax caruanae* Pollonera in Ireland. *Ir. Nat. J.* **16**, 178.

MEAD, A. R. (1961). "The Giant African Snail: a Problem in Economic Malacology." University of Chicago Press, Chicago and London.

MILMAN, P. P. (1951). *Hygromia cinctella* (Draparnaud) at Paignton. *J. Conch.* **23**, 135.

MORDAN, P. B. (1973). "Aspects of the Ecology of Terrestrial Gastropods at Monks Wood NNR and other Woodlands, with Special Reference to the Zonitidae." Ph.D. thesis, University of London.

MORRIS, M. G. (1968). Differences between the invertebrate faunas of grazed and ungrazed chalk grassland. II. The faunas of sample turves. *J. appl. Ecol.* **5**, 601–611.

PAUL, C. R. C. (1963). Snails introduced with bananas. *J. Conch.* **25**, 200.

PRAEGER, R. L. (1896). On the botanical subdivision of Ireland. *Ir. Nat.* **5**, 29–38.

QUICK, H. E. (1943). Land slugs and snails of West Glamorgan. *J. Conch.* **22**, 4–12.

QUICK, H. E. (1960). British Slugs (Pulmonata: Testacellidae, Arionidae, Limacidae). *Bull. Br. Mus. nat. Hist., Zool.* **6**, 103–226.

ROEBUCK, W. D. (1921). Census of the distribution of British land and freshwater Mollusca. *J. Conch.* **16**, 165–217.

RUNHAM, N. W. and HUNTER, P. J. (1970). "Terrestrial Slugs." Hutchinson, London.

SMITH, S. M. (1970). The effect of environment on the distribution of molluscs in Glen Garry and Glen Moriston, Scottish Highlands. *The Conchologist's Newsletter*, March, 1970.

SOUTH, A. (1965). Biology and ecology of *Agriolimax reticulatus* (Müller) and other slugs: spatial distribution. *J. Anim. Ecol.* **34**, 403–417.

SOUTH, A. (1973). Slug damage in Great Britain. *Haliotis* **3**, (in press).

STELFOX, A. W. (1969). *Marpessa laminata* (Montagu) bred in a small box for fifty-four years. *J. Conch.* **27**, 11–12.

STELFOX, A. W. and McMILLAN, N. F. (1966). The distribution in Ireland of very dark forms of *Cochlicella acuta* (Müller). *J. Conch.* **26**, 51.

STRATTON, L. W. (1949). Colonization of land snails. *J. Conch.* **23**, 43.

STRATTON, L. W. (1963). An ecological study. *J. Conch.* **25**, 174–179.

TAYLOR, J. W. (1885). On *Zonites glaber* Studer as a member of the British Fauna. *J. Conch.* **4**, 81.

TAYLOR, J. W. (1894–1921). "Monograph of the Land and Freshwater Mollusca of the British Isles", 3 vols. [plus 3 parts (unfinished)]. Taylor Bros., Leeds.

TAYLOR, J. W. and ROEBUCK, W. D. (1885). Census of the authenticated distribution of British land and freshwater Mollusca. *J. Conch.* **4**, 319–336.

VERDCOURT, B. (1969). A species of *Gulella* Pfr. (Streptaxidae) introduced into Britain. *J. Conch.* **27**, 15–16.

WILLIAMS, J. W. (1889). "Land and Fresh-water Shells". Sonnenschein, London.

WILLIAMS, J. W. (1892). "Land and Fresh-water Shells", 2nd edition Sonnenschein, London.

WILLIAMS, J. W. (1901). "Land and Fresh-water Shells", 3rd edition Sonnenschein, London.

16 | A Century of Change in the Lepidoptera

J. HEATH

Biological Records Centre, The Nature Conservancy,
Monks Wood Experimental Station, Abbots Ripton, Huntingdon

Abstract: The last century has seen the completion of the draining of the fens, the clear felling of many acres of deciduous forest and their replacement with coniferous plantations, the conversion of much lowland heathland to agricultural use as well as an agricultural revolution. These changes together with a general amelioration of the climate in the first half of this century followed by the present deterioration have had a profound effect on the Lepidoptera.

A number of species have become extinct whilst others have declined in numbers or become very localized. In contrast to these losses the same changes in habitat and climate have enabled other species to colonize Britain from mainland Europe or to extend their existing range.

INTRODUCTION

The changes that have taken place in the distribution and abundance of the macro-Lepidoptera during the past 100 years have been due in part to climatic changes and in part to the activities of man (Bretherton, 1951; Mere, 1961).

The climate of Britain showed a steady amelioration during the period 1895–1950, following a twenty-year period of deterioration, and would appear to have been milder by 1950 than at any time since 1840. Thus it would seem that the period 1840–1895 would have tended to produce contractions in the distributional range of Lepidoptera, whereas the converse would have been the case from 1895 to 1950 by which time the climate would seem to have favoured the northward movement of species into Britain from mainland Europe (Lewis, 1950).

An analysis of published assessments of "good" and "bad" years for Lepidoptera was carried out by Beirne (1947). This showed that during the 79 years from 1865 to 1944 there were 40 "bad" seasons, 31 "good" seasons and 8

Systematics Association Special Volume No. 6, "The Changing Flora and Fauna of Britain," edited by D. L. Hawksworth, 1974, pp. 275–292. Academic Press, London and New York.

"average" seasons. Although highly subjective, this analysis showed clearly the period of decline in the 1920s when there were nine successive bad seasons and the period of relative abundance and expansion in range which followed in the 1930s, when there were seven good seasons reported in the ten-year period from 1932 to 1941. The recent decrease in numbers, especially of the butterflies, correlates well with the climatic deterioration now taking place.

Man's activities over the last century have resulted in dramatic changes in land use. For example, although our forests have increased in total acreage the species composition has changed from three million acres of broadleaves and half a million acres of conifers in 1914 to one and three quarter million acres of broadleaves and two and three quarter million acres of conifers in 1971. Similarly over four million acres of species-rich permanent pasture were lost by improvement between 1874 and 1914 and this loss still continues. In eastern England the draining of Whittlesey Mere in 1851 almost completed the destruction of 680 000 acres of Fenland leaving today only a fragmented remnant of a few thousand acres. Large areas of heathland in southern England have also been reclaimed for agriculture. In Dorset for example the acreage was 75 000 in 1811 but only 25 000 in 1960, again in fragmented lots. Similarly in upland Britain many acres of moorland have been afforested (Moore, 1962).

The removal of thousands of miles of hedgerow, the changes in management of the remaining hedgerows, the treatment of roadside verges with herbicides and by mowing and the general mania for rural tidiness have all had their effect. Also the loss of habitats brought about by urban expansion and the construction of motorways together with the human recreational pressure on the countryside will have contributed to change.

Any attempt to correlate these changes with the changes in distribution and abundance of many individual species can only be speculative. Nevertheless such speculation as to what may have happened in the past may provide clues to the solution of future management problems arising in the conservation of endangered species.

EXTINCTIONS AND SPECIES DECLINING
(Tables I and II)

The loss of some species can be attributed directly to specific events. The draining of the fens which was completed in 1851 was responsible for the loss of a number of species confined to that habitat of which *Lycaena dispar* L. (the Large Copper butterfly) (Fig. 1) is perhaps the best known example. At the end of the eighteenth century this butterfly was still found commonly in the fens of Huntingdonshire. Although the larval food plant, *Rumex hydrolapathum*

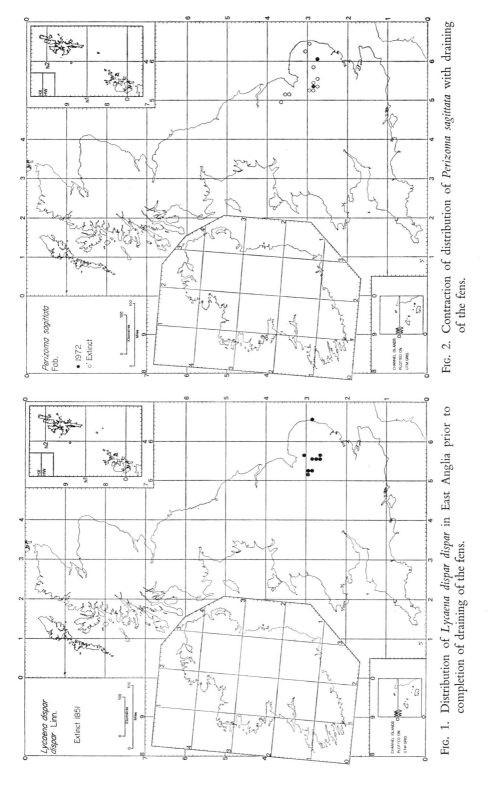

Fig. 2. Contraction of distribution of *Perizoma sagittata* with draining of the fens.

Fig. 1. Distribution of *Lycaena dispar dispar* in East Anglia prior to completion of draining of the fens.

(Great Water Dock) is fairly widespread, the British subspecies of *Lycaena dispar* L. was apparently confined to the fens of East Anglia. This insect, which was collected extensively, finally became extinct in 1851 with the completion of the draining of Whittlesey Mere and the reclamation for agriculture of the surrounding area.

Another purely fen insect was *Laelia coenosa* Hübn. (Reed Tussock) discovered in 1819 at Whittlesey as a larva. It fed on *Cladium* (Fen Sedge) and *Phragmites* (Reed) and was also abundant at Yaxley, Burwell and Wicken but finally became extinct in 1879.

Lymantria dispar L. (Gipsy Moth) was known to be abundant in the Fens and Norfolk Broads in the early 1800s but apparently declined rapidly in the 1850s. The last certain record is from Wennington Wood in 1907. The larvae of the apparently distinct fenland race of this woodland species preferred *Myrica gale* Bog Myrtle) and *Salix repens* (Creeping Willow) to forest trees.

A more widely distributed species associated with the meadows adjacent to the fens, *Trachea atriplicis* L. (Orache Moth), was at one time common in Cambridgeshire and Huntingdonshire, being regularly taken at sugar. The larvae fed on a variety of plants such as *Atriplex* and *Chenopodium*. By the close of the nineteenth century it had become very scarce and was last recorded at Stowmarket in 1915.

Acronicta strigosa D. & S. (Marsh Dagger) is another example of a moth being associated with a habitat adjacent to the fens rather than actually of the fens themselves. This hawthorn feeding species was found in the dense thorn thickets on the edges of Fenland. It occurred in decreasing numbers during the early part of this century and finally seems to have become extinct in 1933 when it was last taken near Cambridge. In recent years there have been unconfirmed records so that it is just possible that this species may still survive. Little is known of the localities in Gloucestershire and Herefordshire where this insect also occurred at the turn of the century.

Perizoma sagittata F. (Marsh Carpet) (Fig. 2), whose larvae feed on *Thalictrum* spp. (Meadow Rue), was at one time known from all the fens of Lincolnshire, Huntingdonshire and Cambridgeshire and still survives at Woodwalton and Redgrave Fens, where it is sometimes not uncommon.

Costaconvexa polygrammata Borkh. (Many-lined Moth) was yet another fen species which formerly occurred at Burwell and Wicken. The larval foodplant was *Galium palustre* (Marsh Bedstraw). The last authentic record of this moth is dated 1879.

Lastly one of the most famous fen insects was *Eugraphe subrosea* Steph. (Rosy Marsh Moth) discovered at Yaxley in 1828. The larvae feed on *Myrica gale*

TABLE I

Species	Recorded range (vice-counties)[a]	Last record (excluding casuals)
(a) Widely distributed species		
Aporia crataegi L.	1, 2, 3, 4, 5, 6, 9, 11, 12, 13, 14, 15, 16, 17, 18, 19, 21, 22, 23, 25, 26, 27, 28, 30, 31, 32, 33, 34, 35, 36, 37, 38, 40, 41, 53	1925 (Kent)
Cyaniris semiargus Rott.	1, 2, 5, 6, 7, 8, 9, 11, 12, 13, 14, 15, 16, 17, 18, 19, 22, 25, 26, 27, 28, 29, 30, 32, 33, 34, 35, 36, 37, 38, 41, 45, 53, 55	1877 (Penarth)
Hecatera dysodea D. & S.	5, 6, 13, 14, 15, 16, 17, 18, 19, 20, 22, 23, 25, 26, 27, 28, 29, 31, 32, 33, 34, 36, 37, 41	1918 (Essex)
(b) Local species		
Lycaena dispar L.	25, 28, 29, 31	1851 (Holme Fen)
Zygaena viciae D. & S.	11	1927 (Wood Fidley)
Costaconvexa polygrammata Borkh.	29	1879 (Wicken Fen)
Leucodonta bicoloria D. & S.	39	1965 (Burn Wood, Staffordshire; still present in Ireland)
Laelia coenosa Hb.	29, 31	1879 (Wicken Fen)
Lymantria dispar L.	27, 28, 29, 31 (elsewhere as casual specimens)	1907 (Wennington Wood, Huntingdonshire)
Eugraphe subrosea Stephs.	31	1851 (Whittlesey Mere)
Acronicta strigosa D. & S.	27, 28, 29, 31, 34, 36, 37	1933 (near Cambridge)
Trachea atriplicis L.	21, 25, 26, 27, 28, 29, 31	1915 (Stowmarket)
(c) Doubtfully extinct		
Phyllodesma ilicifolia L.	4, 39, 40, 64	1913 (Cannock Chase, Staffordshire)
Isturgia limbaria F.	15, 16, 18, 19, 25, 26, 27, 28, 105	1913 (Suffolk)
Lithophane furcifera Hufn.	14, 15, 34, 35, 41	1907 (Bigswier, Monmouthshire)
Apamea pabulatricula Brahm.	20, 25, 26, 27, 28, 40, 41, 53, 54, 55, 56, 57, 58, 61, 62, 63, 64, 65, 70, 73, 76	1919 (Lincolnshire)

[a] For vice-county names and boundaries see pp. 425–428.

(Bog Myrtle). This noctuid became extinct in 1851 with the draining of Whittlesey Mere but was rediscovered in west Wales in 1965 when a large colony was discovered at Borth Bog following a search after a single specimen had been taken the previous year at Harlech. This discovery caused great excitement amongst lepidopterists and aroused much heated discussion in the natural history press on the "pros and cons" of collecting.

TABLE II. Species declining

Species	Extent of decline	Probable reason for decline
Carterocephalus palaemon Pall.	Severe	Change in forestry practice
Hesperia comma L.	Moderate	Reduction of rabbit grazing
Thecla betulae L.	Moderate	Changes in agricultural practice
Lysandra bellargus Rott.	Moderate	Reduction of rabbit grazing
Maculinea arion L.	Severe	Reduction of rabbit grazing
Apatura iris L.	Moderate	Changes in forestry practice
Nymphalis polychloros L.	Severe	Unknown
Argynnis spp.	Moderate in eastern England	Unknown
Euphydryas aurinia Rott.	Moderate	Land drainage
Melanargia galathea L.	Moderate	Changes in agricultural practice
Coenonympha tullia Müll.	Moderate	Land drainage
Eriogaster lanestris L.	Very severe	Disease and changes in hedgerow management
Endromis versicolora L.	Severe	Climatic changes
Perizoma sagittata F.	Severe	Land drainage
Hemaris tityus L.	Very severe	Changes in agricultural practice
H. fuciformis L.	Very severe	Changes in forestry practice
Emmelia trabealis Scop.	Severe	Unknown
Meganola strigula Schiff.	Severe	Climatic changes

Other species which have been affected by land drainage include *Euphydryas aurinia* Rott. (Marsh Fritillary) which frequents wet meadows. The larvae feed gregariously on *Scabiosa succisa* (Devil's Bit). Similarly *Coenonympha tullia* Müll. (Large Heath) has gone from a number of its localities in northern England as a result of the draining of the peat bogs where it used to occur.

The rapid decline and extinction of three widely distributed common species seems to be attributable to changes in agricultural practice. In 1850 *Cyaniris semiargus* Rott. (Mazarine Blue) was widely distributed throughout the southern and eastern counties where it was a common butterfly of pasture

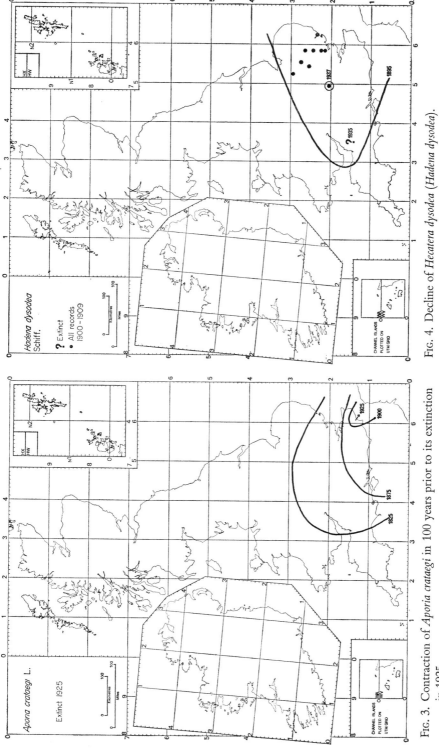

FIG. 3. Contraction of *Aporia crataegi* in 100 years prior to its extinction in 1925.

FIG. 4. Decline of *Hecatera dysodea* (*Hadena dysodea*).

and hayfields, the larvae feeding, presumably, on red clover. By 1875 it had become confined to Glamorgan where it was last seen in 1877. A possible explanation of this sudden decline and extinction is the introduction of new techniques of grassland management. Similarly *Aporia crataegi* L. (Black-veined White) (Fig. 3), a large conspicuous butterfly, was even more common and had a similar distribution. Its gregarious larvae fed on blackthorn as well as cultivated plum and apple. Its decline started about 1850 and by 1900 it was confined to east Kent, finally becoming extinct in 1925. The final extinction of this butterfly could have resulted from the introduction of pest control measures by fruit growers but the initial decline may have been due to virus disease or climatic change, or both. Attempts to reintroduce this species have been singularly unsuccessful.

The third species in this group is *Hecatera dysodea* D. & S. (Small Ranunculus) (Fig. 4) which in 1895 was abundant in south-eastern Britain where its larvae were a pest on lettuce. By 1905 it was restricted to a few localities in East Anglia which, apart from a record of a single specimen from an unknown locality in Somerset in 1935 and four specimens taken at Berkhamstead in 1936–37, are the last records. It is known that the variety of lettuce cultivated was changed at the end of the nineteenth century and this could possibly have caused the spectacular disappearance of this moth.

Eriogaster lanestris L. (Small Eggar) (Fig. 5) has shown a dramatic decline in recent years. Its larvae are gregarious and feed in compact nests on *Crataegus* (Hawthorn). The adult emerges early in the year and the eggs are laid in clusters on the terminal twigs in February and March. Three possible explanations of the decline of this species are firstly, virus or other disease to which gregarious insects are especially prone or secondly climatic change or thirdly destruction of the eggs by modern methods of hedgerow management.

Thecla betulae L. (Brown Hairstreak) is a butterfly which has suffered a similar decline. Its larvae are also dependent on blackthorn—preferring the young growth on hedgerows. These have been removed or severely cut back in many of its old localities.

The disappearance of the rabbit and consequent invasion of scrub has caused a drastic change in the chalk downs and resulted in a number of species becoming much more local. Butterflies which have particularly suffered are *Lysandra bellargus* Rott. (Adonis Blue) and *Hesperia comma* L. (Silver-spotted Skipper). *Melanargia galathea* L. (Marbled White), another calcareous grassland species, has also declined in some areas, such as Huntingdonshire, possibly as a result of changes in roadside verge management.

The decline of the two Bee Hawk-moths, *Hemaris fuciformis* L. and *H. tityus*

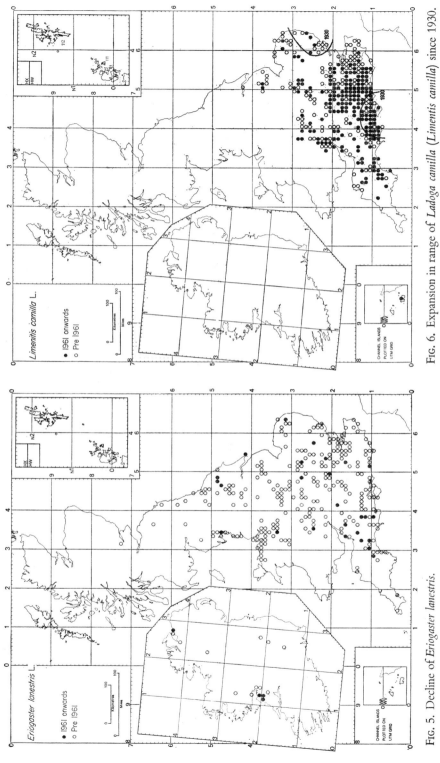

FIG. 6. Expansion in range of *Ladoga camilla* (*Limentis camilla*) since 1930.

FIG. 5. Decline of *Eriogaster lanestris*.

L. is more puzzling. Both species are, as adults, diurnal, visiting the flowers of rhododendron and herbaceous plants in woodland glades early in the day. The larvae feed on *Lonicera* (Honeysuckle) and *Scabiosa succisa* (Devil's Bit) respectively. Could it be that the habits of the lepidopterist have changed—is he perhaps dealing with the catch in his M.V. light trap when he should be out and about observing the flight of *Hemaris*? More probably changes in woodland management have been responsible.

TABLE III. Resident species extending their range

Species	Previous distribution	Present distribution
Ladoga camilla L.	Hampshire, Suffolk in 1930	South-east of line from Bristol Channel to Wash
Polygonia c-album L.	West Midlands in 1920	South of a line from the Mersey to the Humber
Inachis io L.	(General extension northwards)	
Pararge aegeria L.	South and West counties in 1930	Throughout southern England
Hyloicus pinastri L.	Dorset and Suffolk in 1930	Much of south-east England and Thetford Chase
Drepana binaria Hufn.	South of line from Mersey to Humber in 1930	Northwards to Lake District (Lancashire, Westmorland, Cumberland)
Cucullia absinthii L.	Coastal areas of South-west England and Suffolk in 1930	Throughout east and west midlands

Climatic change coupled perhaps with changes in forestry practice may have been the cause of the decline of *Endromis versicolora* L. (Kentish Glory), once a common insect in Tilgate Forest, Sussex; Wyre Forest, Worcestershire, as well as Berkshire, Hampshire and the Highlands of Scotland and now apparently confined to a decreasing number of localities in the Highlands. The larvae of this species feed on birch and the male adults fly actively in sunshine in March and April. It does not seem to have been recorded from Wyre Forest, the last English locality, since 1960.

Four *Argynnis* species, *A. paphia* L. (Silver Washed Fritillary), *A. adippe* L. (High Brown Fritillary), *A. euphrosyne* L. (Pearl-bordered Fritillary) and *A. selene* Schiff. (Small Pearl-bordered Fritillary) were once common throughout the woodlands of Britain. During the past thirty years these butterflies declined in eastern England where they are now almost totally absent. This decline of a whole group of species cannot be wholly explained by changes in woodland

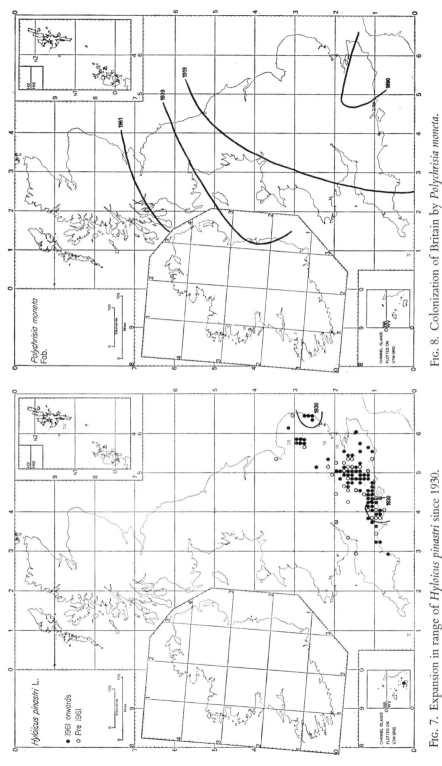

FIG. 8. Colonization of Britain by *Polychrisia moneta*.

FIG. 7. Expansion in range of *Hyloicus pinastri* since 1930.

species-composition as many apparently suitable deciduous woods remain. The larvae of all these species feed on *Viola* spp. which are present in plenty. Is the explanation to be found in micro-climatic changes, increased stocking with game, or perhaps the tidying up of the rides and glades by the forester? These and many other questions concerning the decline of the Lepidoptera remain to be answered.

RESIDENT SPECIES EXTENDING THEIR RANGE
(Table III)

All the changes that have taken place have not however been losses or declines. The exceptionally hot dry summers of the 1930s which coincided with the final stages of the agricultural depression with much land out of cultivation produced an exceptional abundance of many species, some of which had previously declined in the early part of this century. For example, the woodland butterfly *Polygonia c-album* L. (Comma) which, although widespread in the nineteenth century, had contracted its range until it was confined to the west Midlands in 1920, expanded its range during the 1930s over the whole of southern Britain, reaching Lancashire and Yorkshire again by 1950.

Similarly another woodland species, *Ladoga camilla* L. (White Admiral) (Fig. 6), which was confined to Hampshire and Suffolk in 1930, spread throughout southern England in the 1930s and 1940s to reach a line just north of that from the Bristol Channel to the Wash.

During this period *Pararge aegeria* L. (Speckled Wood) also recolonized large areas of southern England from which it had disappeared.

A very similar extension of range occurred during the same period with *Hyloicus pinastri* L. (Pine Hawk Moth) (Fig. 7), due doubtless to the increase in conifer woodland.

Man's activities in creating spoil heaps seems to have encouraged the spread of *Cucullia absinthii* L. (Wormwood Shark). In 1930 this species was confined to the coastal areas of the south-west counties and Suffolk but has in recent years spread throughout the midlands. The larvae feed on *Artemisia vulgaris* and *A. absinthium* (Mugwort) both plants commonly found on waste land.

SPECIES RECENTLY ESTABLISHED OR OVERLOOKED
(Tables IV and V)

These are the colonizers, some of which have spread rapidly whilst others have apparently been unable to do so. *Ennomos autumnaria* Wernb. (Large Thorn) was first noted from east Kent in 1855 and by 1970 had spread to Norfolk and Hampshire. The larvae of this species feed on a variety of deciduous trees and

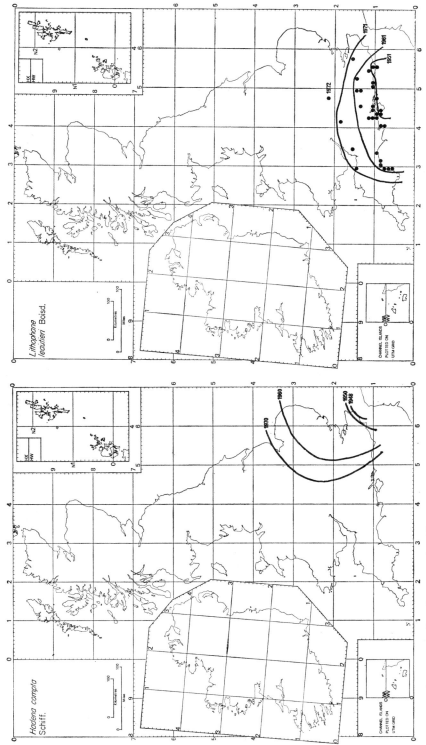

FIG. 10. Spread of *Lithophane leautieri* into southern England.

FIG. 9. Colonization of south-east England by *Hadena compta*.

TABLE IV. Species recently established

(a) *Extending their range*

Species	Original record	Present distribution
Eupithecia phoeniciata Ramb.	1959 (Isles of Scilly)	Southern England to Hampshire
Ennomos autumnaria Wern.	1855 (Kent)	South-east England to Hampshire, Northamptonshire and Norfolk
Hadena compta D. & S.	1948 (Dover)	South-east England to Sussex, Northamptonshire, south Lincolnshire
Mythimna l-album L.	1933 (Lizard)	Along the south coast east to Sussex
M. unipuncta Haw.	1957 Isles of Scilly)	Along south coast east to Dorset
Lithophane leautieri Boisd.	1951 (Isle of Wight and Sussex)	From Sussex to Devon & Somerset and north to Bucks.
Oria musculosa Hübn.	1939 (Salisbury)	Wiltshire, Hampshire, Isle of Wight, Sussex, Somerset, Gloucestershire, Surrey, Berkshire, Northamptonshire, Shropshire
Polychrisia moneta F.	1890 (Kent and Sussex)	Throughout Britain to central Scotland

(b) *Static*

Species	First recorded	Locality
Aplasta ononaria Fuess.	1937 (previously established until 1857)	Folkestone Warren, Sandwich
Thalera fimbrialis Scop.	1950	Dungeness
Cyclophora puppillaria Hübn.	1957	Isles of Scilly
Xanthorhoe biriviata Borkh.	1955	Middlesex
Spargania luctuata D. & S.	1950	Kent, Sussex
Semiothisa signaria Hübn.	1972	Essex
Mythimna obsoleta Hübn.	1957	Tresco, Isles of Scilly
Xanthia ocellaris Borkh.	1920's	Barton Mills, Suffolk
Hydraecia osseola Staud.	1951	Kent, Sussex coast
Archanara geminipuncta Haw.	1957	Tresco, Isles of Scilly
A. spargani Esp.	1957	Tresco, Isles of Scilly
Chilodes maritimus Tausch.	1957	Tresco, Isles of Scilly

(c) *Temporarily established*

Species	Period of establishment	Locality
Calophasia lunula Hufn.	1951–1970(?)	Kent, Sussex
Trigonophora flammea Esp.	1855–1892	Sussex
Sedina buettneri Hering.	1945–1951	Isle of Wight
Catocala fraxini L.	1948–1955	South-east Kent
Minucia lunaris D. & S.	1947–1958	Kent, Sussex

shrubs. *Polychrisia moneta* Fab. (Fig. 8) associated with *Delphinium* had spread from its original locality in Kent in 1890 to southern Scotland by 1961.

A more recent arrival is *Hadena compta* D. & S. (Varied Coronet) (Fig. 9) first recorded as breeding in England on Sweet William (*Dianthus* sp.) at Dover in 1948. In twenty years this species has spread north to Lincolnshire and westward to Northamptonshire and Buckinghamshire.

Lithophane leautieri Boisd. (Blair's Shoulder-knot) (Fig. 10) was first recorded at Freshwater, Isle of Wight in October 1951. Subsequently it was found to be breeding with the larvae feeding on the foliage of *Cupressus macrocarpa*. By 1972 it had spread westwards into Devon and Somerset, eastwards to Kent and north into Buckinghamshire. Although it feeds on Juniper elsewhere in Europe, there is no evidence yet that it is doing so in Britain.

TABLE V. Species overlooked

Species	First recorded	Locality
Zygaena loti D. & S.	1907	Oban
Z. viciae D. & S.	1963	Argyll
Conopia flaviventris Staud.	1925	Hampshire
Eupithecia egenaria H.-S.	1962	Wye Valley
E. extensaria Frey	1874	Yorkshire
E. millefoliata Röss.	1939	Dover, Kent, Sussex
Chloroclystis chloerata Mab.	1971	Surrey
Pelosia obtusa H.-S.	1961	Barton Broad, Norfolk
Eugraphe subrosea Steph.	1965	Morfa Harlech, North Wales
Eriopygodes imbecilla F.	1972	Monmouth
Gortyna borelii Pierret	1968	Essex
Calamia tridens Hufn.	1949	Burren, Co. Clare, Ireland
Schrankia intermedialis Reid	1972	Hertford

Mythimna l-album L. was first recorded as breeding in Britain in Cornwall in 1933 when larvae were found feeding on *Festuca arundinacea*. Since then it has spread steadily along the south coast reaching Sussex in 1960.

TABLE VI. Recently recorded vagrants and adventives

Species	Status	First recorded	Locality
Nymphalis xanthomelas D. & S.	Vagrant	1953	Borough Green, Kent
Drepana curvatula Borkh.	Vagrant	1960	Dover, Kent
Aplocera praeformata Hübn.	Vagrant	1946	Pembroke
Hyles nicaea Prun.	Vagrant	1954	South Devon
Hybocampa milhauseri F.	Vagrant	1966	Aldwick Bay, Sussex
Arctornis l-nigrum Müll.	Vagrant	1946	Sussex
Utetheisa bella L.	Vagrant	1948	Skokholm Island
Graphania dives Phil.	Adventive	1950	Spurn Head, Yorkshire
Tathorhynchus exsiccata Led.	Vagrant	1952	Torquay, South Devon
Cryphia raptricula D. & S.	Vagrant	1953	Arundel, Sussex
Callopistria juventina Stoll.	Vagrant	1959	Lewes, Sussex
Caradrina flavirena Guen.	Vagrant	1967	Totteridge, Herts
Perigea conducta Walker	Vagrant	1958	Fowey, Cornwall
Heliothis nubigera H.-S.	Vagrant	1958	Iwerne Minster, Dorset
Lithacodia deceptoria Scop.	Vagrant	1948	East Kent
Nycteola degenerana Hübn.	Vagrant	1905	New Forest
Charadra deridens Guen.	Adventive	1952	Plumstead, Kent
Chrysodeixis acuta Walker	Vagrant	1955	Horsell, Surrey
C. chalcites Esp.	Vagrant	1943	Torquay, South Devon
Ctenoplusia limbirena Guen.	Vagrant	1947	Swanage, Dorset
Diachrysia orichalcea F.	Vagrant	1943	Torquay, South Devon
Macdunnoughia confusa Steph.	Vagrant	1951	Bradwell-on-Sea, Essex
Plusia biloba Steph.	Vagrant	1954	Aberystwyth, Cardiganshire
Hypena obesalis Treits.	Vagrant	1908	Paignton, South Devon
Plathypena scabra F.	Adventive	1956	Lewisham, Kent
Polypogon tarsicrinalis Knoch	Vagrant	1965	Suffolk
Trisateles emortualis D. & S.	Vagrant	1962	Chilterns

Similarly *Eupithecia phoeniciata* Ramb. another *Cupressus* feeding species first recorded as breeding in the Isles of Scilly in 1959 had spread to Somerset and Dorset by 1961.

Species recently established which have not spread include *Hydraecia osseola* Staud. which was found to be breeding in the Dungeness area in 1953 feeding on *Althaea officinalis*, and *Xanthorhoe biriviata* Borkh. found in 1955 to be established in the Home Counties where its larvae feed on *Impatiens capensis*.

Others recently discovered, such as *Gortyna borelii* Pierret, *Chloroclystis chloerata* Mab. and *Eriopygodes imbecilla* F. have doubtless been overlooked in the past.

From time to time migrant species, for example *Catocala fraxini* L. (Clifden Nonpareil) and *Minucia lunaris* Schiff. (Lunar Double Stripe) in Kent in the late 1940s, *Trigonophora flammea* Esp. (Flame Brocade) in the late 1850s in Sussex

TABLE VII. Species added by taxonomic research

Species	Date recognized in Britain
Lampropteryx otregiata Metcalfe	1917
Epirrita christyi Allen	1906
Diarsia florida Schmidt.	1950
Amphipyra berbera Rungs.	1968
Oligia latruncula D. & S.	1941
O. versicolor Borkh.	1941
Amphipoea crinanensis Burrows	1911
A. fucosa Frey	1911
A. lucens Frey	1911
Archanara neurica Hübn.	1908
Heliothis maritima Grasl.	1958

and *Sedina buettneri* Her. (Blair's Wainscot) in the Isle of Wight in the 1950s, become temporarily established when conditions are favourable (de Worms, 1951, 1963).

In addition to these there have been a large number of vagrants and adventives added to the British list in recent years (Table VI). Finally a few species have been discovered as a result of taxonomic research (Table VII).

DISCUSSION

Whilst it is evident that man's activities have caused or accelerated the decline and extinction of numerous species of Lepidoptera almost all those that have extended their range have probably done so as a result of climatic change. Particularly is this so in the case of those species that have recently colonized Britain.

Examination of the numbers of species established in each decade from 1850 to 1972 shows that between 1850 and 1920 only three new species became established whilst between 1921 and 1960, a period of climatic amelioration, no fewer than 21 species colonized Britain. Since 1961 when the present climatic deterioration started, only one species has been added.

The nomenclature used follows that of Kloet and Hincks (1972).

ACKNOWLEDpEMENTS

I should like to thank Mr M. J. Skelton for his help in abstracting records from the Biological Records Centre data bank and for much helpful criticism and Miss Christine Allen for preparing the maps.

REFERENCES

BIERNE, B. P. (1947). The seasonal abundance of the British Lepidoptera. *Entomologist* **80**, 1–5.

BRETHERTON, R. F. (1951). Our lost butterflies and moths. *Ent. Gaz.* **2**, 211–240.

KLOET, G. S. and HINCKS, W. D. (1972). A check list of British Insects, Second Edition (Revised) Part 2: Lepidoptera. *In* "Handbooks for the Identification of British Insects," **11**(2). Royal Entomological Society, London.

LEWIS, L. F. (1950). Recent Climatic Trend. *Met. mag.* **79**, 189–190.

MERE, R. M. (1961). The recent colonisation of England by new species of macrolepidoptera. *Proc. S. Lond. ent. nat. Hist. Soc.* **1960**, 63–74.

MOORE, N. W. (1962). The heaths of Dorset and their conservation. *J. Ecol.* **50**, 369–391.

SOUTH, R. (1961). "The Moths of the British Isles. Series I & II." Warne, London.

WORMS, C. G. M. DE (1951). New additions to British Macrolepidoptera during the past half century. *Ent. Gaz.* **2**, 153–168.

WORMS, C. G. M. DE (1963). Recent additions to the British Macrolepidoptera. *Ent. Gaz.* **14**, 101–119.

17 | Changes in the British Spider Fauna

E. DUFFEY

Monks Wood Experimental Station, The Nature Conservancy, Abbots Ripton, Huntingdon

Abstract: The number of species of spiders recorded for Britain has doubled during the last 110 years and is now 611. Apart from *Eresus niger* and possibly *Dolomedes plantarius* there is little evidence that native British species have been (or are being) lost by land-use changes although the status of some has declined. Many species thought to be very rare have been shown to be more widely distributed as a result of work during the last 15–20 years. The number of rare species restricted to a single well-defined habitat appears to be comparatively small and those in montane and coastal areas are not likely to be at risk. Rare species of heathlands, grasslands, freshwater marshes and ancient woodlands may be vulnerable, however, and would benefit by protection and appropriate management of the sites where they occur. There is some evidence that the falling water table on East Anglian fens, even when taking place very slowly, results in the local loss of species. The importance and value of man-made habitats in maintaining certain rare species is thought to be considerable but more investigation is necessary. Few species of exotic spiders have succeeded in becoming established by introduction.

INTRODUCTION

The beginning of British arachnology dates from 1861–64 when John Blackwall published his two volume "History of the Spiders of Great Britain and Ireland" which described and illustrated in colour 304 species. Since then there has been an almost continuous succession of outstanding araneologists, mostly amateur, and in 1959 the first Society concerned with the study of this group was formed. This eventually led to the establishment of the British Arachnological Society in 1968 which now has over 370 members.

After Blackwall, the Rev. O. Pickard-Cambridge, whose work on spiders spanned the second half of the nineteenth century and the early years of the

Systematics Association Special Volume No. 6, "The Changing Flora and Fauna of Britain", edited by D. L. Hawksworth, 1974, pp. 293–305. Academic Press, London and New York.

twentieth, raised the total of British species to 510 in 1879 and 532 by 1903. In 1936 W. S. Bristowe recorded 556 species and by 1951–53 G. H. Locket and A. F. Millidge were able to include 574 species in their two volumes, "British Spiders". During the 20 years since this publication, interest in British spiders has increased remarkably and the expansion of field work has now pushed the total to 611 species. The number recorded for Britain has therefore doubled in the last 110 years.

New species, both to Britain and to science, are still being added. Since 1953 nine species new to science have been described and another 31 added to the British list; that is two per year over the last 20 years. Although we can expect new discoveries in the future it is probably true to say that the spider fauna of Britain is better known at present than that of any other country. The least known part of the British Isles is Ireland where araneologists have always been scarce—perhaps connected with the belief that spiders, like toads and snakes, had been banished from Irish soil by St Patrick!

THE PRESENT STATUS OF SOME RARE SPECIES

The history of the British spider fauna since Blackwell is therefore mainly a story of increasing knowledge and there is little evidence of decline or extinction due to changing environmental conditions or loss of habitat. The most famous exception is *Eresus niger*, long regarded as the prize of British araneology and last seen in 1906. Its known localities were few and situated near the south coast so it was clearly on the edge of its range in this country. The best known site, on heathland near Bournemouth, is now covered with a housing estate and intensive work by P. Merrett on surviving heaths nearby has failed to rediscover it.

On the other hand, *Tarentula fabrilis* can still be found in its old haunts on the Bloxworth and Morden Heaths, Dorset, where Pickard-Cambridge discovered it in the 1860s. *Cheiracanthium pennyi* was first described from a specimen taken in 1870 at Wokingham, Berkshire, and for nearly 100 years no fresh specimens were found. In recent years a well established colony has been discovered at a heath in Surrey not very far from the original site. *Lessertiella saxetorum* and *Diplocephalus connatus* are two rare spiders only recorded from the stony valleys of upland rivers. The former was first taken in 1916 and the latter in 1902, both by the Allen River, Northumberland. Fresh specimens were again taken on the original site in the 1960s. *L. saxetorum* is now known from a few other sites where there is a similar habitat but there are numerous examples of rare spiders which are apparently confined to very

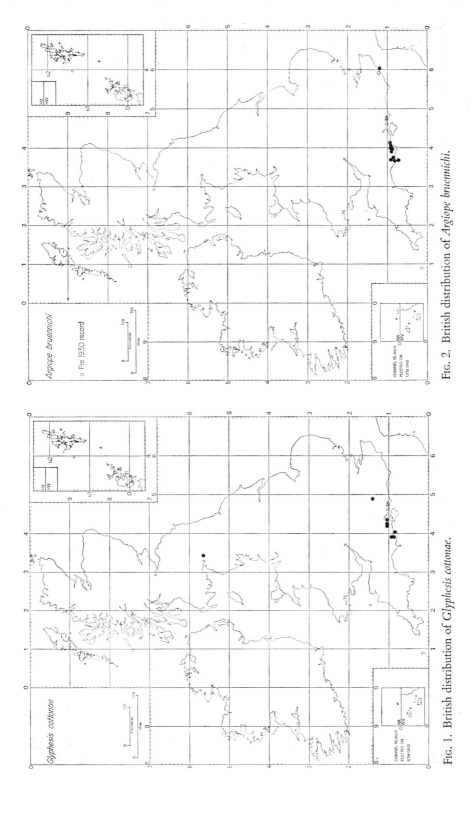

FIG. 2. British distribution of *Argiope bruennichi*.

FIG. 1. British distribution of *Glyphesis cottonae*.

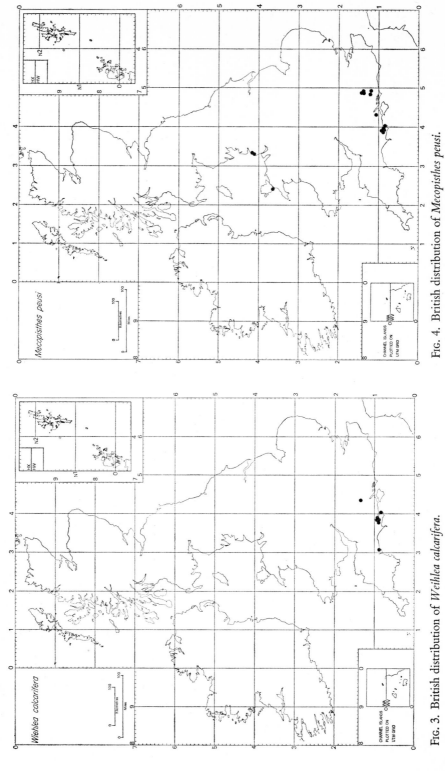

FIG. 4. British distribution of *Mecopisthes peusi.*

FIG. 3. British distribution of *Weihlea calcarifera.*

few places and appear to have been unable to spread elsewhere. *Acanthophyma* (*Lasiargus*) *gowerensis* is known from salt marshes in Gower, Glamorgan, and south-west Ireland, but nowhere else and is unrecorded outside Britain, while *Glyphesis cottonae* occurs commonly in valley bogs in Dorset, Surrey, the New Forest and Cumberland but only in *Sphagnum* (Fig. 1). Other species of localized spiders are *Argiope bruennichi* (Fig. 2), *Weihlea calcarifera* (Fig. 3) and *Mecopisthes*

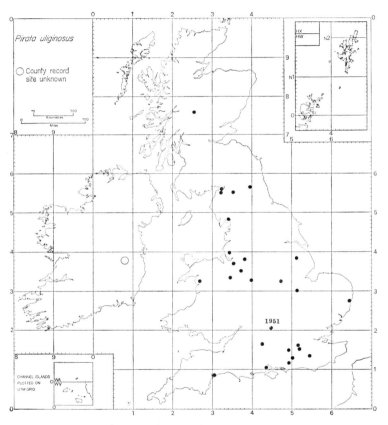

FIG. 5. British distribution of *Pirata uliginosus*.

peusi (*pusillus*) (Fig. 4). A good example is *Zora armillata*, a characteristic spider of Wicken Fen, Cambridgeshire, where it has been known for many years. In recent years two specimens have been taken at Woodwalton Fen, Huntingdonshire, and it has also been discovered on wet heaths in Dorset, but a thorough study of numerous other fens in East Anglia has failed to find it. On the other hand the same survey has shown that a number of species previously thought to be very rare are widely distributed in the fens of Norfolk and Suffolk.

Other species seem to have been overlooked or may even have expanded their range in recent years. *Pirata uliginosus* was first recorded as British in 1951 from a site near Oxford. Today, only 22 years later, it is known from numerous localities all over England, North Wales and from one place in Scotland (Fig. 5). This might look like a genuine example of a species spreading into new habitats. However, an examination of old collections has revealed several specimens which were mistaken for the common *Pirata hygrophilus*. In fact the earliest known record for *P. uliginosus* is dated 1872, again from near Oxford, but its identity was not discovered until recently.

In other cases the true status of the species has not been realized because of inadequate knowledge of the range of habitats occupied. The small erigonid *Thyreostenius parasiticus* was thought to be scarce and associated mainly with artificial habitats until K. Southern found it to be the most numerous spider in dead wood high up in deciduous trees at Wytham, near Oxford. Since then it has been found in this situation in other localities. *Singa hamata* is an orb-web spinner usually associated with northern wet heaths in England but has also been recorded as abundant in a rushy meadow in the south of the country and in a grassy clearing in Monks Wood National Nature Reserve, Huntingdonshire. The last occurrence is particularly surprising because the clearing was made in 1942 when the trees were felled and the land ploughed up to grow potatoes. The attempt at cultivation was a failure, the land abandoned and maintained as grassland when the wood became a National Nature Reserve. The nearest known other sites are in Hampshire to the south and in Shropshire to the north. The rare agelenid *Tetrilus macrophthalmus* has been carefully studied by Crocker (1973) who showed that although it might be well established in the rotten wood and under bark of hollow ancient trees (Plate 1A), its assumed typical habitat, it also occurs under large stones on hilltops in Charnwood Forest, Leicestershire, sometimes in places where trees are absent (Plate 1B). *Clubiona juvenis* is another very rare spider known from only three areas in the British Isles, but all very different and well separated geographically. It was first taken 60 years ago on sand-dunes in County Wicklow, where it still occurs, though in recent years it has been found on a brackish marsh in Poole Harbour, Dorset, and in two freshwater marshes in the Norfolk Broads.

Other species, more widespread in Britain, show a change within their geographical range. *Monocephalus castaneipes* is a characteristic spider of grassy, stony places in upland areas of northern Britain. In the south it was once thought to be very rare but we know that it may occur abundantly in moss on

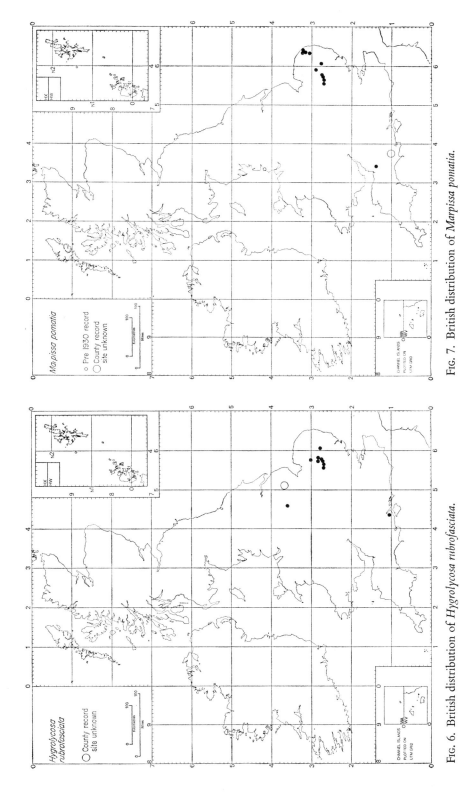

Fig. 7. British distribution of *Marpissa pomatia*.

Fig. 6. British distribution of *Hygrolycosa rubrofasciata*.

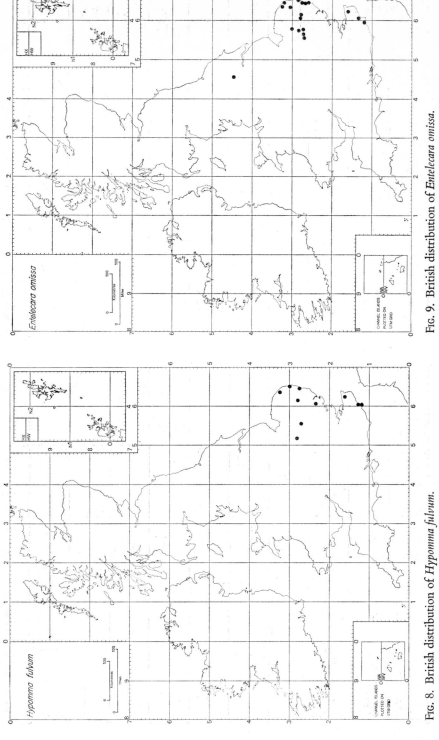

Fig. 9. British distribution of *Entelecara omissa*.

Fig. 8. British distribution of *Hyponma fulvum*.

water levels have been lowered by the proximity of the fen cut-off drainage channel which was completed in 1960 and extends for 27 miles from the Breck to the Wash.

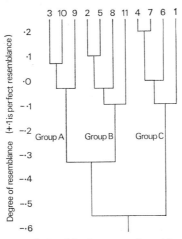

FIG. 10. Dendogram showing relationships between the spider faunas of 11 East Anglian Fens. Group A, Species-poor valley fens; much disturbed. Group B, Species-rich valley fens; little disturbed. Group C, Moderately species-rich isolated fens (analysis based on 8 one-hour recording periods per site).

SOME COMMENTS ON THE FAUNA OF MAN-MADE HABITATS

There is increasing evidence that certain types of artificial habitats are important for the preservation of some very local species. Several are associated almost exclusively with domestic dwellings in this country, although occurring away from buildings in southern Europe. *Tegenaria parietina*, a scarce house spider in the southern half of Britain, is an example of this type of distribution although its close relative, the very familiar *Tegenaria domestica*, is widespread in buildings all over the country. *Pholcus phalangioides* and *Scytodes thoracica* are local species generally found in older houses and *Physocyclus simoni* occurs almost exclusively in cellars. Other species, for example *Euophrys lanigera*, although almost always associated with buildings, are generally found on the outside rather than the inside, in contrast to the preceding species.

Aulonia albimana is a lycosid spider only known from one disused quarry in Somerset where it was first recorded as British by Bristowe in 1936. Extensive searches in many similar areas have failed to discover it. *Pardosa hortensis* is a local or widespread species and often occurs in sparsely vegetated gravel workings. This type of man-made habitat is very vulnerable to destruction either by reclamation, planting or for rubbish dumping. Some of the house spiders mentioned are found in natural caves in southern Europe. A similar example is *Lessertia dentichelis* which has been recorded from cellars, drains and coal mines in this country, but seldom in houses. One of its preferred habitats in Britain is the sewage filter bed where it is sometimes very numerous in the constant environment of moisture, darkness and small temperature fluctuations 2–3 ft below the surface. Another characteristic species of the sewage filter bed spider fauna is *Erigone longipalpis* which elsewhere in Britain is only common on salt marshes although it has occasionally been recorded in freshwater marshes.

INTRODUCTIONS

There is no doubt that a constant flow of exotic spiders reach our shores in imported fruit, timber and other cargoes, as well as via road and air traffic. Every county museum has its collection of "Banana" spiders to show how widespread is this phenomenon. It is strange, therefore, that so few species appear to have established themselves in this country apart from those associated with hot-houses and buildings. There is no well authenticated example of an imported species becoming a successful member of the British fauna under natural conditions, although *Argiope bruennichi* is a possible candidate. It was first recorded in 1922 at Rye, Sussex and later found a short distance away in Kent. However, most present-day records are from southern Dorset and Hampshire (Fig. 2), well separated from the original site. It is also common across the Channel in northern France so could easily have drifted over to the British south coast. Another species, possibly introduced, is *Aulonia albimana* (mentioned above) which since 1936 has apparently been unable to colonize elsewhere from its sole British station.

It is perhaps surprising that a world-wide species such as the Black Widow spider (*Latrodectus mactans*) has not been able to establish indoor colonies in Britain. Very occasionally this spider has been a suspect in the case of human sufferers with mysterious and painful bites but no British-taken specimens are known. However, another spider with a dangerous bite (*Loxosceles laeta*), a native of South and Central America has recently been found to be well-established in the university buildings in Helsinki, Finland (Huhta, 1972).

ACKNOWLEDGEMENTS

I would like to express my thanks to Dr P. Merrett for help and advice in compiling this paper, and to Mr J. Crocker who very kindly allowed me to reproduce his photographs in Plates 1A and B.

REFERENCES

Bristowe, W. S. (1936). "The Comity of Spiders", 2 vols. Ray Society, London.

Bristowe, W. S. (1938). Arachnida. *In* "Victoria County History of the County of Cambridgeshire", pp. 202–212. Oxford University Press, London.

Crocker, J. (1973). The habitat of *Tetrilus macrophthalmus* (Kulcznski) in Leicestershire and Nottinghamshire (Araneae: Agelenidae). *Bull. Br. Arach. Soc.* 2, 117–123.

Duffey, E. (1958). *Dolomedes plantarius. Trans. Norf. Norw. Nat. Soc.* 18, 1–5.

Duffey, E. (1960). A further note on *Dolomedes plantarius* Clerck in the Waveney Valley. *Trans. Norf. Norw. Nat. Soc.* 19, 173–177.

Duffey, E. (1973). A note on comparative invertebrate survey: the spider faunas of Wicken and Woodwalton Fens. *Nature in Cambridgeshire* 16, 13–19.

Huhta, V. (1972). *Loxosceles laeta* (Nicolet) (Araneae, Loxoscelinae), a venomous spider established in a building in Helsinki, Finland, and notes on some other synanthropic spiders. *Ann. Ent. Fenn.* 38, 152–156.

Locket, G. H. and Millidge, A. F. (1951–53). "British Spiders", 2 vols. Ray Society, London.

Pickard-Cambridge, O. (1879). "The Spiders of Dorset". Sherborne.

18 | The British Orthoptera Since 1800

J. A. MARSHALL

Department of Entomology, British Museum (Natural History), London

Abstract: Forty-two species of Orthoptera, of which eleven are established aliens, are currently recorded as breeding in the British Isles. There is no evidence that the number of native species has changed since 1800 (although only six of them were recorded at that time), but at least six of the introduced species have become established here during the last hundred years. Urbanization during this period has removed many areas previously favourable to Orthoptera, and a number of heathland localities have had their Orthoptera depleted as a result of the encroachment of bracken and the trampling effect of the increasingly mobile human population. Some of our rarer species seem to be becoming even more restricted in distribution, and one or two could be in danger of extinction as British insects.

INTRODUCTION

Forty-two species of Orthoptera, of which eleven are established aliens, are currently recorded as breeding in the British Isles. There is no evidence that the number of native species has changed since 1800, although only six were recorded in 1770 and there are five species which have only been recognized during this century. However, the number of introduced species breeding here has certainly increased since 1800, at least six having become established during the last 100 years.

Non-taxonomic knowledge of several British ‘species goes back to the sixteenth century, for they are described by Thomas Moffet [Muffet, Moufet or Mouffet] (1634) in his "Theatrum Insectorum". Although this was not published until 1634, Moffet had prepared the work before his death in 1604, incorporating information from Gesner, Penny and Wotton who had all predeceased him during the sixteenth century.

Systematics Association Special Volume No. 6, "The Changing Flora and Fauna of Britain", edited by D. L. Hawksworth, 1974, pp. 307–322. Academic Press, London and New York.

The earliest lists of British Orthoptera are those of Forster (1770) and Berkenhout between 1769 and 1795, naming nine different species of which six were native and three were established aliens. Forster's stated purpose in publishing his catalogue was to address himself to all "who collect insects" to beg them "to favour him, if possible, with specimens of such insects, as they can spare, and which he is not possessed of", marking in his list those species which he required. A similar plea was made by Curtis (1829–31), in his "Guide to an Arrangement of British Insects" published as a check-list. In this was incorporated information from Donovan's "Natural History of British Insects" (Donovan, 1792–1813), and from Curtis's own "British Entomology" (Curtis, 1825). The number of known species was increased by Curtis to twenty-five natives and four established aliens, although he listed fifty-six species, many synonyms and varieties or mistaken identifications being included. Curtis did however dignify the Orthoptera and Dictyoptera as distinct orders, which both Forster and Berkenhout had included as "Hemiptera" (listing Dermaptera as Coleoptera!). He also distinguished between native and "doubtful native", although he considered the House-cricket (*Acheta domesticus* (L.)) and Common Cockroach (*Blatta orientalis* L.) to be native when they are long-established aliens.

The first work with information on distribution of Orthoptera was that of Stephens (1835) in his "Illustrations of British Entomology", in which he quoted the localities known to him for each species, and added Roesel's Bush-cricket (*Metrioptera roeselii* (Hagenbach)) to the British list.

Shaw (1889–90) followed with his "Synopsis of the British Orthoptera", which he modestly introduced by stating that his "knowledge of the British Orthoptera is but imperfect and scanty"—regretfully there are some instances today where this statement still applies. Shaw gave "analytical tables" to genera and species, at the request of "several entomological friends", and quoted information of localities collected from by several colleagues.

Burr (1897) wrote the first book solely on the British Orthoptera, adding two more established aliens to the British list to make a total of six, the number of known native species remaining at twenty-six.

This number was unchanged with the publication of Lucas's "A Monograph of the British Orthoptera" (Lucas, 1920), but Lucas gave a comprehensive account of each species (including the original description and information on habits and habitat), and lists of all known British localities.

Burr's "British Grasshoppers and their Allies" (Burr, 1936) added four more species, three native and one alien, to the British list, and provided the first "distributional" maps, with the intention of stimulating "interest and collecting in the districts left blank". (It was with the same purpose that Skelton (1973) published a progress report on the Orthoptera distribution maps scheme.)

The system used by Burr to show distribution was only at county level, and was greatly improved upon by Ragge (1965) in his "Grasshoppers, Crickets and Cockroaches of the British Isles". In this work he was able to incorporate much useful information from Kevan's studies on the distribution of the British Orthoptera (Kevan, 1952, 1953, 1954, 1961), and from other recent papers on the biology of various species. In addition to the distribution maps using the Watson-Praeger system of counties and vice-counties (see pp. 425–428), Ragge has provided accounts of habitat, life history, habits and song of each native and established alien species, drawn mainly from his own observations, with identification keys to all species and to the colour varieties of the grasshoppers. With this work the six Orthoptera which have become known or established as British, during this century, were added to the British check-list.

The native species known before 1800, all named as *"Gryllus"*, include *Gryllus campestris* L., *Gryllotalpa gryllotalpa* (L.), *Tetrix subulata* (L.), *Tetrix undulata* (Sowerby) called *"Gryllus bipunctata"*, and *Tettigonia viridissima* L.—from Berkenhout's description, although the name he used was *"Gyrllus verrucivorus"*. The sixth name was *"Gryllus grossus"*, which Berkenhout called the Common Grasshopper, having "a ridge in the form of an X on the Corselet", and which can be taken as referring to *Chorthippus brunneus* (Thunberg), though other common grasshoppers were probably associated with this name.

Lucas (1920) said, ". . . most entomologists think it unnecessary to note the presence of an insect which is considered to be ubiquitous". He was referring to the scarcity of records for the House-cricket, but his statement unfortunately applies to all species with a fairly wide distribution.

As is the case in many other groups, the known distribution of the more common species of Orthoptera tends to represent the distribution of Orthoptera enthusiasts—and of where they go for their holiday collecting trips. For only two species have we any definite information that they have become restricted in their distribution and could be in danger of extinction as British insects.

The Field-cricket (*Gryllus campestris* L.) is one of the insects described in some detail by Moffet (1658). He knew that they sing by "rubbing their wings one against the other", that they live in holes they dig in the ground, and described one ancient method by which children supposedly used to hunt out Field-crickets by putting an ant, "tied about the middle with a hair", into their holes.

However he thought little of this method and described how "sooner and with less labour" the cricket could be enticed out by putting a "small twig or straw" into the hole, and then "draw it out by little and little". Curious though this method may seem, Gilbert White (1779) of Selborne also described how a "pliant stalk of grass, gently insinuated" into the burrow will "quickly bring out the inhabitant"—and this is recognized as the only satisfactory way of capturing the Field-cricket from its burrow.

White also observed that a Field-cricket could be kept "in a paper cage, and set in the sun . . . and thrive, and become so merry and loud as to be irksome in the same room where a person is sitting". A most accurate observation, as the chirp of a Field-cricket kept indoors has a penetrating tone which can give great annoyance.

White observed Field-crickets at Selborne in a nearby "steep abrupt pasture field, interspersed with furze", called the Short Lythe, and though frequent in that area he said that it was by no means a common insect in many other counties. Stephens (1835) considered it to be rare, or at least "seldom captured" though he recorded it from the "vicinity of London and at Windsor, in the New Forest, Devonshire, Cornwall &c". Burr (1897) suggested that it is "rare, but of retiring habits, and probably widely distributed". However, Lucas (1920) was able to give very few records, and suggested that though it was apparently once common in England it had become rare and local. His eleven records were mainly from Hampshire, Surrey and Sussex but included one each from Norfolk and Staffordshire.

Burr (1936) quoted from Hudson's (1903) "Hampshire Days" of an account of an abundant colony of Field-crickets near Hythe, by Southampton Water. This locality was visited by Pickard in 1956 (Ragge, 1956) and specimens were found there. Unfortunately this area has since been built upon and unless the crickets were able to move some distance, the colony probably no longer exists.

The Selborne Short Lythe was examined in 1955 (Pickard, 1956) and no crickets were found, and at another locality where it had once been quite common it has not been found in recent years. In this case, at Frensham Little Pond in Surrey (Ragge, 1956), the habitat has been radically altered by the increasing popularity of the pond as a sailing area and the resultant trampling of the habitat.

In 1956 a large and flourishing colony was discovered near Arundel in Sussex (Ragge, 1956), and more recently the Field-cricket has been found at another locality in West Sussex, on private land where it is being protected (Haes, 1973). These are the only areas where the Field-cricket is now known to breed.

There are still many areas of well-drained grassland or heathland in southern England which are suitable for the Field-cricket, and it is to be hoped that further colonies may be found.

The Mole-cricket (*Gryllotalpa gryllotalpa* (L.)) mentioned by Moffet (1658) as a creature of "moorish and moist ground", seems to have declined during the last 200 years.

Gilbert White (1779) described it as often infesting gardens, "they are unwelcome guests to the gardener . . . if they take to the kitchen quarters, they occasion great damage among the plants and roots, destroying whole beds of cabbages, young legumes, and flowers". Stephens (1835) mentioned it as frequenting "the rich mould of garden grounds", and as occurring "in many places within the metropolitan district". Specimens have been found at various places over the years during the evening or early morning, either flying or having landed in an unlikely area, such as the streets of Exeter (Lucas, 1920).

Several of the more recent records have been reported by the Ministry of Agriculture, through the Mole-cricket damaging root crops (Ragge, 1955). Between 1947 and 1951 five cases were reported from Hampshire and one from Surrey, though there are no known instances since 1951.

Since 1966 the Mole-cricket has been known in a marshy field near Landford in Wiltshire, and in 1970 it was seen in the Southampton area (Ragge, 1973).

The "moorish and moist" habitats which the Mole-cricket favours have been lost from many areas, with the accompanying loss of the Mole-cricket. Since the last war particularly, "a huge backlog of hedging, ditching and draining" has been "wiped off" (Stamp, 1955).

In direct contrast with the two preceding species, there is one native British Orthopteran, the Wart-biter (*Decticus verrucivorus* (L.)), for which there are now more localities known than at any time since 1800.

The Wart-biter was first recorded from near Christchurch, Hampshire, in 1818, and near Rochester, Kent at about the same time. The Christchurch specimens were brown, and were described by Curtis (1825) as "*Acrida bingleii*", a species distinct from *verrucivorus*. Stephens (1829) mentions the Rochester locality, saying he had received fine specimens from Professor Henslow taken from "a field near Rochester in September" where "it once occurred in great plenty"—implying that even a few years later the species was not to be found near Rochester.

To these two localities Lucas (1920) could add only four others in his distribution summary. Single specimens were collected in the New Forest in 1844 and 1891 (the latter supposedly the brown form *bingleii*), one was recorded from Deal, Kent, in 1889 and there were three records from St Margaret's Bay, Kent, for 1886, 1889 and 1900. A new locality had been discovered by Burr at Lydden, near Dover, in 1907, where Porritt found the species rare in 1913.

From 1920 till 1955 there were only two records, from Slepe Heath (near Wareham) and from near Corfe Castle, both in Dorset. Then as Ragge (1955) described, he visited the Isle of Purbeck with the intention of looking for the Wart-biter, and stopped for lunch within hearing of a singing male. More recent visits show the sparse population in this area appears to continue at about the same level. Around this time the species was also discovered in East Sussex (Payne, 1955a), where it is now known to be established over an area of two square miles of chalk downland (Ragge, 1973). In addition, Dolling (1968) rediscovered the species at Burr's 1907 locality where the area is now a Nature Reserve, and a new locality has been found in North Wiltshire (Mason, 1971) on a protected site being carefully watched over. A collection recently presented to the British Museum (Natural History) after the collector's death contained a Wart-biter collected from the Isle of Wight in 1951, where in the area around Ventnor there may have been suitable habitats.

It is to be hoped that this species, for which more breeding localities are now known than would have been thought possible even twenty years ago, may yet be found in more areas in southern England. Although a large insect and apparently obvious, it is possible to search within yards of where it must be and yet not find any trace of it, unless the males are induced to sing by bright sunny weather.

There are three species, considered to be native, which have been discovered in Britain during the last fifty years and which have only a very local distribution:

(1) The occurrence in this country of the Heath Grasshopper (*Chorthippus vagans* (Eversmann)) was first suspected when one specimen labelled "New Forest" was discovered in the British Museum (Natural History) by Uvarov (1922). It was confirmed as a British species when Frazer (1944) discovered two specimens among a grasshopper collection he had made at Wareham, Dorset, in 1934. In 1947 further specimens were found on Sopley Common, Hampshire, an area investigated by Ragge in several years following (Ragge, 1954). He found the population fluctuated considerably in size and density, and though abundant in some years might at other times be easily missed without a thorough search. A collection made from Studland, Dorset, in 1933, and later presented to the British Museum (Natural History) was also found to contain the Heath Grasshopper, a visit by Ragge (1954) to this locality confirming the record.

It has become clear that this species has a restricted distribution over the heaths to the south and east of Poole Harbour, to the north of Hurn, Hampshire, and in the extreme west of the New Forest (Ragge, 1965), although it is possible that it may have been missed in other areas of heathland in east Dorset and south-west Hampshire.

(2) The Scaly Cricket (*Mogoplistes squamiger* Fischer) has been found on several occasions since the first record of 1949 (Bowen, 1950), the last known record being 1967 (Ragge, 1973). All specimens have been found in the same flat, sandy and muddy area of Chesil Beach, Dorset, under or near scattered rocks. The species is native to the north Mediterranean coast, Madeira and the Canary Islands, where it lives in similarly protected habitats. Chesil Beach is within a mile or so of Portland Harbour, and it is possible that the species could have been imported; however, the peculiar habitat does not seem to lend itself to travel and it is equally possible that other localities might be discovered in similarly well sheltered situations on our south coast.

(3) The most recently discovered grasshopper in Britain is *Stenobothrus stigmaticus* (Rambur), the Lesser Mottled Grasshopper, found on the Isle of Man. The first two specimens were collected in 1962, and all subsequent findings have been in the same eastern half of the Langness Peninsula; Burton (1965) failed to find it anywhere else on the island.

The problem posed by the presence of this species on the Continent and on the Isle of Man but nowhere else in Britain (as far as we know), has been discussed fully by Ragge (1963, 1965, 1973). There is an airport within a few miles of the locality, although there are no direct flights to Ronaldsway Airport from the Continent and importation by this method seems most unlikely. Ragge could find no other examples of animals or plants recorded from the Continent and the Isle of Man but from no other part of the British Isles, but there are several animals and plants whose distribution here is confined to the Isle of Man and Ireland (Ragge, 1963). This leads to the possibility that the Lesser Mottled Grasshopper may yet be found in Ireland—a country even less recorded from than Great Britain. It is also possible that the non-occurrence of the species in Great Britain is a result of competition with the Meadow Grasshopper (*Chorthippus parallelus* (Zetterstedt)) which is abundant throughout England, Wales and Scotland, but does not occur in Ireland or on the Isle of Man.

The other native Orthoptera tend to show a gradation from the less common species with a fairly well known distribution, through to the common and widespread species for which far more exact information on distribution is required. The vice-county maps have been of great importance in recording the distribution of Orthoptera up to the present day, though to a certain extent masking deficiencies of exact knowledge. However, within the next year it is hoped that sufficient information will have been collated to enable the

Biological Records Centre to produce valuable maps on the basis of the 10 km National Grid squares, showing a reasonable knowledge of Orthoptera distribution.

The Grey Bush-cricket (*Platycleis denticulata* (Panzer)) has always been known as a local species, confined to maritime districts, but occurring in "all the counties of the south coast of England" (Lucas, 1920). An unusual inland record quoted by Lucas (1920) as "rather strangely, one from the inland county of Derby", it is tempting to associate with the equally odd record of the Great Green Bush-cricket, *Tettigonia viridissima* L. also quoted by Lucas (1920), said to have been "brought to the Museum at Derby by a boy, taken a few miles from Derby about 1896 or 1897". Someone had apparently spent a holiday on the south coast!

The Bog Bush-cricket (*Metrioptera brachyptera* (L.)) is restricted to heathy areas and has a rather patchy known distribution, and, although common where it occurs, there are many counties in central and southern England from which it has not yet been recorded. The rare macropterous form of this species was first recorded here in 1921 (Lucas, 1922), and has only been recorded on six occasions since then, between 1941 and 1970 (Ragge, 1973).

The closely related Roesel's Bush-cricket (*Metrioptera roeselii* (Hagenbach)) has always been considered to be rare in Britain (Stephens, 1835; Burr, 1897), and Lucas (1920) said the species was "confined to the south-east coast of England, and till a few years ago was almost lost to sight". All the records known to Lucas were from about the mouth of the Thames, or on the east coast south of the Humber, all flat, estuarine localities. The known distribution of the species has widened, including a most interesting recent record from the Dovey Estuary in Cardiganshire. This colony was discovered in 1970, and its existence confirmed in 1971 by Ragge (1973) who attempted to investigate the possibilities of the species having been introduced there. He found that much of the land bordering the area is farmed by a firm having large holdings in Lincolnshire, where Roesel's Bush-cricket is known to occur, but unfortunately further information or comment was unforthcoming. The macropterous form of this species has been recorded on several occasions since the earliest record of 1922 (Lucas, 1923).

The Short-winged Cone-head (*Conocephalus dorsalis* (Latreille)) is known principally from the southern counties of England, with a few Welsh records, from coastal areas with rushes, sedges, reeds or sometimes long grass. Recently, brown specimens have been recorded, presenting a strong contrast with the normal mostly green coloration. They were found in the Poole Harbour area, Dorset and Cefni salt marsh, Anglesey. Long-winged forms of this species may occur in normal colonies, and have been recorded since at least 1899 (Lucas, 1920). However, the macropterous form of the Short-winged Cone-head may be confused with the Long-winged Cone-head (*Conocephalus discolor* Thunberg). This species was first collected by Blair in 1931, when he assumed he had "the rare fully-winged form" of the brachypterous species. It was only on being shown a true macropterous *C. dorsalis*, a few years later, that he realized that he had in fact collected *C. discolor* (or *C. fuscus* as it was then known) (Blair, 1936).

The Long-winged Cone-head has since been recorded in Sussex (Menzies, 1946) and several southern coastal regions, seldom far from tidal water. In 1969 at the Chapman's Pool colony in Dorset, and in 1970 at Telscombe Cliffs in Sussex, unusually long-winged forms were found (Ragge, 1973), known previously only from the continent.

The most common bush-cricket in the southern half of England is the Dark Bush-cricket (*Pholidoptera griseoaptera* (Degeer)), which is found in almost any habitat which provides cover without obscuring too much light. It is not, however, an easy insect to catch—Burr (1907) suggested that "this insect was formerly regarded as a prize owing to its retiring disposition and great activity", assisted by its unfriendly way of living in bramble patches and nettle beds, "but those familiar with its stridulation cannot fail to recognize its presence in almost every roadside hedge . . . in the southern counties". Its single chirp, repeated at variable intervals, may be heard at any time of the day or evening.

A song which is familiar to many holiday-makers in the south-west is that of the Great Green Bush-cricket (*Tettigonia viridissima* L.) the only bush-cricket known to Forster and Berkenhout before 1800. This species Stephens (1835) said was very common, and that he had "frequently taken it in Battersea field,". The only area near to London where the species can still be found is probably Erith Marshes.

The largest native grasshopper is *Stethophyma grossum* (L.), the Large Marsh Grasshopper, known in England from only a few southern counties and East Anglia, and from western and southern Ireland. Stephens (1835) said it was "found in those marshy districts about Camberwell, Deptford etc. near London", but these habitats have long since changed in their suitability.

The Lesser Marsh Grasshopper (*Chorthippus albomarginatus* (Degeer)) is now recorded from almost all coastal vice-counties of southern England, as far north as Lancashire and south-eastern Yorkshire, and from Co. Clare in Ireland and Caernarvonshire in Wales. These scattered records suggest the distribution is much wider than our present knowledge indicates, particularly as many more records have recently been collected in the Cambridgeshire area by M. J. Skelton of the Biological Records Centre at Monks Wood.

The Common Field Grasshopper (*Chorthippus brunneus* (Thunberg)) is probably the grasshopper seen most often, because of its occurrence near buildings, on roadsides and cultivated land, and Lucas (1920) suggested that it "may be fairly looked upon as *the* British grasshopper". It is known from every vice-county in England and Wales, most of Scotland, the Isle of Man and Ireland. The rare green variety has now been recorded seven times (Ragge, 1973), from localities in England, Wales and Ireland. The Common Field Grasshopper is not recorded from as far north in Scotland as either the Meadow Grasshopper (*Chorthippus parallelus* (Zetterstedt)) or the Common Green Grasshopper (*Omocestus viridulus* (L.)) which both prefer a rather more moist grassland than does the Common Field Grasshopper. The Meadow Grasshopper can be found on all forms of grassland that are not mown or heavily grazed, including marshland, and heathland that is not too dry, and the macropterous form of this species is now taken frequently.

The Common Green Grasshopper, the Mottled Grasshopper (*Myrmeleotettix maculatus* (Thunberg)) and the Common Groundhopper (*Tetrix undulata* (Sowerby)) all show a similar distribution pattern to that of the Common Field Grasshopper.

The remaining native Orthoptera are confined mainly to southern England and Wales, some with a more restricted distribution which may be related to a particular habitat. The occurrence of the Rufous Grasshopper (*Gomphocerippus rufus* (L.)) is closely related to its liking for limestone vegetation, particularly on steep, south-facing slopes, and the Stripe-winged Grasshopper (*Stenobothrus lineatus* (Panzer)) prefers dry, limestone or sandy areas.

ESTABLISHED ALIENS

The introduced species recorded as British in 1770 (Forster, 1770) are the House-cricket, *Acheta domesticus* (L.), the Common or Oriental Cockroach (*Blatta orientalis* L.) and the German Cockroach (*Blattella germanica* (L.)). Of these both the House-cricket and the Common Cockroach are known to have been established here by the sixteenth century or earlier, as they were mentioned as British insects in Moffet's (1634, 1658) "Theatrum Insectorum".

The Common Cockroach was mentioned in some detail by Moffet (1634), where he described and illustrated male and female as two of three different species of "Moths called Blattae". He has been quoted many times as saying the cockroach was common in London wine-cellars, and in the original Latin it is easy to pick out the relevant words, "Londini apud nos in cellis vinarijs". However, the meaning of the sentence of which these words are part is only made clear on referring to the English translation by John Rowland, published in 1658. From this it is obvious that the insect found "in London in Wine-cellars, and dark dungeons" was the third sort of "Moth". This species is described as "so unsavory, and carries with it such a stinking smell", and from the drawing is obviously a Cellar Beetle (*Blaps* sp.). Moffet goes on to say that "the other species are more frequent in Bake-houses, and warm places"—where the Common Cockroach is usually found.

Gilbert White in his "Observations on Insects", noted the occurrence of *Blatta orientalis* twice, in 1790 and 1792, describing the invasions made by them from neighbouring houses during a hot season, through the open windows. He wondered how the females entered, as he had assumed the males could fly in though the brachypterous females could not, when in fact both sexes would have walked up the walls.

Although Curtis (1829) considered the Common Cockroach to be a native species, Stephens (1835) recognized that it is "not an aboriginal native", although occurring "in houses, especially in London and in maritime commercial towns, in utter profusion, swarming in myriads". Shaw (1889) mentioned that "it is not only generally distributed in houses, but is also found out of doors", a point also mentioned by Burr (1897), where he said the species "has now spread all over the country", including the Orkneys. Lucas (1912) related a letter from Burr describing both the Common and the German Cockroach (*Blattella germanica* (L.)), "swarming within a rubbish-heap in a brick-yard" during January 1911 when the "country was iron-bound in a black frost". He did add that though the weather was cold, "the fermentation in the large heap of ashes and refuse produced much heat". It has thus been known for some time that the Common Cockroach could survive out-of-doors in this country throughout the winter, providing some artificial warmth and protection were available. However there have recently been reported (Beatson and Dripps, 1972) three cases of outdoor colonies of the Common Cockroach, with cover but no extra warmth available. In these cases the cockroaches appear to be breeding out-of-doors, and only adults have been occasionally discovered in nearby houses, presumably foraging for food.

The House-cricket (*Acheta domesticus* (L.)) may have been introduced into Britain as early as the Crusades (Payne, 1955), but was certainly known here by the sixteenth century (Moffet, 1634). Gilbert White (1779) described the species as residing "altogether within our dwellings, intruding itself upon our notice whether we will or no . . . they are particularly fond of baker's ovens".

The occurrence of the familiar "cricket on the hearth" has been greatly reduced during this century, associated with improved standards of hygiene, although several authors (Burr, 1897; Lucas, 1920; Ragge, 1965) have suggested that some reduction in the records for the House-cricket can be accounted for by competition with the Common Cockroach.

The House-cricket now occurs commonly on rubbish dumps as well as indoors, and in warm weather will readily fly, as Gilbert White observed.

The American or Ship Cockroach (*Periplaneta americana* (L.)) has been recorded here since 1829 (Curtis, 1829–31). It was said by Stephens (1835) to be found in "warehouses and outbuildings, by the side of the Thames, especially below London Bridge". Shaw (1889) recorded that the Ship Cockroach was fast establishing itself, and was also able to report the arrival of the Australian Cockroach (*P. australasiae* (Fab.)) in Belfast. Both of these species are common in ports and large coastal towns, and are frequently introduced into the country in consignments of imported goods. A related species *P. brunnea* Burmeister, the Brown Cockroach, has twice established breeding colonies at London Airport but has on each occasion been eradicated (Bills, 1965, 1967).

The Brown-banded Cockroach (*Supella supellectilium* (Serville)) has established itself here in recent years, in houses and offices, though very few records are known so far.

The Surinam Cockroach (*Pycnoscelus surinamensis* (L.)) and the Greenhouse Camel-cricket (*Tachycines asynamorus* Adelung) are two species that have established breeding colonies here in heated greenhouses since the turn of the century. Both are Asian in origin and can only survive in artificially heated habitats.

The final three of our eleven introduced and established species are Stick-insects, one of Indian origin and two from New Zealand.

The Laboratory Stick-insect (*Carausius morosus* Brunner) has been widely cultured in Europe for many years, probably originating from eggs collected in southern India between 1898 and 1911 but none later than this; the origin of the species is discussed by Ragge (1973). The species has only been recorded as establishing itself in greenhouses in south Devon and Surrey, but it is reared widely and prolifically and escapes—and throw-outs—may often occur.

The first record of *Acanthoxyla prasina* (Westwood), the Prickly Stick-insect, is from Paignton, Devonshire in 1908 (Kirby, 1910) though the species was first correctly named by Uvarov (1944) when he described the specimens discovered at Tresco Abbey, Isles of Scilly in 1943 and 1944. The Smooth Stick-insect (*Clitarchus hookeri* White) was discovered in Tresco Abbey Gardens in 1949 (Uvarov, 1950). The origin of both these species is probably in the importation of a large number of live plants from New Zealand, in 1907 and 1909, some of which were also distributed in Paignton. New Zealand plants were also introduced to south-west Ireland, where the Smooth Stick-insect has been found in South Kerry.

The Prickly Stick-insect was found in large numbers in Paignton in 1950 (Rivers, 1953), and was observed there in 1962 although not more recently. One of the original habitats, a Japanese Cedar tree, has been chopped down but many other of the original exotic trees and shrubs still occur there, and the species is probably still in existence (Ragge, 1973).

CASUAL INTRODUCTIONS

Many foreign species of Orthoptera have been introduced into this country on fruit, vegetables and other imported goods, and a few other species have managed to fly here. Most of these are of little importance because they only occur rarely, and so far have not established breeding colonies.

The major source of foreign introductions is with bananas, from the Canaries and West Africa and particularly the West Indies. Several West Indian cock-roaches reach us so frequently that they have been given common names (Ragge, 1965) and one West Indian and one West African cockroach are known only from imported specimens (Gurney, 1965). The Madeiran Cockroach (*Leucophaea maderae* (Fab.)) has been imported occasionally since before 1829, when Curtis listed the species as a "doubtful native". The species was also known to Moffet (1634): he illustrated both dorsal and ventral views (among the Cicadas!) but did not mention the origin of the specimen. Stephens (1835) reported specimens of *Blaberus giganteus* (L.) found in the West India Docks, and recognized them as imported.

Several bush-crickets are imported now mainly from the West Indies though until the mid-1960s a West African species, the Prickly Bush-cricket, (*Cos-moderus maculatus* (Kirby)) was frequently imported.

Stray specimens of the Migratory Locust (*Locusta migratoria* L.) occasionally reach us, and have done so for many years with records from Yorkshire to Devonshire and Cornwall, from at least 1842 (Burr, 1897). The Desert Locust (*Schistocerca gregaria* (Forskal)) also reaches this country occasionally but neither species arrives in swarm proportions. The Migratory Locust and the very large Egyptian Grasshopper, *Anacridium aegyptium* (L.), are also occasionally imported with goods from Mediterranean countries.

During the last few years the number of imported specimens has gradually declined, until now it is quite a rare event to find a cockroach or bush-cricket among the imported foodstuffs at a warehouse or greengrocers. The importers are presumably well satisfied with this state of affairs, however disappointing it may be for those interested by the accidental importation of live exotic Orthoptera.

REFERENCES

BEATSON, S. H. and DRIPPS, J. S. (1972). Long term survival of cockroaches out of doors. *Envir. Hlth* **80**, 340–341.

BERKENHOUT, J. (1895). "Synopsis of the Natural History of Great-Britain and Ireland." Cadell, London.

BILLS, G. T. (1965). The Occurrence of *Periplaneta brunnea* (Burm.) (Dictyoptera, Blattidae) in an International Airport in Britain. *J. Stored Prod. Res.* **1**, 203–204.

BILLS, G. T. (1967). The second established colony of *Periplaneta brunnea* (Burm.) (Dictyoptera, Blattidae) in an airport in Britain. *Entomologist's mon. Mag.* **102**, 130.

BLAIR, K. G. (1936). *Conocephalus fuscus* Fab., a grasshopper new to Britain. *Entomologist's mon. Mag.* **72**, 273–274.

BOWEN, H. J. M. (1950). *Mogoplistes squamiger* Fisch. (Orth., Gryllidae) in Dorset. *Entomologist's mon. Mag.* **86**, 81.

BURR, M. (1897). "British Orthoptera." The Economic and Educational Museum, Huddersfield.

BURR, M. (1907). Orthoptera in East Kent in 1907. *Entomologist's Rec.* **19**, 252–254.

BURR, M. (1936). "British Grasshoppers and their Allies." Allen, London.

BURTON, J. F. (1965). Notes on the Orthoptera of the Isle of Man with special reference to *Stenobothrus stigmaticus* (Rambur) (Acrididae). *Entomologist's mon. Mag.* **100**, 193–197.

CURTIS, J. (1825). "British Entomology," **2**. Sherwood, Gilbert and Piper, London.

CURTIS, J. (1829–31). "A Guide to an Arrangement of British Insects." Westley and Davis, London.

DOLLING, W. R. (1968). *Decticus verrucivorus* (Linnaeus) in (Orthoptera, Tettigoniidae) in Kent. *Entomologist* **101**, 168.

DONOVAN, E. (1792–1813). "The Natural History of British Insects." Rivington, London.

FORSTER, J. R. (1770). "A Catalogue of British Insects." Eyres, Warrington.

FRAZER, W. R. (1944). First authentic record of *Chorthippus vagans* Ev. (Orthopt., Acrididae) in Britain. *J. Soc. Brit. Ent.* **2**(6), 224.

FRAZER, W. R. (1949). *Chorthippus vagans* von Eversm. (Orthopt., Acrididae) in Hampshire and Dorset. *J. Soc. Brit. Ent.* **3**(2), 59–60.

GURNEY, A. B. (1965). Two new cockroaches of the genera *Pelmatosilpha* and *Henschoutedenia*, with a key to the West Indian species of *Pelmatosilpha* (Dictyoptera: Blattaria). *Proc. R. ent. Soc. Lond.* (B) **34** (1–2), 5–11.

HAES, E. C. M. (1973). The distribution of native Saltatoria in Sussex (1965–70). *Entomologist's Gaz.* **24**, 29–46.

HUDSON, W. H. (1903). "Hampshire Days." Longman, London.

KEVAN, D. K. McE. (1952). A summary of the recorded distribution of British Orthopteroids. *Trans. Soc. Br. Ent.* **11**, 165–180.

KEVAN, D. K. McE. (1953). Notes on the distribution of British Orthopteroids. *J. Soc. Br. Ent.* **4**, 119–122.

KEVAN, D. K. McE. (1954). Further notes on the distribution of British Orthopteroids. *J. Soc. Br. Ent.* **5**, 65–71.

KEVAN, D. K. McE. (1961). A revised summary of the known distribution of British Orthopteroids. *Trans. Soc. Br. Ent.* **14**, 187–205.

KIRBY, W. F. (1910). An undetermined species of stick-insect found in Devonshire. *Zoologist, ser.* IV, **14**, 197–198.

Lucas, W. J. (1912). British Orthoptera in 1911. *Entomologist* **45**, 114–117.

Lucas, W. J. (1920). "A monograph of the British Orthoptera." Ray Society, London.

Lucas, W. J. (1922). Notes on British Orthoptera in 1921. *Entomologist* **55**, 200–203.

Lucas, W. J. (1923). Notes on British Orthoptera in 1922. *Entomologist* **56**, 104–107.

Mason, J. L. (1971). *Decticus verrucivorus* (L.), the Wart-biter (Orth., Tettigoniidae), new to North Wiltshire (V.C.7), *Entomologist's mon. Mag.* **107**, 126.

Menzies, I. S. (1946). *Conocephalus fuscus* F. (Orth., Tettigoniidae) in Sussex. *Entomologist's mon. Mag.* **82**, 39.

Moffet [as Moufet], T. (1634). "*Theatrum Insectorum*". Cotes, London.

Moffet [as Moufet], T. (1658). "The Theater of Insects: or, Lesser Living Creatures." [Translation by J. Rowland.] E.C., London.

Payne, R. M. (1955). The Home of the House Cricket, *Acheta domesticus* (L.) (Orth., Gryllidae). *Entomologist's mon. Mag.* **91**, 263.

Payne, R. M. (1955). *Decticus verrucivorus* (L.) (Orth., Tettigoniidae) in Sussex. *Entomologist's mon. Mag.* **91**, 263.

Pickard, B. C. (1956). *Gryllus campestris* L. in Britain (Orth., Gryllidae). *Entomologist* **89**, 200.

Ragge, D. R. (1954). The distribution of *Chorthippus vagans* (Eversmann) in Dorset and Hampshire (Orth., Acrididae). *Entomologist* **87**, 230–231.

Ragge, D. R. (1955). Recent records of the Mole-Cricket from Hampshire and Surrey. *Entomologist's Rec.* **67**(6), 161.

Ragge, D. R. (1955a). Rediscovery of *Decticus verrucivorus* (L.) in Dorset (Orth., Tettigoniidae). *Entomologist* **88**, 260–261.

Ragge, D. R. (1956). Some notes on the Field-cricket, *Gryllus campestris* L. (Orth., Gryllidae). *Entomologist* **89**, 300–301.

Ragge, D. R. (1963). First record of the grasshopper *Stenobothrus stigmaticus* (Rambur) (Acrididae) in the British Isles, with other new distribution records and notes on the origin of the British Orthoptera. *Entomologigist* **96**, 211–217, 2 figs.

Ragge, D. R. (1965). "Grasshoppers, crickets and cockroaches of the British Isles." Warne, London.

Ragge, D. R. (1973). The British Orthoptera: A Supplement. *Entomologist's Gaz.* **24**, 227–245.

Rivers, C. F. (1953). A New Zealand stick-insect in South Devon. *Bull. amat. Ent. Soc.* **12**, 92–94, 1 fig.

Shaw, E. (1889–90). Synopsis of the British Orthoptera. *Entomologist's mon. Mag.* **25**, 354–359, 365–372, 409–421, 450–455; **26**, 56–64, 94–797, 167–176.

Skelton, M. J. (1973). Orthoptera distribution maps scheme. Progress report. April 1973. *Entomologist's Gaz.* **24**, 223–226.

Stamp, L. D. (1955). "Man and the Land." Collins, London.

Stephens, J. F. (1835–37). "Illustrations of British Entomology", **6**. Baldwin and Cradock, London.

Taylor, E. (1954). A further record of *Mogoplistes squamiger* Fisch. (Orth., Gryllidae) in Dorset. *Entomologist's mon. Mag.* **90**, 300.

Uvarov, B. P. (1922). A grasshopper new to Britain. *Entomologist's mon. Mag.* **58**, 211.

UVAROV, B. P. (1944). A New Zealand Phasmid (Orthoptera) established in the British Isles. *Proc. R. ent. Soc. Lond.*, **B**, **13**, 94–96.

UVAROV, B. P. (1950). A Second New Zealand Stick-insect (Phasmatodea) established in the British Isles. *Proc. R. ent. Soc. Lond.*, **B**, **19**, 174–175.

WHITE, G. (1834). "The Natural History of Selborne; with Observations on Various Parts of Nature, and the Naturalist's Calendar." Fraser, Edinburgh.

19 | Changes in the British Coleopterous Fauna

P. M. HAMMOND

Department of Entomology, British Museum (Natural History), London

Abstract: Changes in the composition of the British coleopterous fauna since the last glaciation are briefly discussed; special significance is attached to the man-made factors involved and to the effects of forest clearance in particular. Sources of data for the study of recent change are reviewed and the difficulties encountered in recognizing and documenting genuine change are stressed. Changes in range and abundance, probable extinctions and successful colonists are considered in the context of recent environmental change, much of it man-made. Evidence of change in the last two centuries is also drawn from data concerning a limited area, the county of Essex. Conservation measures best suited to the maintenance of a rich and varied coleopterous fauna are briefly outlined. A more systematic approach in the gathering of distributional and other data is advocated. Much of the change in the coleopterous fauna in the period since 1800 is regarded as a continuation of changes begun at least 5000 years ago. Man-made alterations in the environment are considered to be the principal agents of faunistic change throughout historical time.

INTRODUCTION

As the aims of this Symposium were to "consider changes . . . in Britain since systematic recording of the British fauna and flora began" the present review of the coleopterous fauna is bound largely to beg the question. It may be argued that systematic recording of this group of insects is only now beginning; our knowledge of the present-day fauna is an inadequate basis for any discussion of past change. The composition of the fauna is still incompletely understood in taxonomic terms and, even in the case of species that have been well understood for one or two centuries, our knowledge of their present-day distribution, frequency, abundance and habits is slight. As by no means the least popular of insect groups with entomologists in the post-Linnaean era, many observations on British Coleoptera are to be found in the literature, but these data are mostly scattered and uncollated. There are very few satisfactory accounts

Systematics Association Special Volume No. 6, "The Changing Flora and Fauna of Britain", edited by D. L. Hawksworth, 1974, pp. 323–369. Academic Press, London and New York.

of the present status of individual species and very few convenient reviews, bibliographies, or reliable summaries of data which can be used for reference.

During the period for which published accounts are available (*ca.* 1800 to the present) considerable change in the British coleopterous fauna is likely to have taken place but evidence drawn from published work concerning the nature of this change is largely anecdotal in the context of the fauna as a whole. Major changes which are likely to have taken place during this period are more readily identified in terms of known environmental changes and on the basis of what is known of the present-day occurrence and habits of species of Coleoptera, than by utilizing the sparse data concerning the occurrence in time of these species provided by published records and historical collections. Now that fossil assemblages of Coleoptera have been described from a number of post-glacial British sites the composition of the coleopterous fauna of the early part of the post-Linnaean period is scarcely more amenable to direct study than the faunas of earlier periods. The relative availability of reliable records concerning the presence and abundance of species in different periods, therefore, can provide only an arbitrary delimitation of the present discussion.

In these circumstances a review of likely changes in the coleopterous fauna within the whole of the post-glacial period is considered to be the most useful contribution. This is attempted here as much on the basis of known environmental changes and the apparent requirements of species found in Britain today as on a comparison of the available records for different points in time. However, in the very recent period for which a greater number of entomological collections and published records are available there is relatively good evidence of the change in status of certain species. Changes in such individual species are discussed on the basis of the direct evidence. It should be remembered, however, that changes in individual species may not be at all indicative of general trends or illustrative of major changes in the fauna as a whole.

In general, an attempt has been made to provide information and documentation which may serve as useful reference material in future discussions of change in the British coleopterous fauna although the large size of this fauna makes a very thorough collation of these data impracticable here. Discussions of change will have a much sounder basis when the taxonomy of the species found in Britain is more fully understood and when more extensive and reliable data concerning habits and distribution are available.

COLEOPTERA AS INDICATORS OF ENVIRONMENTAL CHANGE

Coleoptera may be counted on most scores as a highly successful group of organisms. My own estimates suggest that the approximately 370 000 currently recognized species represent about a third only of extant species. The British fauna, however, is an impoverished one, composed of 3690 species (plus 21 casuals) at the last published count (Kloet and Hincks, 1945). Subsequent studies have extended this list and it seems likely that about 4000 species have been established members of the British fauna during at least part of the post-glacial period. Many of the features of interest and value in studies of faunal history and short-term population changes are exhibited by species of Coleoptera which occur in Britain. Although most species reproduce sexually, some are parthenogenetic. Some species are good fliers, some are quite wingless and some wing-dimorphic. The powers of rapid dispersal of some are very consider-able, of others very limited. Some are ecological "specialists" and some "generalists", some are predacious, some phytophagous, many are scavengers, some are parasitic and some are otherwise closely associated with other animal species. Some are greatly favoured by the activities of man and some have be-come pests. Almost every niche has its species.

The special value of Coleoptera in studies of environmental change is due not only to their ecological characteristics but to the nature of the fossil record. Fossil remains of vascular plants, particularly of pollen, have for some time been much used in interpretations of past climatic changes, but recent studies (Coope, 1967, 1969, 1970a, b; Coope and Brophy, 1972; Coope *et al.*, 1971) have begun to demonstrate how fossil assemblages of Coleoptera may be used to pinpoint the timing of changes with more precision and act as indicators of short-term oscillations. This sensitivity is due to the greater speed with which a coleopterous fauna will respond to climatic change when compared with many vascular plants. Such promptness of response is equalled or surpassed by other groups of organisms but fossil assemblages of Coleoptera are peculiarly suited to a role as climatic indicators. Abundant remains are more frequently present in Pleistocene deposits than those of Mollusca or other groups in which remains are readily identifiable; remains of other insect groups which are well represented in deposits are generally poorly preserved or are, at least at present, more difficult to identify at the specific level. Many of the characteristics which lend significance to the coleopterous fauna as indicators of past climate change apply also to interpretations of changes in the environ-ment created by man.

GENERAL TRENDS THROUGHOUT THE POST-GLACIAL PERIOD

Recent changes in the British coleopterous fauna can be seen largely as part of a continuing process extending from the end of the last (Weichselian) glaciation until the present day. Although the effects of man have become more acute in certain respects since the industrial revolution, and more particularly in the present century, modern changes can scarcely be understood except in the context of the last 9000 years or so.

Only a few general studies of the history of the British coleopterous fauna have been attempted. Although Sharp (1899) speculated on the derivation of the British fauna and other authors much more briefly considered the question, until very recent years only Sainte-Claire Deville (1930) and Lindroth (1935) had focused any close attention on the subject. Other work by Lindroth (1931, 1957, 1960, 1972; Holdhaus and Lindroth, 1939), although not dealing specifically with the British fauna, contains much that is relevant.

The increased attention paid in recent years to Quaternary fossil insect assemblages has produced extremely valuable data, relating to changes in the British coleopterous fauna. More particularly, the studies of the past fifteen years or so, mostly centred on the Department of Geology at Birmingham University, have resulted in a considerable increase in our knowledge of the British fauna at various points during the latter half of the Pleistocene period. The nature and implications of these studies have been summarized by Coope 1970b) who provides a good bibliography of studies in this field. Additional papers concerned with this topic but not listed by Coope (1970b) are those of Allen (1969d), A. Bell (1922), F. G. Bell *et al.* (1972), Bradley (1958), Buckland and Kenward (1973), Celoria (1970), Coope (1970a), Coope and Brophy (1972), Coope *et al.* (1971), Duffey (1968), Gaunt *et al.* (1970), Osborne (1971a, b, 1972), Pearson (1961a, b, 1964), Shotton (1965) and Stafford (1971).

Although the evidence that is now accumulating from studies of assemblages of fossil insects provides objective evidence concerning the occurrence of beetle species in time, the study of modern faunas still has an important part to play in assessing past faunal change. Simplistic extrapolations from present-day, often extremely artificial, distribution of beetle species in northern Europe can no longer be accepted, but critical appraisal of the ecology of these species can help in the creation of a theoretical framework for discussion of faunistic change. In fact, for the post-glacial period, the sum of evidence drawn from observations on the occurrence of species today is still more likely to lead to useful hypotheses than is the presently known fossil record. The validity of such hypotheses, however, will increasingly be amenable to test as more fossil data becomes available.

Most Pleistocene faunas so far investigated date from the period of the Weichselian glaciation or from previous glacial or interglacial periods. These faunas are of considerable indirect interest in any assessments of post-glacial changes as they demonstrate the general nature of responses to climatic changes in environments not influenced by man. The relatively few post-glacial faunas that have been studied are, naturally, of more direct interest. Post-glacial changes in the coleopterous fauna are likely to correspond fairly well with the general picture sketched in terms of the vascular plant flora by Godwin (1956) although the timing and precise nature of responses to climatic and other changes will differ.

The final retreat of the ice enabled the spread northwards of species of Coleoptera already present in southern Britian, followed by other species entering (no doubt mostly re-entering) Britain from the south. Eventually a number of cold-adapted species were (almost literally) pushed out of the north of Scotland. Coope (1969) provides a list of 53 species recorded from the mid-Weichselian interstadial which are now extinct in the British Isles. The majority of these are cold-adapted and many of them are likely to have been present in the British Isles immediately before and during the final retreat of the ice. A few relatively cold-adapted species maintained a presence in part of northern Britain even through the post-glacial climatic optimum and are still represented in Britain by relict populations. Some of these species, for example *Phyllodecta polaris* Schneider (see Morris, 1970) have been discovered here only in recent years.

Before any major afforestation a large percentage, possibly as much as half, of the present-day coleopterous fauna was already established in the British Isles. The spread of pine forest and later successions of forest types led to the immigration and spread within the British Isles of another important section of the present-day fauna. The apparent failure of spruce to enter the British flora at this time and the failure of beech to dominate any woodland areas in the south, at least until much later (Pennington, 1969), may well explain some very recent changes in the coleopterous fauna as well as the inability of some forest species to colonize Britain during the period of maximum forest cover. The eventual restriction, in the main, of pine forest to the northern parts of Britain produced a corresponding restriction of the many species of Coleoptera closely associated with pine. It is possible that some such species, if at all warm-adapted, were lost from the British fauna during the retreat of pine forest towards the north; evidence from the fossil record and elsewhere on this point is so far insufficient. The fossil record does already show, however, that some Coleoptera associated with pine were still present in the southern parts of Britain in periods considerably post-dating the local rise to dominance of deciduous woodland.

The timing of the severance of land connection with continental Europe is of some significance, especially in terms of flightless species or others with poor powers of natural dispersal. However, this break, at about 5500 B.C. (Pennington, 1969) post-dates the major changes in environment mentioned already, and is unlikely to have excluded many species from the British fauna. The many "lacunae" in the British coleopterous fauna (see Sainte-Claire Deville, 1930) are undoubtedly due largely to causes other than delayed post-glacial immigration prevented by a sea barrier. The much earlier severance of land connections between Britain and Ireland is probably responsible in much greater measure for the numerous lacunae of the Irish fauna. The paucity of woodland species in Ireland, for instance, is difficult to explain entirely in terms of local extinction or climate. That certain woodland species should be very local or rare in Ireland is not surprising in view of massive deforestation, but of approximately 650 "woodland species" of Coleoptera found in the British Isles, at least two-thirds are unknown from Ireland.

Little direct evidence is available regarding the fate, during the periods of almost complete forest cover, of species favouring "open" ground. Many of these species are likely to have been restricted to a few suitable sites, mostly on the coast. A few may have been totally eliminated from the British Isles and possibly reintroduced following the clearing of forests by man. The extent to which the fauna of open-ground species present in the British Isles before the post-glacial development of forest can be equated with the "culture-steppe" faunas of later, man-made environments cannot be argued at length here. However, one may suggest that this early "steppe" element in the British fauna cannot be precisely equated with a true steppe fauna (which exists under different climatic conditions) and probably even less with a British culture-steppe fauna, although a certain correspondence between this pre-forest fauna and that of man-made open spaces is to be expected.

MAN–MADE ENVIRONMENTAL CHANGES

Changes in the environment brought about by the early activities of man were many and varied. Since at least 5000 years ago such changes have had a pronounced effect on the British coleopterous fauna; in fact, throughout the period 5000 B.P. to the present human influence on the coleopterous fauna is likely to have considerably exceeded that of climate. One particular change— forest clearance—wrought by man from almost the earliest days of his colonization of the British Isles in post-glacial times has been of the greatest significance. The clearing of land by early Neolithic man is likely to have been on a fairly small scale but had an accumulative effect on the coleopterous fauna which continues today. The effect of forest clearance on the British flora and fauna has been the subject of much study, and possible effects on the coleopterous fauna have been discussed by Osborne (1965). Clearly, such a change in the environment will involve both retractions and extensions of range of species.

The effect of fragmentation of forest areas on forest-dependent species, at least of certain kinds, appears to be out of proportion to the actual amount of forest cleared, because of the relatively poor powers of dispersal of many species. It is the species associated, in the larval stage at least, with the wood of mature or dead trees and with arboricolous fungi that are hardest hit. Such species may suffer local extinction throughout an area in which forest has been fragmented. Species of the forest floor and those dependent on leaves, seeds and flowers are apparently able to persist more successfully in small pockets of forest or are able to recolonize such areas. Although the decline of forest species has been an almost continuous process from the time of neolithic man, it may be possible to pinpoint certain periods of particularly drastic change. Of fourteen species identified from British post-glacial sites but unknown as British in the recent period of entomological collections and records, twelve are closely associated with woodlands, many of them with the wood of mature trees (Table I). As more post-glacial assemblages are investigated this list may be expected to lengthen considerably. The twelve woodland species now presumed extinct in the British isles were all present as recently as 3500 years ago. Some extinctions of forest species may have occurred before this time but I suspect that a major series of extinctions may have been produced in the period beginning about 2500 years ago. These local extinctions still continue but the major changes in composition of the British forest fauna are likely to have been achieved by the Middle Ages. It should be mentioned that the extinction in the British Isles of certain woodland species may be due not only to the effects of man; climatic changes, notably oceanization of climate, may also have played a significant, if subsidiary, role.

In terms of the general composition of the British coleopterous fauna, the decline of forest species is overshadowed by changes of a quite different kind produced by the progressive clearance of the forests by man. The very beginnings of encroachment on the afforested areas gave opportunities for the spread of species which favour open ground. The creation of "culture-steppe" is likely to have evoked an immediate response in such species, pre-dating significant decline of forest species. The rapidity with which culture-favoured species can invade and colonize areas of newly cleared woodland is demonstrated by the work of Lindroth (1957, 1963) in North America. Although only indirectly applicable to the British situation, Lindroth's work probably furnishes the most useful basis, at the present time, for predicating the nature

TABLE I. Coleoptera from deposits dated post-9000 B.P. but not known in Britain today

Species	Usual habitat	Age of deposit	Source
Rhysodes sulcatus F.	Rotten wood	*ca.* 3000 B.P.	Buckland and Kenward (1973)
Agabus wasastjernae Sahlb.	Acid pools in coniferous woodland	Pollen Zone VIIb (Bronze Age)	Osborne (1972)
Porthmidius austriacus Sch.	Rotten wood	Pollen Zone VIIb (Bronze Age)	Osborne (1972)
Isorhipis melasoides (Lap.)	Dead wood	*ca.* 3000 B.P.	Buckland and Kenward (1973)
Prostomis mandibularis (F.)	Rotten wood	*ca.* 3000 B.P.	Buckland and Kenward (1973)
Zimioma grossa (L.)	Rotten wood	post 3000 B.P.	Buckland and Kenward (1973)
Mycetina cruciata Schall.	Fungi on rotten wood	*ca.* 3000 B.P.	Buckland and Kenward (1973)
Dermestes lanarius Ill.	Skins and other material of animal origin	Late Bronze Age	Osborne (1969)
Aphodius quadriguttatus (Hbst.)	Dung in sandy places	Late Bronze Age	Osborne (1969)
Cerambyx cerdo L.	Dead wood	*ca.* 4000 B.P.	Duffey (1968)
Hesperophanes fasciculatus (Fald.)	Wood	Roman	Osborne (1971b)
Eremotes elongatus (Gyll.)	Dead wood	Pollen Zone VIIb (Bronze Age)	Osborne (1972)
E. strangulatus Ferriss	Dead wood	Pollen Zone VIIb (Bronze Age)	Osborne (1972)
Pycnomerus terebrans (Ol.)	Dead wood	Pollen Zone VIIb (Bronze Age)	Osborne (1972)

of change in the British coleopterous fauna brought about by the earliest production of culture-steppe.

The work of Godwin (1956) and others suggests that a large proportion of present-day ruderal and weedy species of vascular plants were to be found in the British Isles immediately prior to any major environmental changes effected by Neolithic man. The fossil record of Coleoptera is not yet sufficient to accurately assess what percentage of culture-favoured beetles were present in the British Isles during the same period. It may be noted, however, that only about twelve culture-favoured species are so far recorded from post-glacial sites prior to about 8000 B.P., while some forty of these species are known from Bronze Age or Roman sites. Omitting from consideration those that are more directly synanthropic (i.e. dependent on Man's dwellings, or other buildings and constructions, on his stored products, his crops or rejectamenta) rather more than 300 species of Coleoptera found in the British Isles today may be regarded as markedly favoured by forest clearance. Most of these are predacious, generally ground-dwelling forms, of the families Carabidae, Staphylinidae, Silphidae, etc., or are phytophagous species associated with weed or ruderal vascular plants. Any movements of species into Britain at this time must be seen in the context of changes in continental Europe. It is likely that some species previously confined largely to true steppe conditions entered Britain following an advance across much of Europe from the east. Although other explanations are possible, some relatively recent immigrants to the British Isles may represent delayed movements of the same kind. It is notable that more than half of the British species, regarded as culture-favoured here, now occur as introductions in North America (Lindroth, 1957, etc.). At the same time it is clear that the list of species established in North America culture-steppe following introductions from Europe cannot be equated in a precise way with those species favoured by the early creation of culture-steppe in the British Isles. Apart from their ability to exploit man-made environments other factors discussed by Lindroth (1957) operated to determine which species of Coleoptera were transported by man to North America and were suited for colonization there. It should also be noted that culture-steppe in different regions may occur under a variety of geological and climatic conditions and that these factors appear to exert a considerable influence on the composition of a given culture-steppe fauna. Some species of Carabidae regarded as typical of woodland in much of Scandinavia appear to be much more capable in the British Isles of colonizing culture-steppe. Even within southern Britain some species rarely taken in the east except in woodland appear in much more open habitats in the west. As an example of the individuality of the culture-steppe of any given area, the distribution of the ground-beetle (*Pterostichus madidus*(F.)) is of interest. Although eurytopic it is a distinctly culture-favoured species and is probably the most familiar of "back-garden" beetles in Britain where it is generally distributed and often abundant. However, it is strictly confined to Western Europe (absent from Scandinavia and Central Europe) and has not become established in North America, although likely to have been imported there. Other culture-favoured European species, such as *Carabus auratus* L. appear, conversely, to be debarred from permanent membership of the British fauna. In each case, climatic factors are probably of major importance.

Another group of species much favoured by the activities of man is that characteristic of the accumulations of vegetable deposited, often as refuse, in the course of many human activities. This type of habitat is represented in the British Isles today by hay and straw stacks, some rubbish dumps, garden refuse and compost heaps, hot-beds, manure heaps, piles of cut grass or other vegetation. The sludge drying beds and filter beds of sewage works are habitats which in some respects are of a similar type.

In the most recent historical period a number of species characteristic of such man-made habitats are well-known as adventives, both in Europe (Horion, 1949) and elsewhere. Extrapolations suggest that a number of "plant-refuse species" present in the oldest entomological collections or known to early entomologists also first entered the British fauna, at an earlier date, as a consequence of man's activities in making accumulations of plant material. The movement of such species into the British Isles is likely to have begun following the first agriculture-based settlements of man; the process continues today. The documented recent spread of many plant refuse species from continent to continent demonstrates how frequently they may be accidentally transported by man; early immigrants to the British Isles may have arrived in this way or, perhaps more frequently, by natural means once established in adjacent areas of continental Europe. As well as providing a niche for immigrant species, the provision of man-made heaps of plant matter is certain to have enabled extensions of range and increases in abundance of certain species already present in the British Isles.

The identification of those species that are likely to have first found a place in the British fauna because of their close association with man-made plant refuse habitats requires the consideration of several types of evidence. Recognition of the different species associations to be found in the many variants of the plant-refuse type of habitat is of the greatest importance. Although a number of species, mostly members of the families Staphylinidae, Hydrophilidae and Histeridae, which are predators (and in some cases parasites) of dipterous larvae and small insects such as Collembola, are to be found in many types of habitat and are not confined to one particular variant of the plant refuse habitat, it is misleading to lump all such habitats together when examining species associations. Man-made plant refuse heaps (themselves divisible into several types) frequently contain species in unnatural association, species typical of or otherwise confined to differing natural habitats: decaying macrofungi, carrion, dung of herbivorous mammals, the nests of birds or mammals, etc. Species which exploit the plant refuse habitats made by man but which are likely to have been present in the British Isles before such habitats existed include most British species of *Omalium*, *Tachinus* and *Proteinus* as well as many species of Histeridae, Hydrophilidae and Staphylinidae of the genera *Aleochara*, *Anotylus*, *Atheta* and *Philonthus*. Some of these species, e.g. *Tachinus rufipes* (Deg.), have become established in recent times in North America or elsewhere as the result of accidental introduction by man. However, as with the species of culture-steppe, not all the species thus exported from northern Europe owe their original presence in the British Isles to the activities of man. As well as providing great concentrations of potential food, man-made plant refuse heaps of certain types contain micro-climates of a special kind. In particular, fermentation processes contribute to a high temperature. Species found, at least in the larval stage, only in such habitats and which are never fully exposed to the rigours of the British climate are almost certain to owe their introduction to man, Other species that are very largely confined to synanthropic situations of the plant refuse type are also likely to fall into this category. Data concerning the phylogenetic relationships, behaviour and habitat preferences outside the British Isles of such species is of great assistance in estimations of their geographical origins. It is notable that many of the species belong to groups which are otherwise restricted to the tropics and that they occur in warmer regions in habitats uninfluenced by man. Likely origins for most of the earlier immigrants to the British Isles and other parts of northern Europe are the Mediterranean area and the nearer parts of Africa and Asia. A number of the later immigrants, particularly those of the past 200–300 years, have come from much further afield.

However, the behaviour of certain recently introduced species in the British Isles demonstrates that by no means all adventive species found in man-made plant refuse habitats are confined in newly colonized areas to such situations. This is particularly true of species originating from areas with climates, at least in respect to certain features of temperature, not differing greatly from that of the British Isles. Species of this type, a number of which are fungus-feeders, are often to be found primitively in habitats of a very specialized kind but their ability to utilize man-made plant refuse habitats greatly facilitates their spread and colonization of new areas. Some species thus favoured by the activities of man are likely to have found a place in the British fauna as early immigrants.

Excluding species that are known or are very likely to have been introduced to the British fauna during the past 150 years or so (to be considered later), some 80 well established British species probably first entered the British Isles because of their ability to exploit man-made accumulations of plant material. I know of no records of any of these species as members of fossil assemblages from British sites of pre-Roman date but several (e.g. *Oxytelus sculptus* Grav., *Lithocharis ochracea* (Grav.), *Leucoparyphus silphoides* (L.) are recorded from one Roman site (Osborne, 1971b). The estimation of the past history in the British Isles of individual plant refuse species requires the presentation of a great body of data. It is hoped that such data can be included in a fuller discussion of this topic elsewhere.

Other environmental changes brought about by man prior to the Industrial Revolution, and in some cases prehistorically, also created habitats enabling species to enter the British Isles for the first time or to extend their ranges. The most striking examples are those species, many of them pests today, which inhabit and often feed on stored food and other products. Some such species will breed in the British Isles only inside buildings or stores and are clearly of exotic origin. Early immigrants of this type, as in the case of plant refuse species, are likely to have originated from the warmer parts of the Palaearctic region or adjacent areas in the Old World. An extensive literature is available on this topic and it will not be considered further here except to note that some species are now known to have been present in Britain at an earlier date than had previously been generally supposed. Osborne (1971b) records five species characteristic of grain stores from a Roman site in Warwickshire.

Most of the species intimately associated with larger animals are likely to have been present in the British Isles before widespread human activities, but the beginnings of animal husbandry and the later adoption of open pastures for the rearing of sheep and cattle cannot have been without effect on the many beetle species associated with the dung of herbivorous mammals. Although the occurrence of many dung-inhabiting species (*Aphodius*, *Cercyon*, *Sphaeridium*, etc.) transported from Europe to North America in recent times with livestock (Lindroth, 1957) demonstrates the ease with which such species are spread by man, the overall composition of the British dung-fauna is unlikely to have been greatly changed by early human activities. However, at least one probable extinction and one dramatic decline seem to be indicated from fossil evidence, between the Bronze Age and the present day (Osborne, 1969); such changes may, of course, be largely due to climatic factors.

Irrigation, drainage, mining and numerous other human activities all led to early environmental changes of significance to the beetle fauna. It is clear that most of the environmental changes brought about by man which are of great significance in the history of the British coleopterous fauna began very early. Many of the most crucial changes had already taken place more than 2000 years ago.

SOURCES OF DATA IN THE STUDY OF RECENT CHANGE

Evidence of change in the coleopterous fauna during the past 200–300 years can be obtained from insect collections and published records, but the earliest collections and published works contain data of only limited use. However, old collections and their documentation undoubtedly deserve more intensive and more critical study if progress is to be made in reconstructing the very recent history of the fauna. A few useful accounts of old collections are already available, for instance that of Walker (1932) on the Dale collection. In more recent years, Allen (1969, etc.) has contributed several critical evaluations of the identity and status of doubtfully British species represented in old collections. The collections of many early British entomologists have been lost or dispersed, but, in the case of Coleoptera, a few notable collections survive largely intact, foremost among them that of J. Stephens at the British Museum (Natural History). The collections of C. C. Babington (University Museum, Cambridge), Dale (University Museum, Oxford), Kirby and Marsham (both British Museum (Natural History)) and others also contain early material of interest. An earlier but little-known collection, that of Leonard Plukenet, a well-known botanist of his period, has recently been studied by the present writer. This collection, dating from 1696–97, contains 103 species of Coleoptera from the London area and provides some information on the local fauna of the period. The collections of early entomologists are highly selective in nature and are generally poor in the types of data (regarding distribution and abundance) most useful in the study of faunal history. However, the data which can be gleaned may complement those derived from the "natural" assemblages of archaeological sites, which are themselves selected but in quite different a manner.

In relation to their number, published works on British Coleoptera provide a disappointingly small amount of the kind of information needed in assessment of faunal change. There has been no dearth of competent systematists in Britain who have worked with Coleoptera and no great lack of interest in the group by ecologists and others concerned with species as members of communities, but professional taxonomists in the present century have taken relatively little interest in the local fauna and have contributed little to the systematic gathering of relevant data. A much greater involvement with the British fauna has been exhibited by several generations of amateur coleopterists, manifesting itself in much expert collecting and many published observations. However, lack of appropriate organization has meant that few really useful summaries of their data have emerged. Recent summaries by continental coleopterists of the biology and distribution of North and Central European Coleoptera (Horion, 1956, etc.; Lindroth, 1960) are of a kind not yet attempted on any scale in Britain.

The gathering of distributional data with regard to Coleoptera in the British Isles is still very much in its infancy. The few large-scale works incorporating such data base summaries on very gross units of distribution: the Irish "provinces" of Johnson and Halbert (1902) or counties and larger areas in Britain (Fowler, 1887–91). Later authors (e.g. Balfour-Browne, 1940, 1950, 1958; Moore, 1957), have largely used counties or vice-counties (see pp. 425–428). Despite the small number of competent and interested individuals who are able or willing to contribute data, following the early lead of British botanists, British coleopterists have recently begun to gather distributional data on the basis of 10 km square units. About 25 preliminary distribution maps of this kind have already been published (Holland, 1972; Hammond, 1971, etc.) and a further 75 have been prepared, of which some 25 are in press. The most intensively recorded species so far mapped is known from only 366 squares and the real information content of such maps compared with alternative forms of presentation (e.g. maps based on vice-counties; 50 km squares) has yet to be fully evaluated. However, the value of data-gathering schemes cannot be assessed purely in terms of cost-effectiveness in production of preliminary distribution maps. More important features are the provision of a framework for the long-term accumulation of distributional data which would otherwise be lost, encouragement of interest in the group, and the provision of a vehicle for the accumulation of badly needed data on usual habitats, seasonal occurrence, etc. Several data-gathering schemes concerned with Coleoptera are now in operation or about to be launched in conjunction with the Biological Records Centre of the Nature Conservancy and it is hoped that this important basis for any discussions of faunal change will be placed on a firm footing.

CHANGES IN THE COLEOPTEROUS FAUNA IN THE POST-LINNAEAN PERIOD

Understanding of the composition of the fauna and the number of species recognized has grown steadily since the time of the first major works on

British Coleoptera (Stephens, 1828–32). Impetus has been provided at various points by the publication of monographs, check-lists and key-works, notably those of Fowler (1886–91). The period since the most recent catalogue of British species, that of Kloet and Hincks (1945), is not atypical and may be used to illustrate the rate at which knowledge of the fauna grows and the difficulty of distinguishing real from apparent change in a situation where our knowledge of the composition of the fauna is constantly being revised.

I have noted published records for the British Isles, since 1945, of 193 species not included in Kloet and Hinck's check-list, omitting previously known but misidentified species, those which are probably of only casual occurrence and those of very local occurrence in indoor situations. In the same period I have noted newly recognized and published synonymies, corrections of misidentifications, etc., which together reduce the number of species recognized in Britain by 72. In the 28-year period since 1945 the net increase in species recorded from the British Isles is thus 121, raising the total of 3690 given by Kloet and Hincks (1945) to 3811. However, a number of these species should be regarded as "doubtful", either of casual occurrence only or probably extinct, and should be added to the 21 species so classed by Kloet and Hincks (*op. cit.*). If other necessary additions and deletions known to me are considered a figure of approximately 3730 is arrived at for established British species and an additional 70 species are placed in the "doubtful" category.

Not all of the 193 species first noted for the British Isles since 1945 can be regarded as genuine additions to the British fauna; a number are certainly or very probably synonymous with species already known in the British Isles or records for them are based on misidentifications. However, there is little doubt, on taxonomic or other grounds, that about 145 of these are good species which are established here. Most of these additions represent an increase in taxonomic understanding of the groups in question (especially Clambidae, Cryptophagidae, Piliidae, Staphylinidae, etc.), and the same is true of most of the deletions. The use of new collecting techniques and investigation of previously unworked areas are also responsible for a certain number of additions. Utilizing the criteria proposed by Lindroth (1957) for regarding a species as an introduction, it is clear that about 20 of these species are genuinely new to the British fauna, although not necessarily since 1945, and have not merely been previously overlooked. A further ten species are probable introductions and several others, for which the evidence is not at all conclusive, may fall into this category.

That the majority of the species first recorded for the British Isles in the post-1945 period are long-standing members of the British fauna is indicated, not only by circumstantial evidence, but also by the presence of British specimens in old collections. Circumstantial evidence indicating a long British history for some species not (to my knowledge) yet known from old British collections is, nevertheless, very great. Species such as *Coeliodes nigritarsis* Hartm., *Gyrophaena pseudonana* Strand, *Hypocyptus hanseni* Palm, *Nebria nivalis* Payk., *Ostoma ferrugineum* L., *Phyllodecta polaris* Schneider, *Rhizophagus parvulus* Payk., *Simplocaria maculosa* Er., *Stenelmis canaliculata* Gyll., etc., have all been taken in circumstances which suggest populations of long standing in Britain. These species have no known synathropic tendencies and, in view of their known distribution in continental Europe, might be expected to occur in the British Isles. Some of them are represented by populations of the relict type.

The recent British history of some other species not present in old British collections is not so clear. Certain species, especially those whose northern limit of distribution in Europe is at about the level of southern Britain, may possibly occur in Britain as a result of recent, possibly natural, extensions of range, e.g. *Axinotarsis marginalis* Lap., but the majority of these too are more likely to have been previously overlooked, often because of extreme localization or retiring habits, e.g. *Eucinetus meridionalis* (Cast.), *Hypocoelus cariniceps* Reitter, *H. olexai* Palm, *Lagria atripes* (Muls. & Guille.), etc. A particularly interesting case is presented by three species, *Omophron limbatum* F., *Heterocerus hispidulus* Kies. and *Dyschirius obscurus* Gyll., recently found to be well established in the same locality on the Sussex coast (Farrow and Lewis, 1971; Allen, 1971). *O. limbatum*, a particularly distinctive species, and *H. hispidulus* were previously unknown from Britain and are not present (to my knowledge) in any old collections; *D. obscurus* was not previously known from this part of the coast. The ocurrence of these three species together on a part of the south coast, where beetles had previously been much collected, is certainly suggestive of an importation, but the continental distribution of all three is such that they may have been long established in this country and merely overlooked. In this instance, as in a number of others, no conclusive evidence is presently available.

1. Factors in Recent Change

The activities of man have undoubtedly been the principal agents of change in the British coleopterous fauna since 1800. Climatic factors are likely to have played a part in adjustments of range, some of which may have involved extinctions or the immigration of species new to the fauna. Changes in the distribution or abundance of certain species are likely to have resulted from changes within populations of such species which are not a direct expression of environmental change.

As changes in individual species may be illustrative of the nature of change in the fauna as a whole, examples may assist in the identification of the principal factors involved.

2. *Contractions of Range, Declines in Abundance and Extinctions*

As the available evidence for the decline of species is both negative and largely anecdotal such changes are particularly difficult to prove. However, there are certainly a number of species recorded from parts of the British Isles Isles in the nineteenth century but for which the same areas are without subsequent records. Unfortunately, there is little or no reliable information concerning the frequency or abundance of species at the time of the old records and it is difficult to estimate whether records based on one or only a few individuals are representative of established breeding populations.

An accumulation of anecdotal evidence suggests a marked and continuing decline of species typical of deciduous forests, especially these species associated with mature trees or dead wood. Species dependent on old forests ("Urwaldtiere"), particularly members of the families Cerambycidae, Cleridae, Colydiidae, Cucujidae and Elateridae, appear to have continued the decline begun in prehistoric times. As such species were already represented in the British Isles 200 years ago by essentially relict populations in widely separated localities, the elimination of a species from one locality may have represented a considerable, and often permanent (see Buckland and Kenward, 1973) contraction of range. However, the persistence of some species in confined areas is well known and is illustrated by several instances of distinctive species thought to be extinct in the British Isles only to be rediscovered, after a considerable interval of time, often in the localities where originally found. *Lacon querceus* (Hbst.), for example, a large and highly distinctive species, had not been found in Britain for more than a century when rediscovered in Windsor Forest by Allen (1936). Intensive collecting in the Windsor Forest area by Donisthorpe and others has produced a number of re-discoveries of Urwaldtiere (Donisthorpe, 1936–38). Other notable rediscoveries of forest species include *Diaperis boleti* (L.), *Globicornis nigripes* (F.) (Allen, 1947; Woodroffe, 1971), *Platydema violaceum* (F.) and *Uleiota planata* (L.) (Welch, 1964). In the same category belong species previously undetected although undoubtedly present and first discovered only in recent years. *Ostoma ferrugineum* (L.) is the outstanding example among the Urwaldtiere. Other discoveries of species associated with forest conditions or with woodland trees include *Epuraea biguttata* (Thunb.), *Ernoporus caucasicus*

M

Lind. and *Leperesinus orni* Fuchs (Allen, 1970), *Eucinetus meridionalis* (Cast.), *Hypocoelus* spp. (Allen, 1969*a*) and *Rhizophagus parvulus* Payk. (Johnson, 1963). One of these, *E. caucasicus*, is known from a British fossil assemblage (Kelly and Osborne, 1964). A number of old forest species and other species associated with mature trees, probably upwards of 20 in all, are likely to have become finally extinct in the British Isles during the period since 1800. Probably to be included in this category are *Plagionotus arcuatus* (L.), *Platycerus caraboides* (L.), *Mycetophagus fulvicollis* F., *Rhynchites auratus* (Scop.) and *R. bacchus* (L.) (Allen, 1962), *Strangalia attenuata* (L.), *Leptura virens* (L.) and *Corymbites cruciatus* (L.) (Allen, 1969a, b). The decline of many species of Cerambycidae is documented by Kaufmann (1946, 1948). The altered ratio of woodland to open countryside since 1800 must have been detrimental to all woodland-favouring Coleoptera, but apart from the old-forest species (most properly regarded as relics in the British fauna), many woodland species, especially those characteristic of the ground layer and soil, appear to persist wherever their habitats remain. Some, including a number of phytophagous species, have maintained populations in hedgerows and gardens. As certain species which are rarely found outside of woodland in the east may be found in the open in the west, the destruction of woodland habitats may be expected to have had a differential effect on such species according to the part of the country.

The relict populations of those species restricted, in the British Isles, to a few localities in the north are likely to have been isolated from continental populations for at least 10 000 years, and as such represent important biological capital. In some cases morphological peculiarities of members of British populations, (e.g. in *Eudectus whitei* Sharp) have led some coleopterists to claim specific distinctness for the British forms. Many of these northern relics may have suffered declines or elimination of some populations in recent historical times, but most of them occur in habitats (high moorland, mountain tops, etc.) apparently relatively free from human interference. However, at least two species, *Stenus glacialis* Heer and *Agonum sahlbergi* Chd., have not been found in the British Isles for more than 50 years. Populations of the majority of these cold-demanding northern relict species appear to be hanging on in the north of the British Isles despite the small areas they occupy; survival of relict populations of warmth-demanding species, isolated since the post-glacial climatic optimum, is also illustrated by the several species of this type still to be found in parts of eastern Scotland (Crowson, 1967) although apparently separated, in many cases, by hundreds of miles from the nearest populations to the south.

Lowland heath and chalk downland, both largely man-made in origin, have been increasingly utilized by man for new purposes, principally arable farming, plantations of conifers and building, during the past 200 years. New practices in animal husbandry and the activities of rabbits have also played a part in promoting change in such habitats. Where local habitat destruction is complete the many species characteristic of these habitats are eliminated and become increasingly localized in occurrence. Although few of the chalk down-land species appear to be threatened with total extinction in the British Isles some of the species typical of lowland heath and other sandy situations appear to have suffered severe declines, not just of a local nature, during the present century. Ground-beetle species of the genera *Amara* and *Harpalus*, at least those largely confined to open, sandy situations are a case in point. Habitat destruction may not be the only factor involved. Species such as *Amara fulva* (Deg.), *A. lucida* (Dufts.) and *Harpalus sabulicola* (Panz.), all found in open, sandy places and often commoner near the coast appear to have suffered dramatically in the past 50 years. Sandy coastal situations, with many species of Coleoptera in common with inland sandy heath, have also been much affected by man's activities. The development of seaside resorts, begun on a large scale in Victorian times, and other new uses of parts of the coast-line have frequently involved radical changes in areas with a characteristic fauna. Sand-dune systems, cliffs of sand or clay and some chalky or rocky cliffs have all suffered. A number of the species of Coleoptera to be found in such habitats are at the northern edge of their range in the British Isles; holiday resorts and other developments on the south coast of Britain have led to extreme localization of some species and to the possible extinction of others. The "cosmetic" treatment accorded to some chalk, sand and gravel pits in recent years has also led to the local elimination of already highly localized species which are to be found in such situations.

Much of the draining of fens on a large-scale preceded the period for which we have records and collections of Coleoptera but a continued contraction of range is indicated for most fen-favouring species during the period since 1800. The only aquatic species to have become probably extinct in this period, such as *Agabus striolatus* (Gyll.), *Rantus aberratus* (Gemm. and Har.) and *Hydaticus stagnalis* (F.), are principally associated with fenland situations. A number of species are known in the present century from only one or two small areas of East Anglian fen; some of these may already be extinct while others should be regarded as endangered relicts.

A marked decline in certain dung-feeding Coleoptera during the past 100 years has been suggested by a number of authors (e.g. Johnson, 1962) and certain species of Scarabaeidae (e.g. *Onthophagus nutans* (F.), *O. fracticornis* (Preyssler)) may have become extinct in the British Isles during this period.

Only a minority of dung-feeding species appear to be so affected and most of these apparently favour sandy soil into which they dig burrows. The factors involved are difficult to identify but may include the conversion of sandy pastures to other uses, the feeding of cattle and horses in richer pastures and a decline in the number of sheep grazed on heathland and other sandy areas.

Increasingly hygienic care of many buildings and areas of human habitation is likely to have had a variety of effects. Among species not regarded as pests, a general decline in those which may be regarded as typical of old cellars, most of these not native species, e.g. *Blaps* spp. (Allen, 1972), *Sphodrus leucophthalmus* (L.) and small mould-feeding species such as *Orthoperus atomarius* (Heer) appears to be indicated.

Other activities of man which have undoubtedly led to the reduction of populations of many species are largely centred around changes in agricultural practices and increased urbanization. Changes in animal husbandry, the draining of ponds and ditches, removal of hedges, the use of insecticides and herbicides, etc., are all likely to have had an influence in fragmenting populations. Insecticides are likely to have had a considerable effect on predatory as well as phytophagous species which are directly or indirectly exposed. Apparent declines of many of the larger predacious species of beetle in agricultural areas may be due, at least in part, to the effect of insecticides. A fall in frequency and abundance appears to be particularly marked in genera such as *Carabus*, *Silpha* and *Staphylinus*. Although a variety of factors may be involved, the dramatic decline of such species as *Anisodactylus binotatus* (F.) and *Zabrus tenebrioides* (Goeze), which are not restricted to habitats much encroached upon in recent years, may be due in part to insecticides and herbicides. As noted by Lindroth (1972), the contraction of range of species following the industrial revolution in northern Europe is perhaps most marked in vertebrate animals and vascular plants. Unless an already localized type of habitat is destroyed most insects do not appear to be affected so severely by industrial development. Aquatic species are undoubtedly adversely affected or even eliminated locally by pollutants in fresh water but as probably all aquatic species are also to be found in the many unpolluted waters such pollutants have only a local effect. Any direct effect of air-borne pollutants on beetles is likely to be small but severe pollution of this type undoubtedly influences at least certain species through the now well-documented effects on plants (see Chapters 3, 4). That many ground-living, and by no means necessarily culture-favoured species, are able to flourish in conditions of heavy air pollution is demonstrated by the number of species of this type found in industrial areas wherever parks or other spaces provide suitable habitats. In many parts of the London suburbs where air pollution is moderately severe, gardens contain an extremely rich and varied Coleoptera fauna. Despite extreme artificiality many such gardens, from the point of view of the beetle fauna, appear to represent a good substitute for naturally occurring mixed deciduous woodland. Almost half of the British Coleoptera fauna have been taken by careful and persistent collecting over a period of years in one suburban garden at Blackheath by A. A. Allen. Any effects of pollutants in favouring the development of melanic varieties of beetle species are contentious. The polymorphic ladybird *Adalia bipunctata* (L.) may possibly fall into this category but the case remains very much to be proved.

Most of the factors involved in recent declines of species of Coleoptera have exerted their effects more severely in the southern parts of the British Isles than elsewhere. As many species are apparently restricted by climatic factors to these areas, those under the greatest pressure from modern environmental changes are threatened with extinction in the British Isles. Although short-term climatic changes and a variety of man-made environmental changes are also involved, the destruction of special habitats is probably the principle factor in the decline to extinction of species in the south of Britain.

3. Expansions in Range and Increase in Abundance

Except very locally, or over a period of a very few years, increases in abundance are difficult to prove as adequate data are lacking. Equally, records of a species from new areas are by no means necessarily testimony to an expansion of range. Only when patterns or trends based on a number of records become apparent can we have any confidence that these represent genuine changes. In the case of apparent changes in range we are also faced with the problem of distinguishing dispersal from colonization (see Lindroth, 1972). Some species frequently disperse to areas which are not colonized and "good" years when dispersal is particularly extensive may give an impression of expansion although the breeding range of the species has remained more or less constant. On the other hand, however, a change in the range through which a species is found as a migrant may be produced by a change in its breeding range. The absence of recent British records for such species as *Calosoma sycophanta* (L.) and *Polyphylla fullo* (L.), known from Britain in the nineteenth century as occasional migrants, may be due to a recent contraction of their continental ranges.

A general increase in abundance and sometimes an expansion of range in the British Isles of many culture-favoured species would have been expected to have occurred in the period since 1800. For example, the spread of some weeds and cultivated plants has enabled comparable extensions in range of species associated with them. At least a relative increase in abundance is likely to have taken place, especially in southern Britain, in those species best adapted to urban and intensively cultivated conditions. Most notable expansions of range within the British Isles are also likely to be attributable to man-made environmental changes. In many cases the creation of suitable habitats in areas where none previously existed has paved the way for the colonization of these areas by species previously absent. For example, many species characteristic of coniferous forests and known in the British Isles 100 years ago only from Scotland, have been noted during the present century from southern Britain and in some cases also from Ireland. Some examples of such expansions of range are given in an earlier survey of changes in the British beetle fauna (Blair, 1935). To such species as *Asemum striatum* (L.) may be added *Abdera triguttata* (Gyll.), *Brachonyx pineti* (Payk.) (Morley, 1941), *Nudobius lentus* (Grav.) (Atty, 1964, etc.), *Ips suturalis* Gyll. (Donisthorpe, 1933), *Pissodes* spp., *Phloeostiba lapponica* (Zett.) (recently found near Brandon, West Suffolk) and *Rhinomacer attelaboides* F. The Breckland area of East Anglia is particularly rich in species previously unknown from the south. Some other species favouring coniferous forest but not so markedly restricted in their previous distributions have become more widespread and abundant, especially in the south. Examples in this category are to be found in the genera *Rhizophagus*, *Pityophagus*, *Epuraea*, *Pissodes* and *Hylobius* and in the families Anobiidae, Cerambycidae, Coccinellidae and Scolytidae. Although in terms of the British Isles expansions of range of species associated with conifers are largely from north to south it is likely that many new southern populations are not derived from those in the north but are of continental origin.

The recorded range of some species not associated with coniferous trees has also extended southwards in recent years but it must remain doubtful whether new records of such species as *Acrulia inflata* (Gyll.) and *Atheta eremita* (Rye) (Hammond and Bacchus, 1972) represent genuine expansions of range. Apart from certain pronouncedly synanthropic species, some of which are probably relatively recent colonists in the British Isles and still in the process of expansion here (e.g. *Oligota parva* Kraatz), it is also doubtful whether many of the species for which new records increase the known range northwards are genuinely spreading. Some apparent increases in abundance or expansions of range of species typical of deciduous woodland, e.g. *Diplocoelus fagi* Guér., *Pediacus depressus* (Hbst.), *P. dermestoides* (F.) and *Silvanus bidentatus* (F.) are, however, difficult to explain.

Many minor environmental changes brought about by man are likely to bring new species into an area. If the recent increase in number of records of two species associated with burnt wood, *Acritus homoeopathicus* Woll. and *Pterostichus angustatus* (Dufts.), are indicative of a genuine change in their status then man-made factors are likely to be involved. Completely new habitats are sometimes produced which unexpectedly provide suitable habitats for beetle species. The concentrations of chicken dung resulting from battery methods of chicken farming have provided a habitat in which *Alphitobius diaperinus* (Panz.) has spread widely in the British Isles during the present century. The factors involved in the rapid spread and increase in abundance of some species are difficult to identify. The weevil *Barypithes pellucidus* (Boh.), which has exhibited a dramatic expansion of range in the past 50 years, might be considered a case in point, although, as perhaps in some other cases, this species may be a relative newcomer to the British fauna.

An expansion of range of several adventive species, some at least fairly long established as members of the British fauna, appears to be based on changes in habitat or extensions of habitat range. *Alphitophagus bifasciatus* (Say), for example, was known to nineteenth century British entomologists only as an indoor species, most usually found in grain stores. In recent years this species has been taken increasingly frequently in outdoor situations, often of the man-made plant refuse type (compost, etc.) but also on several occasions in fungi and under bark (e.g. Allen, 1948). An apparent change-over to outdoor habitats is also exhibited by *Ahasverus advena* (Waltl) (e.g. Donisthorpe, 1936), *Carcinops pumilio* (Er.), *Carpophilus sexpustulatus* (L.) and *C. marginellus* Motschulsky. Most of the species exhibiting these extensions of habitat range are relative newcomers to the British fauna and apparent changes in habitat preference may often be due to no more than the "settling down" characteristic of many introduced species. In some cases, however (e.g. *Alphitophagus bifasciatus*), an endogenous change in the species itself may be involved.

4. Recent Colonists

The task of recognizing which species of Coleoptera have become established members of the British fauna during the period since 1800 is not easy, but, utilizing the five criteria for assessing immigrant status proposed by Lindroth (1957), with modifications to suit the British situation, most recent colonists can probably be identified. Lindroth's first (historical) criterion is of great value in a region as well studied as the British Isles. The approximate location and date of establishment of the earliest British colonies of the larger and more distinctive species may often be adduced from the historical record. In some cases the subsequent spread of such species can be monitored to a certain extent as it takes place. A second (geographical) criterion is applicable, in a different form from that used by Lindroth, in suggesting recent immigration of species newly discovered in the British Isles but otherwise known only from relatively distant parts of the world. The identification as likely immigrants of species dependent on man-made habitats is probably the most useful application of the third (ecological) criterion, although, except where such habitats derive from human activities only of very recent date, this criterion alone cannot tell us at what point in time successful colonization took place. The fourth (biological) criterion is of limited application but can provide conclusive evidence of exotic origin of species specifically associated with another organism which is itself known to be introduced. Infraspecific variation as utilized in Lindroth's fifth (taxonomic) criterion is also of limited application although it may be of some potential value in identifying immigrant populations of species represented already in the British Isles by isolated, and possibly morphologically distinct, populations. Another, equally "taxonomic" criterion is, however, of value when a suspected immigrant species, unknown from any other country, can be shown to belong to a species group peculiar to another continent or to distant parts of the Eurasian land mass.

Horion (1949) grouped adventive species of Coleoptera in Europe in three categories: those associated with stored products, those found in plant refuse, and phytophagous species. However, not all successful immigrants in the British Isles can be fitted into these three classes and additional classifications of newcomers to the British fauna may profitably be constructed. Geographical origin, feeding habits, method of entry, degree of success in colonization and environmental changes (if any) necessary to such success, as well as the nature of habitats occupied, all form useful bases for categorization.

Species first established in the British Isles in post-Linnaean times have entered the British fauna for a variety of reasons. Some have probably entered and become established in this country as part of a general expansion of range in northern or western Europe and exhibit the same kind of tendencies as those expanding their range from centres within the British Isles. Such expansions of range to include part of the British Isles may involve climatic factors, factors endogenous to the species themselves and man-made environmental changes, but in the original area of distribution and in the recipient area. In these cases, the mode of immigration of species from Europe often remains problematic; it is difficult to establish which species have travelled to Britain by natural means of dispersal (by air, rafting on the sea, etc.) and which have been brought accidentally by human transport. Most of the successful colonists from Europe are good fliers and some are certainly able to cross from adjacent parts of continental Europe when conditions of wind and temperature are suitable (e.g. Hurst, 1970). The nature of dispersal flights and the differential ability of species to withstand exposure to sea water are important to any discussions of the possible routes of arrival of immigrants from Europe.

Immigrant species from distant continents and many of those from parts of Eurasia undoubtedly rely on accidental transport with man, mostly by ship, for their arrival in the British Isles. Some species from far afield, establishing themselves first in parts of Europe, may finally enter Britain without human assistance.

The list of immigrant species which are likely to have first established breeding populations in the British Isles since 1800 is a long one. Documentation of immigrant status and British records in a collated form are unavailable for the majority of these species, but as a collation of much relevant data will be presented elsewhere a summary only will be given here.

Species which are more or less confined to indoor situations are well represented among recent immigrants. Many of these remain localized, often more or less restricted to the vicinity of their point of entry, in warehouses and stores, while some, although continually reimported, establish only temporary colonies. The most successful species of this type, often pests, spread widely and become sufficiently abundant to be fairly frequently found outdoors, or even become established in some outdoor situations, for example *Dermestes peruvianus* Lap. and *D. haemorrhoidalis* Küster (Allen, 1960; Bezant, 1963). Recent colonists of habitats largely isolated from the rigours of the British climate, in dwellings, stores, hot-houses, etc., are numerous and frequently of considerable economic significance. However, as such species are likely to mix with and impinge to only a small extent on our native flora and fauna, they are omitted from further discussion here.

Some of the most successful of recent colonists are listed in Table II. All of these species have established themselves, since 1800, in outdoor situations where, to a greater or lesser extent, they mingle with native species of Coleoptera and have spread widely within the country.

A high proportion of the species listed originate from distant parts of the world and are likely to have been brought to the British Isles accidentally by man. Some are clearly highly "potent" species which are often in the process of general expansion but the success of others in colonizing the British Isles following a chance introduction can be interpreted only in terms of adaptation to specialized niches, previously unfilled in the British Isles.

Many of the most successful colonists from distant parts are plant refuse species of the type which have flourished in the British Isles in man-made habitats since pre-Roman times. A number of these, species such as *Philonthus*

TABLE II. Particularly successful colonists among species of Coleoptera first established in the British Isles in the period since A.D. 1800

	Date of first British record	Country of region of origin	Present range	Distribution in the British Isles (1973)	Usual habitat and food
Amara montivaga (Sturm)	1948	W. Palaearctic	W. Palaearctic	Scattered, mostly S. England	Open, sandy and chalky places; predacious
Anthicus tobias Marseul	1935	?Asia Minor	Palaearctic (?)+	Scattered, England only	Rubbish dumps, manure heaps, etc.; predacious
Aridius bifasciatus (Reitter)	1949	Australia	Australia, New Zealand, Britain (?)+	General in most of England; also in Scotland	Almost ubiquitous; a mould-feeder
A. nodifer (Westwood)	1839	Australia	Almost cosmopolitan	Throughout the British Isles	Almost ubiquitous; a mould-feeder
Atheta immigrans Easton	1968	?New Zealand	Britain(?)+	Scattered; coast of Scotland and S. England	Sea-shore; probably predacious
Atomaria lewisi Reitter	1938	?E. Asia	Almost cosmopolitan	Throughout the British Isles	Varied; probably a mould-feeder
Bohemiellina paradoxa Machulka	1952	?E. Asia	S.E. Asia, Europe (?)+	Scattered, England only	Manure heaps, etc.; probably predacious
Brachypterolus vestitus (Kies.)	ca.1929	S. Europe	Europe (?)+	Widespread in S. England	Feeds on flowers of *Antirrhinum majus*
Caenoscelis subdeplanata Bris.	1954	W. Palaearctic	W. Palaearctic	Scattered, England only	Fungoid wood, compost, etc.; probably a mould-feeder
Carpelimus incongruus Steel	1968	New Zealand	New Zealand and Britain	S.E. England	In wet sand and chalky soil; probably feeds on algae
Cercyon laminatus Sharp	1959	E. Asia	Palaearctic and Oriental regions	Scattered, England only	Compost, etc.; predacious

Species	Year				Notes
Ceuthorhynchus turbatus Schultze	1949	W. Palaearctic	W. Palaearctic	S. England	Feeds on Cardaria draba
Cis bilamellatus Wood	1884	Australia	Australia and Britain	General in England also in S. Scotland	Feeds on Piptoporus betulinus
Corticeus fraxini (Kug.)	1922	W. Palaearctic	W. Palaearctic	S. England	Under bark of coniferous trees; predacious
Cryptopleurum subtile Sharp	1966	E. Asia	Palaearctic and Oriental regions	Scattered, England and Wales only	Compost, etc.; predacious
Euconnus murielae Last	1942	?E. Asia	W. Europe (?)+	Scattered, England only	Rotting straw, etc.; probably predacious
Harmonia quadripunctata (Pont.)	1937	W. Palaearctic	W. Palaearctic	Widespread in England	Associated with coniferous trees; predacious
Leistus rufomarginatus Dufts.	1942	N. Europe	N. Europe	Widespread in S. England	Deciduous woodland; predacious
Lithocharis nigriceps Kraatz	1955	E. Asia	Palaearctic and Oriental (?)+	Widespread in England	Plant refuse of various types; predacious
Mecinus janthinus Germar	1948	Europe	Europe	S. England	Feeds on Linaria vulgaris
Philonthus parcus Sharp	1961	E. Asia	Palaearctic, Oriental and Australia (?)+	Widespread in England	Compost, etc.; predacious
P. rectangulus Sharp	1921	E. Asia	Almost cosmopolitan	Throughout the British Isles	Compost, dung, etc.; predacious
Plegaderus vulneratus Panz.	1962	Europe	Europe	Widespread in England	Under bark of coniferous trees; predacious
Pragensiella marchii Dodero	1936	?Asia	Palaearctic (?)+	Widespread in England	Compost, etc.; probably predacious
Stenopelmus rufinasus Gyll.	1921	N. America	Nearctic and Palaearctic	Widespread in S. England	Feeds on Azolla filiculoides
Teropalpus unicolor (Sharp)	1900	?New Zealand	New Zealand, Australia, S. Africa, England	S. England	Sea-shore; probably feeds on algae

parcus Sharp, *P. rectangulus* Sharp, *Lithocharis nigriceps* Kraatz, *Bohemiellina paradoxa* Machulka, *Pragensiella marchii* (Dod.), *Euconnus murielae* Last, *Cercyon laminatus* Sharp and *Cryptopleurum subtile* Sharp, are predators, mostly feeding on dipterous larvae or small arthropods such as Collembola. Although these species have all undoubtedly spread from country to country because of their synanthropic tendencies, not all of them are confined in the British Isles to man-made habitats; *Philonthus rectangulus* is frequently found in the dung of herbivorous mammals and *Lithocharis nigriceps* may be found in a variety of natural accumulations of plant refuse as well as in compost, etc. Although some of these species have not previously been identified as adventives in this country they are all likely to originate from the same general area in Eastern Asia. Many of the other plant refuse species are probable mould feeders. Species of this type are well suited to accidental transport by man but many are not successful colonists or, as in the case of *Aridius australicus* (Belon) and *Metoph-thalmus serripennis* (Broun), achieve only a limited success in establishing themselves outdoors. However, three probably mould-feeding species have been strikingly successful. *Atomaria lewisi* Reitter and *Aridius nodifer* (Westwood) are today found in almost all temperate regions of the world, the latter species also in parts of the tropics; both are now generally abundant throughout the British Isles. *Aridius bifasciatus* (Reitter) has achieved a striking success in the short period since the first recorded British capture in 1949 (Fig. 1). Although first described from Germany (in imported Australian tobacco) and also known as an accidental import in Belgium and France, Britain remains the only country in the Northern Hemisphere where *A. bifasciatus* is known to be established. In at least the southern half of Britian this species is now almost ubiquitous. Unlike many other mould-feeding species it is not restricted to indoor situations, to man-made accumulations of mouldy vegetable matter or to specialized habitats such as wood in a certain stage of decay, and its abundance, in terms of biomass, is perhaps unlikely to be matched in southern England by many recent immigrants, animal or plant. A few recent colonists of the plant refuse type of habitat in Britain originate from Europe or adjacent parts of Africa and Asia. *Caenoscelis subdeplanata* Bris., a species until this century found only in association with fungoid wood, appears to have spread westwards through Europe with an accompanying extension of habitat range. The case of *Tachinus flavolimbatus* Pand., found in Britain principally in synanthropic habitats such as compost, demonstrates how all the criteria for recognizing a species as an introduction should be examined. Its occurrence in man-made plant refuse habitats and the relatively recent date of the first recorded British captures, 1939 (Steel, 1961), are suggestive of an immigrant. However, great superficial similarity to a common native species, the relative scarcity of *T. flavolimbatus* in Britain and its distribution abroad (Atlantic European north to northern France) suggest that it may have been overlooked here. Although some recent expansion of range is possible, British specimens dating from the nineteenth century are now known and colonization of Britain in the period since 1800 appears unlikely.

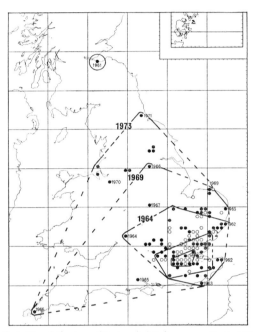

FIG. 1. Records of *Aridius bifasciatus* (Reitter) (Coleoptera, Lathridiidae) in the British Isles.

The lines represent maximum recorded dispersal at five different dates: 1954, 1959, 1964, 1969, 1973. Open circles represent records for dates later than that of the enclosing line.

More colonists occupying plant refuse habitats are to be expected in the British Isles and it may not be too rash to predict that many of them, like those of recent years, will come from Eastern Asia, Australia and New Zealand.

Several immigrant species from far afield, although undoubtedly transported to the British Isles by man, have become established in habitats largely un-influenced by man. The spread of such species within the British Isles is likely to have been almost exclusively by their own powers of dispersal. An example is *Cis bilamellatus* Wood, one of our few immigrant species whose spread has been the subject of some study (Paviour-Smith, 1960). The chance importation of this species from Australia, possibly in herbarium specimens, quite unpredictably led to successful colonization, in the fungus *Piptoporus betulinus* (syn. *Polyporus betulinus*), of much of Britain. *Teropalpus unicolor* (Sharp) (almost certainly of New Zealand origin) and *Carpelimus incongruus* Steel (now identified as a New Zealand species) are both equally unpredictable but successful colonists. Another probable immigrant, *Atheta immigrans* Easton, recently described from Scotland and also known in England, has not been recognized outside the British Isles. The sea-shore habitat favoured by this species provides few clues as to its origin but, as very similar species are known from New Zealand, that country may again be the source. The unpredictable nature of colonization by non-synanthropic species of this type is further demonstrated by their apparent lack of success elsewhere in the Northern Hemisphere. Although well established in Britain for more than 70 years, *Cis bilamellatus* and *Teropalpus unicolor*, like the more recently imported *Carpelimus incongruus* and *Atheta immigrans*, remain unknown from continental Europe.

Two other immigrants from other continents are phytophagous species which undoubtedly owe their introduction to importation with their host plants. *Stenopelmus rufinasus* Gyll. has become widespread in southern Britain and elsewhere in western Europe during the present century and is probably to be found in most places where its food plant, *Azolla filiculoides*, is established. The size of populations may be very great: Ashe (1939), for example, provides a "conservative" estimate of 500 000 individuals on one small *Azolla*-covered pond approximately 15 yd in diameter. *Syagrius intrudens*, Waterh., a less successful colonist, probably of Australian origin, has been known in the British Isles since 1903 on exotic ferns, and more recently, has been taken on bracken (Philp, 1963).

Many wood-feeding and other Coleoptera are accidentally imported to the British Isles with timber but few of the species from other continents are very successful in establishing themselves in the wild here. Many which are regularly taken, for example *Buprestis aurulenta* L. from North America (Shaw, 1961), do not appear to breed here. Some establish themselves only in timberyards whilst others, such as *Pycnomerus fuliginosus* Er., *Lyctus sinensis* Lesne and *Paratillus carus* Newman (from Australia and a predator of *Lyctus* spp.), have at least a limited success as breeding species in the wild. Species entering Britain with imported timber from southern Europe, for example *Aulonium ruficorne* (Ol.) (Allen, 1965), generally appear to meet with a similar lack of success. One wood-boring Cossonine weevil, *Euophryum confine* Broun (from New Zealand), is exceptional. Although most commonly found in the timber of buildings it can also be taken in dead and dying wood outdoors and has spread to most parts of the British Isles in the past 30 years. It is commonly observed that few New World organisms are successful colonists of the Old World. Certainly, few British Coleoptera appear to be of New World origin. Apart from *Stenopelmus rufinasus*, the few North American species to have gained a foothold in the British Isles in post-Linnaean times are associated with timber, for example *Epiphanis cornutus* (Esch.), *Lyctus planicollis* Lec. and *L. cavicollis* Lec., but are not generally well established in the wild. A further possible exception is a Nearctic species of *Psammobius*, recently found on sand-dunes in Lancashire, but it is not yet known whether this capture represents a breeding population.

Many of the species first established in the British Isles since 1800 and derived from northern or western Europe are associated with coniferous trees. Although some species of this type were probably imported to the British Isles fairly frequently during the nineteenth century, in logs, often those used as pit-props, or even in those washed up on the shore (Bartlett, 1921), successful colonization is largely a feature of the last 50 years or so. An important factor may be the extensive planting of conifers, including spruce and other species not native to this country, in southern Britain during the present century. The creation of suitable habitats for the first time in parts of Britain, within their normal climatic range may have provided the conditions for colonization by species associated with coniferous trees. The distinctive lady-bird, *Harmonia quadripunctata* (Pont.) is a recent colonist of this type. Elsewhere in Europe the range of this species towards the north is limited, reaching, in Scandinavia, only the very southern tip of Sweden. No recent expansion of range to the north has been noted but this species has spread westwards through Holland and Belgium in this century, probably due to planting of conifers in those countries. The earliest British specimens known to me were taken in the vicinity of conifer plantations in West Suffolk in 1937. East Anglia is a likely centre from which this species has subsequently spread (Fig. 2) through much of southern and eastern England. Immigration of this species to Britain without the aid of human transport is a possibility as dispersal flights are well known both in Britain and continental Europe. Numbers have been noted in tidal drift on the coast of East Anglia and a single individual found on a sea wall in south Essex during the summer of 1961, some miles from any coniferous trees, may have been a migrant from Europe. If *H. quadripunctata* has spread by natural means to Britain then the ready access of coniferous plantations from the south-east coast as well as their occurrence in an appropriate climatic belt has probably contributed to successful colonization. Other Coleoptera of North European origin to have successfully invaded conifer plantations in southern Britain during the present century include *Corticeus fraxini* (Kug.), *Plegaderus vulneratus* Panz. and several species of Scolytidae.

Aulonium trisulcum (Geoff.) is perhaps the only species intimately associated with a species of deciduous tree to have established itself, although possibly not for the first time, in Britain during the past 100 years following a spread from other parts of northern Europe. However, even in this case,

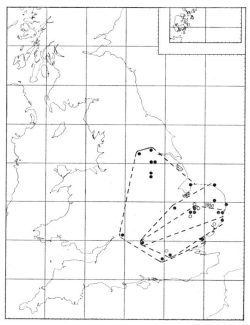

Fig. 2. Records of *Harmonia quadripunctata* (Pont.) (Coleoptera, Coccinellidae) in the British Isles.

The lines represent maximum recorded dispersal at three different dates: 1950, 1960, 1973. Open circles represent records for dates later than that of the enclosing line.

although a predator of elm-frequenting Scolytidae and apparently favoured by the spread of Dutch elm disease (see Chapter 6), *A. trisulcum* appears to have met with only limited success in establishing itself here (Elton, 1971). Recent colonists from neighbouring parts of Europe appear to include also some phytophagous species associated with herbaceous plants. Although its host plant, *Linaria vulgaris*, is a native and there appears, in terms of its known distribution in continental Europe, no special reason why *Mecinus janthinus* Germar should not have been a long-established member of the

British fauna, the sequence of recent records of this distinctive species from southern England is suggestive of recent colonization. Genuinely delayed post-glacial immigration of an otherwise normal type could provide an explanation, but endogenous changes in the species itself or short-term climatic changes are likely to be involved. *Brachypterolus vestitus* (Kies.) and *Ceuthorhynchus turbatus* Schultze have both become established in Britain following the intro-duction of their host plants, *Antirrhinum majus* and *Cardaria draba*, respectively, but, in the first case at least, have followed their hosts with some delay.

Successful colonization by two species of Carabidae, which appear to be twentieth-century immigrants in the British Isles, cannot be explained readily in terms of recent man-made environmental changes. The spread of *Leistus rufomarginatus* Dufts. during the early part of this century westwards from well defined centres in central and northern Europe has been documented by several continental authors. Following an expansion of range to include Holland and north-west France and after apparently arriving in England in the south-east, this species has spread through much of southern and eastern England. As no spread of this species has been noted in other parts of its range, for example to the north in Scandinavia, climatic factors appear unlikely to be involved. Endogenous developments in populations of the species may have initiated its expansion. As *L. rufomarginatus* is an inhabitant of leaf litter, typically in beech woodland, the relatively late development of beech as a dominant woodland tree in Britain and north-west France may have contributed to a delay in fully occupying its potential range to the west.

Although its British history remains somewhat enigmatic another species of ground-beetle, *Amara montivaga* (Sturm), is also likely to be a recent immigrant. Although not recorded from the British Isles until some 20 years ago (Allen, 1950), it is now known from many parts of the British Isles, including Scotland and Ireland (Speight, 1972). The earliest British specimens known (and which I have examined) were taken in County Kerry in 1948. As old specimens of *A. montivaga* have not been discovered in British collections and as it has a history of expansion in Scandinavia during the past 100 years, a recent increase of range to include the British Isles is likely. However, as almost simultaneous introduction to England, Ireland and Scotland is hard to envisage, this species has probably been established here somewhat longer than the 25 years since the first known captures. Typical of open, sandy places, the recent spread of *A. montivaga* may be akin to the much earlier expansion of "culture-steppe" species into western Europe, but, once again endogenous factors are likely to be involved.

In terms of the total number of species comprising the British coleopterous fauna, extinctions are probably largely balanced by newcomers. Although many of the incoming species are of the potentially cosmopolitan type and a poor substitute for the Urwaldtiere and other species which are eliminated as a result of man-made environmental changes, some of them occupy niches in close association with native species and add to the diversity of the communities which they invade. The effects of immigrant species on the native flora and fauna are difficult to assess. The most successful of the mould-feeders are likely to have a large, but as yet unmeasured influence. The most abundant species in heaps of plant refuse may have some effect on competitors. Although past exaggerated claims regarding adverse effects on native species as a direct result of competition with related species introduced from elsewhere renders any such hypotheses contentious, one successful recent colonist may have produced effects of this type. A growing body of anecdotal evidence, both from this country and from various parts of Europe (e.g. Coiffait, 1954) suggests that the success of *Lithocharis nigriceps* Kraatz as an adventive has been at the expense of its already established relative *L. ochracea* (Grav.).

EVIDENCE OF LOCAL CHANGE

Some indication of the nature of changes in the British coleopterous fauna may be obtained by extrapolation from observations in smaller areas. The changes that have taken place, over the past 50, 100 or occasionally more years, in certain sites of special scientific interest (e.g. Monks Wood, Huntingdon-shire and Wicken Fen, Cambridgeshire) are documented to some degree. Such small areas do not, however, necessarily provide evidence that can be applied with any confidence to the country as a whole. Data concerning areas of larger size (e.g. vice-county or county) may be of greater applicability but, although the coleopterous faunas of some counties are relatively well-known, there are few collated data of value in assessing change. Johnson (1962) discussed apparent changes in the Scarabaeid fauna of the Lancashire–Cheshire area during the preceding 100 years and noted that 12 of 51 species taken in the area prior to 1908 had not been found there since. He attributed the apparent decline or loss of these species from the local fauna to the effects of industrialization and the development of coastal holiday resorts. However, there are very few useful attempts to gather and interpret information of this type. My own studies in the county of Essex have provided some data relating to the total coleopterous fauna of a kind which is not presently available for others area of equivalent size.

Data concerning the 2572 species recorded from Essex have been collated in three date-classes: pre-1904, 1904–49 and post-1949. 200–300 additional species might be expected to occur if carefully searched for, and a comparable number in other southern counties with a coastline. The data gathered are inevitably uneven, both in the nature and number of records available for the three different date-classes. However, the number of species recorded for each period (in the region of 2000) is approximately the same (see Table III). That only 1432 species are common to all three date-classes indicates the serious deficiencies in these data as a basis for estimating genuine change. Known

TABLE III. Numbers of species of Coleoptera taken in the county of Essex in three different date classes (including totals for some selected families)

Family	Approximate number of British species	Essex total	Post-1949	1904–49	Pre-1904
Carabidae	337	232	173	188	182
Dytiscidae	107	76	61	53	57
Hydrophilidae	80	70	57	54	53
Ptiliidae	61	26	19	11	7
Scarabaeidae	77	49	35	38	44
Coccinellidae	46	34	26	29	28
Ciidae	22	16	8	15	12
Anobiidae	28	15	6	14	11
Curculionidae	418	243	157	191	191
Chrysomelidae	253	174	109	133	142
Silphidae	20	17	11	12	16
TOTAL FOR ALL FAMILIES	3750	2572	1802	2120	1966

biases in these data readily explain some of the differences in records for the different periods. A large proportion of the 606 species known from the county but lacking pre-1904 records are members of "difficult" groups, little studied or poorly understood in the nineteenth century; many were not then generally recognized as specific entities. However, some 25 of these species are recent immigrants to the British Isles and a few others are likely to be immigrants to the county from elsewhere in Britain during this period, unlike the majority which have merely been overlooked. It is likely that most of the species recorded during the period 1904–49 (many of these also from the period pre-1904) but not since, still occur in the county but have not been revealed by the selective collecting carried out during the post-1949 period. However, I would regard some 35 of these as likely to be now extinct in Essex. The 244 species on record for the county but not found since 1904 probably include a much higher proportion which have been lost from the local fauna. My estimate is that as many as 110 of these no longer occur in the county or are represented by only temporary populations. It should be remembered, however, that many of these species were already scarce in the nineteenth century; in some cases the most recent records are well over a century old.

Evidence for these assertions cannot be presented in detail here, but Table II gives data concerning a representative selection of families. No great change in the proportional representation of some of them (e.g. Carabidae, Dytiscidae, Hydrophilidae and Coccinellidae) in the three date-classes is indicated. Of course, an equal number of species recorded from each period does not necessarily mean that the *same* species are involved in each total. In the Carabidae, for example, several species are exclusive to the pre-1904 period and several have only been taken since. As members of the family Carabidae have received a uniformly high level of attention during the three periods they probably provide the best source of examples that demonstrate genuine change. Most members of this family are rather unspecific predators but many have fairly precise habitat requirements in terms of micro-climate and physical and chemical properties of the substratum. Few are culture dependant. A marked decline appears to be indicated for a number of species associated with open, sandy, or chalky ground or sandy coastal situations. No species of this type provide any evidence of local increases, but those that appear to have become more scarce, more local, or in some cases, possibly to have disappeared from the county during the period since about 1850 include *Amara lucida* (Dufts.), *A. fulva* (Deg.), *Bembidion ephippium* (Marsh.), *Broscus cephalotes* (L.), *Calathus ambiguus* (Payk.), *C. mollis* (Marsh.), *Dyschirius arenosus* Steph., *Harpalus sabulicola* (Panz.), *H. obscurus* (F.), *H. azureus* (F.), *H. rupicola* Sturm, *Masoreus wetterhali* (Gyll.), *Trechus fulvus* Dej. and *Zabrus tenebrioides* (Goeze). Species associated with muddy salt marsh habitats and with most inland marsh or waterside situations in Essex appear to have fared moderately well. Some species which favour damp if not marshy situations, for example *Clivina fossor* (L.), *Bembidion bruxellense* Wesmael and *Pterostichus strenuus* (Panz.), may have become more generally distributed and abundant within the county during the present century. Such species are probably somewhat culture-favoured. Species more strictly confined to water-side or marsh habitats include a few that may have increased the number and size of their populations, for example *Bembidion assimile* Gyll., *B. dentellum* (Thunb.), *B. doris* (Panz.), *B. lunulatum* (Fourc.) and *Trichocellus placidus* (Gyll.). Species favouring "disturbed" ground (arable fields, gardens, etc.), such as *Pterostichus madidus* F., may also have exhibited an increase in abundance. In comparison with other large species of this family, most of which appear to have undergone a decline in the twentieth century, some culture-favoured species such as *Carabus nemoralis* Müller appear to have maintained their population levels moderately well. There is little evidence of general decline in any of the few species which (in Essex at least) are largely confined to woodland although where woodland has been clear-felled such species have naturally been lost from these areas. It may be that predators inhabiting the woodland floor are less affected by fragmentation of habitat and by modern forestry management practices than species dependent on habitats (e.g. rotten wood) which are relatively rare within much modern managed woodland. One species associated with burnt wood, often that of conifers, *Pterostichus angustatus* (Dufts.) is known in Essex only from the post-1949 period and may be a recent immigrant to the county.

The rise in number of recorded species of Ptiliidae (Table III) (also in Clambidae, Helodidae, Scraptiidae and the smaller Staphylinidae) is illustrative of the increased attention paid to small and obscure species in recent years. Rises in the number of species of some other families (e.g. Trixagidae—2 species recorded pre-1904 and 4 species since; Silvanidae—3 species recorded pre-1904 and 5 species since) are more difficult to explain.

The decline in numbers of phytophagous species is only partially explained by the greater attention paid to such species in earlier periods. In particular, a decline in the species of Chrysomelidae, Apionidae, Attelabidae and Curculionidae associated with woodland, sandy heathland and coastal habitats is indicated. The decline in the number of recorded species of Cerambycidae is even more marked. Many of these are associated with mature woodland and the number of records of even many of the commoner species for localities outside of the main wooded areas (Epping Forest, etc.) is very low for the post-1949 period. Other species associated with mature woodland trees (e.g. some Cleridae, some Melyridae, Anobiidae, etc.) or more particularly associated with woodland fungi or fungoid wood (e.g. Ciidae, Leiodidae, Mycetophagidae, Tetratomidae) also seem to have fared badly.

A fall in recorded captures of species of Silphidae is not simply explained, as species which may have suffered declines have different feeding habits and are found in a variety of habitats. A notable decline in the number of records of several species of burying-beetle (*Nicrophorus*) cannot be attributed entirely to biases in the records. This decline is most marked in areas which are now "open" countryside (i.e. with few copses or hedgerows) and a decrease in the amount of carrion available may provide at least a partial explanation.

Most of the species that appear to have become more localized, less abundant or even locally extinct in Essex during the past century or so are "ecological specialists", often associated with biotopes of already localized occurrence. Much of the Essex coast and countryside has been subject to considerable man-made change during this period and many of the apparent declines in species of Coleoptera are readily explicable in terms of the destruction of their specialized habitats. Woodland and old forest species have probably suffered greatly. An increasing number of such species are restricted to the Epping Forest area, although they were once found more widely. Some species previously known from only the Epping area appear to be extinct in the county; in some cases these species were formerly found in localities around Wanstead, Leyton and Woodford where woodland, once part of Epping Forest, has been felled in recent times. The general effect of deforestation in southern England is also illuminated by a comparison of Essex with counties to the immediate south and west. Apart from Epping Forest and adjacent areas, Essex woodland is extremely fragmented. Woodland and old parkland occupy only some $2\frac{1}{2}\%$ of the surface area of the county. Making due allowances for other differences (soil type, etc.), the greater area and less fragmented nature of woodland in Kent, Surrey, Sussex and Berkshire are likely to be principal factors contributing to a richer woodland fauna in those counties. Not only are many woodland species of Coleoptera less localized in the more extensively wooded counties but a number of "old forest" species totally absent from Essex are found. A number of species becoming increasingly scarce in Essex, or in some cases eliminated from the local fauna, are associated with coastal habitats which have been much affected by human activity in the past one hundred years or so. Species associated with sandy cliffs, shingle banks and other coastal habitats of local ocurrence in Essex, and previously known principally from the north-east of the county (between Harwich and Mersea Island), and from the environs of Southend and Westcliff-on-Sea, have suffered particularly. Both of the coastal areas in question have been severely affected by the development of holiday resorts. Recent changes in the coleopterous fauna of Essex illustrate some of the effects on terrestrial invertebrates of urbanization, more intensive arable farming and other alterations in land use typical of much of southern Britain during the period since 1800. Increasingly large areas of the country have been occupied by a more or less uniform fauna of species favoured by or resistant to these man-made environmental changes. Some of the dominant components of this uniform fauna may be identified by considering the frequency of records during the post-1949 period. The number of 10 km squares within the county from which there are records for this most recent period is employed here as an arbitrary but convenient measure of frequency. Fifty species, listed in Table IV, have been recorded from more than 20 each. Despite the many sampling biases, this list may be of some value in distinguishing species which (at the 10 km square level) are uniformly distributed from those which may be abundant but are more local. It is these uniformly distributed culture-species *in extremis* which are an increasingly dominant feature of the coleopterous fauna in many parts of the British Isles.

TABLE IV. Species of Coleoptera taken from 20 or more different 10 km squares within the county of Essex in the period since 1949

Adalia decempuncta (L.)	*Helophorus brevipalpis* Bedel
Agonum dorsale (Pont.)	*Leistus ferrugineus* (L.)
A. ruficorne (Goeze)	*Lema melanopa* (L.)
Amara familiaris (Dufts.)	*Megasternum obscurum* (Marsh.)
Anotylus rugosus (F.)	*Meligethes aeneus* (F.)
A. sculpturatus (Grav.)	*M. viridescens* (F.)
Apion dichroum Bedel	*Notiophilus biguttatus* (F.)
Aridius bifasciatus (Reitter)	*Propylea quattuordecimguttata* (L.)
Bembidion guttula (F.)	*Proteinus ovalis* Steph.
B. lampros (Hbst.)	*Psyllobora vigintiduopunctata* (L.)
B. lunulatum (Fourc.)	*Pterostichus madidus* (F.)
B. obtusum Serv.	*P. strenuus* (Panz.)
Ceuthorhynchus contractus (Marsh.)	*Rhyzobius litura* (L.)
C. pollinarius (Forst.)	*Rugilus orbiculatus* (Payk.)
Chaetocnema concinna (Marsh.)	*Sepedophilus marshami* (Steph.)
Cidnorhinus quadrimaculatus (L.)	*Sitona lineatus* (L.)
Coccidula rufa (Hbst.)	*Stenus clavicornis* (Scop.)
Coccinella septempunctata L.	*Stilbus testaceus* (Panz.)
C. undecimpunctata L.	*Sunius propinquus* (Bris.)
Corticarina gibbosa (Hbst.)	*Tachinus rufipes* (Deg.)
Demetrias atricapillus (L.)	*Tachyporus chrysomelinus* (L.)
Dromius linearis (L.)	*T. hypnorum* (F.)
D. melanocephalus Dej.	*T. nitidulus* (F.)
Drusilla canaliculata (F.)	*T. obtusus* (L.)
Harpalus affinis (Schrank)	*Xantholinus linearis* (Ol.)

CONSERVATION

Relict populations in the north of the British Isles have considerable claim to conservation as they contain much biological information which is irretrievably lost once the populations are finally extinguished. Many groups of organisms are represented by populations, which have been isolated from those in continental Europe since the period immediately following the last glaciation in similar, often the same, situations. Habitat conservation in relevant areas (certain

Scottish Highland peaks, parts of the Pennines, etc.) will at least delay their extinction. Other northern species of Coleoptera are likely to remain in most of the areas which they presently occupy, as long as such areas (moorland, etc.) continue to be relatively free from human interference.

It is clearly the southern parts of the British Isles that exhibit the greatest recent changes in the composition of their coleopterous fauna. Although the number of species comprising this southern fauna has probably remained relatively constant throughout the historical period, the rise to dominance of culture and other synanthropic species and the spread of new immigrants is leading to an increased uniformity. Species that have not responded well to man-made environmental changes are dependent on a variety of habitats. Not all of these can be readily conserved, but as many of the species are found in a few principal types of habitat, some attempt at conservation of these is possible. Efforts to preserve variety in the British coleopterous fauna should include conservation of long-established deciduous woodland, fenland, lowland heath, chalk downland and areas of the coast-line which include sand-dunes and other localized features.

In deciduous woodland the species most adversely affected by man's activities appear to be those dependent on mature trees or on decaying wood and associated fungi. Conservation and appropriate management, much on the lines suggested by Chalmers-Hunt (1969), is to be recommended. A good succession of trees at all stages of growth must be assured while, at the same time, old wood and other decaying matter must be left *in situ*. In comparison with areas of high population density elsewhere in Europe, the southern parts of Britain are relatively rich in conserved areas of ancient forest and parkland (see Sainte-Claire Deville, 1930); in coleopterous terms, continued protection of these habitats is, perhaps, the most important single act of conservation to be advocated. At least some woodland species are able to take advantage of newly afforested areas (including coniferous plantations) and such changes in land use are not to be totally deplored from this point of view. However, conservation of existing faunas in lowland heath or downland, which is frequently utilized for new plantations, may be of primary concern.

Although most species characteristic of "wet-land" are widely distributed in the British Isles, a number are more or less restricted to the ancient fens of East Anglia. It is to be hoped that further sites of special coleopterological interest will be added to those, such as Wicken Fen and Wood Walton Fen, which are already conserved. Habitat conservation at many coastal sites is likely to be difficult because of increasing utilization of coastal areas for re-creational and other purposes. However, as the special fauna of sand-dunes and cliffs of clay or sand are particularly susceptible to man-made environmental changes, some effort should be made to protect habitats of this kind.

Many of the practices involved in present-day intensive farming are likely to exert a profound effect on species of Coleoptera. Activities such as widespread draining of marshes, filling in of ponds or gravel-pits, removal of hedges and a generally increased utilization of previously untilled land for crops, are bound to produce the further localization of many species whose species habitats are thus fragmented or destroyed. Conservation of any features adding to habitat diversity in agricultural areas may help, not only in the preservation of a rich local fauna, but in the maintenance over larger areas of species adversely affected by fragmentation of populations.

NEEDS FOR THE FUTURE

Any assessment of the post-glacial history or of recent changes in the British coleopterous fauna is very much restricted by the lack of relevant data in collated forms. Future work in this field may be best served by a more systematic approach to the gathering and storage of available data concerning the distribution and habits of British species. However, to accelerate a general growth in understanding of the coleopterous fauna and thus provide data for analysis of change, a variety of developments are to be advocated. In some areas critical taxonomic work is still urgently needed to provide a base from which other investigations can proceed. Identification manuals, incorporating new taxonomic conclusions, are essential to a broader participation in the process of gathering information. In the case of some families, handbooks published by the Royal Entomological Society already fulfil this need. However, only the minority of species have so far been dealt with in this series; an accelerated rate of production of new parts and revised editions of those already out of date are to be desired. A completely new fauna of the type produced by Fowler (1887–91) is not to be expected and is by no means essential. What is needed rather, as an accompaniment to manuals containing only the data most essential for accurate identification, is a British equivalent to the work of Horion (1956, etc.). On this basis of new key-works and appropriate recording schemes coleopterists should be in a position to provide monographic treatments of sectors of the British fauna.

As fresh taxonomic and other work changes so rapidly our understanding of the British fauna, check-lists which are nomenclatorially sound and true representations of taxonomic knowledge of the fauna at the date of preparation are of great importance. A new check-list of British Coleoptera is now in the final stages of preparation. However, such works cannot be expected to retain currency for long and there may be a case for periodic, even annual, supplements.

Present schemes for gathering distributional and other data, which omit the greater part of our fauna, need to be extended in order to provide, eventually, a repository for information on all the British species of Coleoptera. As with other groups of organisms, the gathering of records from Ireland and other neglected areas must be encouraged. Special recording schemes concerned with certain categories of pests are urgently needed. A "stored products" scheme and a "forestry" scheme would be most appropriately organized by entomologists involved professionally in these fields. More complete records of casual importations would also be of great value. Only when the fullest data concerning imported species are collated can the factors involved in successful colonization by the minority be fully evaluated. More attention to successful colonists is to be recommended including more energetic attempts to monitor their spread. Ideally, detailed studies of expanding populations and their effects on the established flora and fauna should also be made. A considerable contribution could be made by a study, on the lines of Lindroth's (1957) work on North America and Europe, of faunal exchange between Australia, New Zealand and the British Isles.

At the local level, coleopterological surveys of areas where much recent change has taken place and resurveys of previously investigated sites could provide much data of interest. The effects of known man-made changes such as the use of pesticides or the removal of hedgerows can best be elucidated by detailed ecological studies in small areas. An increased knowledge of the variety of species associations to be found may add considerably to our understanding of the factors involved in change. Much useful information on long-term changes may be derived from assemblages of coleopterous remains. It is to be hoped that the fauna of archaeological sites and assemblages of modern origin will receive increasing attention.

CONCLUSIONS

Recent changes in the British coleopterous fauna may be regarded as largely due to a continuation of processes of environmental change which began several thousand years ago. Man's contribution to these processes has been considerable since the earliest human settlements in the British Isles, and was of major significance long before the industrial revolution and the era of entomological collections and records. However, this most recent period, one of a rapidly expanding human population, has seen an accentuation of some existing trends.

Alterations in land use, which bring about the further localization or extinction of species with special habitat requirements, have been responsible for the major changes in overall composition of the British fauna. Habitat destruction by man has probably been the principal factor leading to extinctions and retractions of range. However, the British fauna has long been characterized by a high proportion of species able to adapt to a variety of conditions, and many of these re-adjust to habitats newly remoulded by man. Man-made habitats are also often those which are occupied by new immigrants. Successful colonists originating from distant continents are a particular feature of the past 200 years. Some types of environmental change are largely peculiar to the recent period: widespread air pollution may have had little effect on the majority of species; pollution of aquatic habitats has had a pronounced but local effect; and the effects of pesticides are likely to be complex, but possibly deleterious to many species.

Change, both in composition of the fauna and in the relative abundance and range of individual species, appears to be greatest in the southern parts of Britain, where culture-favouring species become steadily more abundant and generally distributed. While many of the already scarce and localized species at the northern limits of their range in southern Britain become scarcer still or extinct, relict species in the northern parts of the British Isles appear to be less adversely affected.

The influence of known climatic change is difficult to assess, on the basis of available data, but compared to the effects of human activity is likely to be of secondary importance.

REFERENCES

ALLEN, A. A. (1936). *Adelocera quercea* Herbst (Col., Elateridae) established as British. *Entomologist's mon. Mag.* **72**, 267–269.

ALLEN, A. A. (1947). *Globicornis nigripes* F. (Col., Dermestidae) in Britain: corrections and additions. *Entomologist's mon. Mag.* **83**, 171.

ALLEN, A. A. (1948). A Note on the Habitat of *Alphitophagus bifasciatus* Say, etc. Col., Tenebrionidae). *Entomologist's mon. Mag.* **84**, 46.

ALLEN, A. A. (1950). Two species of Carabidae (Col.) new to Britain. *Entomologist's mon. Mag.* **86**, 89–92.

ALLEN, A. A. (1956). *Pediacus depressus* Hbst. (Col., Cucujidae) in Berks. and Sussex; with a summary of its British history. *Entomologist's mon. Mag.* **92**, 212.

ALLEN, A. A. (1960). Is *Dermestes peruvianus* Lap. (Col., Derestidae) becoming naturalized in Britain? *Entomologist's mon. Mag.* **96**, 34.

ALLEN, A. A. (1962). A reputed twentieth-century(?) occurrence of *Rhynchites bacchus* L. (Col., Curculionidae) in Kent. *Entomologist's mon. Mag.* **98**, 50.

ALLEN, A. A. (1965). The status of *Aulonium ruficorne* Ol. (Col., Colydiidae) and *Platysoma oblongum* F. (Col., Histeridae); with a few suggestions as to the treatment of imported species in faunal lists. *Entomologist's mon. Mag.* **100**, 278.

ALLEN, A. A. (1969a). Notes on some British Serricorn Coleoptera, with adjustments to the list. 1. *Sternoxia*. *Entomologist's mon. Mag.* **104**, 208–216.

ALLEN, A. A. (1969b). Notes on the genus *Cerambyx* (Col.) in Britain, and on the British status of two other Cerambycids. *Entomologist's mon. Mag.* **104**, 216.

ALLEN, A. A. (1970). *Ernoporus caucasicus* Lind. and *Leperisinus orni* Fuchs (Col., Scolytidae) in Britain. *Entomologist's mon. Mag.* **105**, 245–249.

ALLEN, A. A. (1971). Notes on *Omophron limbatum* F. (Col., Carabidae) in Britain. *Entomologist's mon. Mag.* **106**, 221–223.

ALLEN, A. A. (1972). On the present-day incidence and habitats of *Blaps mucronata* Latr. (Col., Tenebrionidae). *Entomologist's mon. Mag.* **107**, 192.

ALVEY, R. C. (1968). Beetle remains: A Roman Well at Bunny, Nottinghamshire. *Trans. Thoronton Soc. Notts.* **71**, 5–10.

ATTY, D. B. (1964). *Nudobius lentus* Grav. (Col., Staphylinidae) in Gloucestershire. *Entomologist's mon. Mag.* **100**, 93.

BALFOUR-BROWNE, W. A. F. (1940, 1950, 1958). "British Water Beetles", 3 vols. Ray Society, London.

BARTLETT, C. (1921). A note on Coleoptera in drift pine-logs. *Entomologist's mon. Mag.* **57**, 15.

BELL, A. (1922). On the Pleistocene and Later Tertiary British Insects. *Ann. Rep. Yorks. Phil. Soc.* **1921**.

BELL, F. G., COOPE, G. R., RICE, R. J. and RILEY, T. H. (1972). Mid-Weichselian Fossil-Bearing Deposits at Syston, Leicestershire. *Proc. Geol. Ass.* **83**, 197–211.

BEZANT, E. T. (1963). The occurrence of *Dermestes peruvianus* La Porte and *Dermestes haemorrhoidalis* Küster (Col., Dermestidae) in Britain. *Entomologist's mon. Mag.* **99**, 30–31.

BRADLEY, P. L. (1958). An assemblage of Arthropod Remains from a Roman Occupation Site at St. Albans. *Nature, Lond.* **131**, 435–436.

BUCKLAND, P. C. and KENWARD, H. K. (1973). Thorne Moor: a Palaeo-ecological Study of a Bronze Age Site. *Nature, Lond.* **241**, 406–407.

CELORIA, F. (1970). Insects and archaeology. *Science and Archaeology* **1**, 15–19.

CHALMERS-HUNT, J. M. (1969). Insect Conservation in Mixed Woodland and Ancient Parkland. *Entomologist's Rec. J. Var.* **81**(6), 156–163.

CLARK, J. T. (1967). The distribution of *Lucanus cervus* (L.) (Col., Lucanidae) in Britain. *Entomologist's mon. Mag.* **102**, 199–204.

COIFFAIT, H. (1954). Note sur trois espèces de staphylinides en voie d'expansion et sur une espèce en voie de régression. *Vie Milieu* **4**, 75–78.

COOPE, G. R. (1967). The value of Quaternary insect faunas in the interpretation of ancient ecology and climate. *Proc. VII Congr. I.N.Q.U.A. (Boulder 1965)*, **7**, 359–380.

COOPE, G. R. (1969). The response of Coleoptera to gross thermal changes during the Mid Weichselian interstadial. *Mitt. int. Verein. theor. angew. Limnol.* **17**, 173–183.

COOPE, G. R. (1970a). Climatic interpretations of Late Weichselian Coleoptera from the British Isles. *Rev. Géogr. phys. Géol. dyn.* **12**, 149–155.

COOPE, G. R. (1970b). Interpretations of Quaternary Insect Fossils. *A. Rev. Ent.* **15**, 97–120.

COOPE, G. R. and BROPHY, J. A. (1972). Late Glacial environmental changes indicated by a coleopteran succession from North Wales. *Boreas* **1**, 97–142.

COOPE, G. R., MORGAN, A. and OSBORNE, P. J. (1971). Fossil Coleoptera as indicators of climatic fluctuations during the Last Glaciation in Britain. *Palaeogeogr. Palaeoclimat. Palaeoecol. (Amsterdam)* **10**, 87–101.

COOPE, G. R. and OSBORNE, P. J. (1968). Report on the coleopterous fauna of the Roman Well at Barnsley Park, Gloucestershire. *Trans. Bristol Glos. arch. Soc.* **86**, 84–87.

CROWSON, R. A. (1967). Refuges for warmth-loving species along the Scottish south coast. *Entomologist's mon. Mag.* **102**, 245–246.

DONISTHORPE, H. (1933). *Ips (Tomicus) suturalis* Gyll. in Windsor Forest. *Entomologist's mon. Mag.* **69**, 105–106.

DONISTHORPE, H. (1936). *Cathartus (Ahasverus) advena* Waltl in the open. *Entomologist's mon. Mag.* **82**, 228.

DONISTHORPE, H. (1936–38). A Preliminary List of Coleoptera of Windsor Forest. *Entomologist's mon. Mag.* **72**, 210–219, 260–267; **73**, 20–28, 70–77, 110–125, 167–176, 237–246, 273–274; **74**, 21–27, 67–77, 115–126, 166–171, 212–222, 259–270.

DUFFEY, E. A. J. (1969). The status of *Cerambyx* L. (Col., Cerambycidae) in Britain. *Entomologist's Gaz.* **19**, 164–166.

ELTON, C. S. (1971). *Aulonium trisulcum* Fourc. (Col., Colydiidae) in Wytham Woods, Berkshire; with remarks on its status as an invader. *Entomologist's mon. Mag.* **106**, 190–192.

FARROW, R. A. and LEWIS, E. S. (1971). *Omophron limbatum* (F.) (Col., Carabidae) an addition (or restoration?) to the British List. *Entomologist's mon. Mag.* **106**, 219–221.

FOWLER, W. W. (1886–91). "The Coleoptera of the British Islands", 5 vols. Reeve, London.

GAUNT, G. D., COOPE, G. R. and FRANKS, J. W. (1970). Quaternary deposits at the Oxbow opencast coal site in the Aire valley, Yorkshire. *Proc. Yorks. geol. Soc.* **38**(2), 9, 175–200.

GODWIN, H. (1956). "The History of the British Flora." Cambridge University Press.

HAMMOND, P. M. (1971). Notes on British Staphylinidae 2.—On the British species of *Platystethus* Mannerheim, with one species new to Britain. *Entomologist's mon. Mag.* **107**, 93–111.

HAMMOND, P. M. and BACCHUS, M. E. (1972). *Atheta* (s.str.) *strandiella* Brundin (Col., Staphylinidae) new to the British Isles, with notes on other British species of the subgenus. *Entomologist's mon. Mag.* **107**, 153–157.

HOLDHAUS, K. and LINDROTH, C. H. (1939). Die Europaischen Koleopteren mit Boreal-piner Verbreitung. *Annln. naturh. Mus. Wien*, **50**, 123–293.

HOLLAND, D. G. (1972). A Key to the larvae, pupae and adults of the British Species of Elminthidae. *Freshwater Biol. Ass., Sci. Publ.* **26**, 1–58.

HORION, A. (1949). Adventivarten aus faulenden Pflanzenstoffen, besonders aus Komposthaufen. *Kol. Zeitschr.* **1**, 203–215.

HORION, A. (1956). "Faunistik der Mitteleuropäischen Käfer, 5 Heteromera". Tutzing.

HURST, G. W. (1970). Can the Colorado potato beetle fly from France to England? *Entomologist's mon. Mag.* **105**, 269–272.

Johnson, C. (1962). The Scarabaeoid (Coleoptera) Fauna of Lancashire and Cheshire and its apparent changes over the last 100 Years. *Entomologist* **95**, 155–165.

Johnson, C. (1963). *Rhizophagus parvulus* Payk. (Col., Rhizophagidae): an addition to the British list. *Entomologist's mon. Mag.* **98**, 231.

Kaufmann, R. R. U. (1946). On some doubtful or rare Longicornia included in the new check-list of British insects. *Entomologist's mon. Mag.* **92**, 181–185.

Kaufmann, R. R. U. (1948). Notes on the distribution of the British Longicorn Coleoptera. *Entomologist's mon. Mag.* **84**, 66–85.

Kloet, G. S. and Hinkcs, W. D. (1945). "A Check List of British Insects." Privately printed, Stockport.

Lindroth, C. H. (1931). Die Insektenfauna Islands und Ihre Probleme. *Zool. Bidr. Upsal.* **13**, 103–599.

Lindroth, C. H. (1935). The Boreo-British Coleoptera. A study of the faunistical connections between the British Isles and Scandinavia. *Zoogeografica* **2**, 579–634.

Lindroth, C. H. (1957). "The Faunal Connection Between Europe and North America." Wiley, Stockholm and New York.

Lindroth, C. H. (1960a). On *Agonum sahlbergi* Chd. (Col., Carabidae). *Entomologist's mon. Mag.* **96**, 44–47.

Lindroth, C. H. (ed.) (1960b). "Catalogus Coleopterorum Fennoscandiae et Daniae." Entomologiska Sällskapet, Lund.

Lindroth, C. H. (1963). The fauna history of Newfoundland. *Opusc. ent. Suppl.* **23**, 1–12.

Lindroth, C. H. (1972). Changes in the Fennoscandian Ground-beetle fauna (Coleoptera, Carabidae) during the twentieth century. *Ann. Zool. Fenn.* **9**, 49–64.

Lloyd, R. W. (1953). *Ostoma ferrugineum* L. (Col., Clavicornia, Ostomidae) new to Britian. *Entomologist's mon. Mag.* **89**, 251.

Moore, B. P. (1957). The British Carabidae (Coleoptera), Part 2. *Ent. Gazette* **8**, 170–180.

Morley, C. (1941). Another Scots Beetle in Suffolk. *Trans. Suffolk Nat. Soc.* **4**, 249.

Morris, M. G. (1963). *Ceuthorhynchus turbatus* Schultze (Col., Curculionidae) in Cambridgeshire and Huntingdonshire. *Entomologist's mon. Mag.* **98**, 188.

Morris, M. G. (1966). *Ceuthorhynchus unguicularis* C. G. Thomson (Col., Curculionidae). New to the British Isles, from the Suffolk Breckland and the Burren, Co. Clare. *Entomologist's mon. Mag.* **101**, 279–286.

Morris, M. G. (1968). *Ceuthorhynchus unguicularis* Thomson (Col., Curculionidae) in Wiltshire and Suffolk. *Entomologist's mon. Mag.* **104**, 45.

Morris, M. G. (1970). *Phyllodecta polaris* Schneider (Col., Chrysomelidae) new to the British Isles from Wester Ross and Inverness-shire, Scotland. *Entomologist's mon. Mag.* **106**, 48–53.

Osborne, P. J. (1965). The effect of forest clearance on the distribution of the British insect fauna. *Proc. Intern. Congr. Entomol., 12th, London* **173**, 37–55.

Osborne, P. J. (1969). An insect fauna of Late Bronze Age date from Wilsford, Wiltshire. *J. Anim. Ecol.* **38**, 555–566.

Osborne, P. J. (1971a). Excavations at Fishborne 1961–1969. The Insect Fauna from the Roman Harbour. *Rep. Res. Comm. Soc. Antiquaries Lond.* **27**, 393–396.

Osborne, P. J. (1971b). An insect fauna from the Roman site at Alcester, Warwickshire, *Britannia* **2**, 156–165.

OSBORNE, P. J. (1972). Insect faunas of Late Devensian and Flandrian age from Church Stretton, Shropshire. *Phil. Trans. R. Soc.*, B, **263**, 327–367.

PAVIOUR-SMITH, K. (1960). The invasion of Britain by *Cis bilamellatus* Fowler (Coleoptera: Ciidae). *Proc. R. ent. Soc. Lond.*, A, **35**, 145–155.

PEARSON, R. G. (1961a). "The Coleoptera from some Late-Quaternary deposits and their significance for zoogeography." Ph.D. thesis, University of Cambridge.

PEARSON, R. G. (1961b). *Chlaenius (Chlaeniellus) tristis* Sch. (*holosericeus* F.) (Col., Carabidae) from a deposit of the last interglacial at Selsey Bill. *Entomologist's mon. Mag.* **97**, 86–87.

PEARSON, R. G. (1964). *Diachila* cf. *artica* Gyll. (Col.) from the full-glacial deposits at Barnwell Station, Cambridge. *Entomologist's mon. Mag.* **99**, 200.

PENNINGTON, W. (1969). "The History of British Vegetation." English Universities Press, London.

PHILP, E. G. (1963). *Syagrius intrudens* Waterh. (Col., Curculionidae) recorded from Kent. *Entomologist's mon. Mag.* **99**, 79.

SAINTE-CLAIRE DEVILLE, J. (1930). Quelques aspects du peuplement des Îles britanniques. (Coléoptères). *Société de Biogéographie*, **3**, *Contribution à l'étude du peuplement des Îles britanniques*, 99–150. Society for Biogeography.

SHARP, W. E. (1899). Some Speculations on the Derivation of our British Coleoptera. *Proc. Trans. Lpool biol. Soc.* **13**, 163–184.

SHAW, M. W. (1961). The golden Buprestid *Buprestis aurulenta* L. (Col., Buprestidae) in Britain. *Entomologist's mon. Mag.* **97**, 97–98.

SHOTTON, F. W. (1965). Movements of Insect Populations in the British Pleistocene. *Geol. Soc. Am. Papers* **84**, 17–33.

SPEIGHT, M. (1970). "Archaeoentomology". [Unpublished mimeographed report.]

SPEIGHT, M. C. D. (1972). Ground beetles (Col., Carabidae) from the Bourn Vincent National Park. *Ir. Nat. J.* **17**, 226–230.

STAFFORD, F. (1971). Insects of a medieval burial. *Science and Archaeology* **7**, 6–10.

STEEL, W. O. (1961). *Tachinus flavolimbatus* Pandelle, a Staphylinid (Coleoptera) new to Britain. *Entomologist* **94**, 77–78.

STEPHENS, J. (1828–32). "Illustrations of British Entomology. *Mandibulata*", 5 vols. Baldwin and Cradock, London.

WALKER, J. J. (1932). The Dale Collection of British Coleoptera. *Entomologist's mon. Mag.* **68**, 21–28, 71–75, 105–108.

WELCH, R. C. (1964). *Uleiota planata* (L.) (Col., Cucujidae) breeding in Berkshire. *Entomologist's mon. Mag.* **99**, 213.

WOODROFFE, G. E. (1971). *Globicornis nigripes* F. (Col., Dermestidae) in Buckinghamshire. *Entomologist's mon. Mag.* **106**, 148.

WORMS, BARON DE (1961). A Ladybird plague on the East Coast. *Trans. Suffolk Nat. Soc.* **12**, 55.

20 | Changes in the British Dipterous Fauna

K. G. V. SMITH

Department of Entomology, British Museum (Natural History), London

Abstract: The faunistic history of Diptera is poorly documented owing to difficult taxonomy, scattered literature and paucity of students, and is sometimes a reflection of the activities of taxonomists rather than of the flies themselves.

Diptera probably originated in swamps during the Carboniferous period and estimates from fossils indicate that they constituted 3%, 5%, 27% and 10% of the Permian, Mesozoic, Tertiary and recent insect faunas, respectively. Thus, flies have passed their peak and are now declining. Climatic changes and man's effect on the environment have caused resultant changes in the Diptera fauna.

Many synanthropic Diptera declined, at least in numbers, as man's standard of hygiene improved and were probably present in vast numbers in the nineteenth century.

Some blood-sucking species are declining because their hosts are vanishing; man is destroying them by chemical means and draining their breeding grounds.

Other flies dependent on the primordial type of swampy habitat decline as land is drained.

The success of predaceous and parasitic families depends largely on that of their hosts; where the relationship is intricate (e.g. host specific) their future may be doubtful. Highly adaptable species, particularly those intimately associated with man, his buildings, his animals and plants and their waste products should succeed.

INTRODUCTION

The Diptera probably originated in the swampy conditions of the Carboniferous period and from known fossils Carpenter (1930) has estimated that Diptera constituted 3% of the Permian insect fauna, 5% of the Mesozoic, 27% of the Tertiary and 10% of recent insects. This indicates that, as an order of insects, flies have passed their peak and are now declining. There are some 80 000 species of flies distributed throughout the world. About 5700 of these species are known to occur in Britain.

Systematics Association Special Volume No. 6, "The Changing Flora and Fauna of Britain", edited by D. L. Hawksworth, 1974, pp. 371–391. Academic Press, London and New York.

Among British insects the Diptera are not as well documented as other groups, such as the Lepidoptera, Coleoptera and Orthoptera. The order has always been regarded as difficult taxonomically and apart from the early, rather unsatisfactory, work of Walker (1851–56) there has been no comprehensive monograph of the British species. A useful work to students early this century was that of Wingate (1906) which was primarily a list of Durham Diptera but contained keys to most of the common British species, with the exception of the difficult families Cecidomyiidae and Mycetophilidae. This pioneer work was a great stimulus to the study of the order. Fortunately among the very few serious students of the Diptera there have been some entomological giants such as G. H. Verrall, J. E. Collin and F. W. Edwards. Verrall commenced a projected series of volumes on the British Diptera of which only two volumes appeared (Verrall, 1901, 1909) and his nephew J. E. Collin produced a third volume on Empididae in 1961. Verrall and Collin were wealthy amateurs and thus able to devote much time to field work and published papers sometimes adding a hundred species at a time to the British list. Collin published 212 papers on British Diptera and added some 650 Diptera to the British List, including over 200 new to science. Edwards was a professional Dipterist at the British Museum (Natural History), London, but showed much of the enthusiasm of the amateur, particularly for field work, and was a prodigious worker, producing 408 papers between 1907 and 1941 which brought order into the difficult suborder Nematocera and earned for him Fellowship of the Royal Society.

It only became possible to consider seriously the distribution and faunistic history of the Diptera after the careful revisionary work of these three great entomologists had been published.

Fortunately there are some Diptera large enough and attractive enough to gain the attention of general entomologists and naturalists and some of these are recognizable in the plates of the early works on British insects. Such flies include the family Syrphidae and the larger Brachycera and it is from these groups that most of my examples illustrating changes in the British fauna are drawn.

Individual studies of Diptera distribution are few but most authors publishing revisions indicate distribution of known records which draws out data on occurrence and distribution in the form of short notes in the British entomological journals. There are few detailed studies of individual species so that information on occurrence and distribution can only be gleaned from a study of the lists given in the "Victoria County History" and individually published local lists, the identifications in which may need careful scrutiny.

The series of handbooks published by the Royal Entomological Society (see reference list) has stimulated the study of the Diptera, as I am sure it has of most other orders.

Over the years the occurrence of Hippoboscid ectoparasites on birds and mammals has been fairly regularly recorded and more recently bat parasites have received a similar treatment. Cave research groups have also recorded the occurrence of cave-dwelling Diptera. In 1972 the Nature Conservancy's Biological Records Centre at Monks Wood co-operated in the establishment of a cranefly (Tipulidae) recording scheme.

THE DISTRIBUTION AND ABUNDANCE OF THE ESTABLISHED FAUNA

1. Geographical factors

Oldroyd (1970) refers to the dipterous fauna of the British Isles as "passive" rather than "active", in that it consists largely of a small and fortuitous sample of the western European fauna rather than a coherent assembly produced by the positive character of the environment. There are two distinct elements in the established fauna; those that reached us before the sea separated Britain from Europe, and those that have invaded us since. As with other insects there are migrant Diptera (mainly Syrphidae) which reinforce established species or which appear each year but do not survive the winter. These migrant species include *Metasyrphus corollae* (F.), *Syrphus vitripennis* Meigen, *S. ribesii* (L.), *Scaeva pyrastri* (L.), *Episyrphus balteatus* (Degeer), *Platycheirus albimanus* (F.), *P. manicatus* (Meigen), *Eristalis tenax* (L.), *Sphaerophoria scripta* (L.), *S. menthastri* (L.), *Syritta pipiens* (L.) and *Volucella zonaria* Poda.

Other rare British Diptera which may be migrants are *Doros conopseus* F. (Syrphidae) and *Stomorrhina lunata* F. (Calliphoridae), (see Wainwright, 1944; Goffe, 1945). Possibly some of the so-called "reputed British species" are really occasional migrants.

Some migrant species may eventually establish themselves as will be shown to be the case with the syrphid *Volucella zonaria* Poda. Small Diptera may be dispersed by air currents or even caught up in the wind-borne gossamer threads of spiders.

It is significant that the proportions of apparently endemic species in the British Isles are highest among lesser known orders such as Diptera and Hymenoptera. As one would expect of an island there is probably an endemic element in our dipterous fauna, whether at the species level or below it, but many if not all of our so-called endemic species have subsequently been found to occur on the continent. This has happened and indeed is still happening as a result of taxonomic errors. These errors can now be largely corrected or avoided because freer loaning and exchanging of specimens and easier travel has facilitated access to type material. Much revisionary work of a high standard is now being published on the continent and this has had a considerable repercussion on the interpretation of our dipterous fauna. Many British taxonomists have tended to be rather insular in the past and, as in other spheres of activity, we are now thinking European.

It is worth emphasizing the reality and effect of this insular outlook which I believe has had a considerable influence on the assessment of our fauna among taxonomically difficult groups. During the First World War the American dipterist C. H. Curran collected flies whilst serving in Britain and on his return published a list of species (Curran, 1920). Most of these were common but J. E. Collin (1920) was quick to point out that no less than six species were additional to the British List. Similarly the Canadian dipterist J. R. Vockeroth during visits to this country has found several species not on our list. There is no real mystery about this—it is simply that they have identified their captures with "European" literature rather than with existing keys to purely British species, where the characters of closely related taxa may be inadvertently obscured.

Within the British Isles the asilid *Epitriptus cowini* (Hobby) was thought to be endemic in the Isle of Man, but has since been found in Ireland by Chandler and may yet be found on the continent thus constituting part of the Lusitanian element in our fauna along with such species as *Haematopota bigoti* Gobert (Tabanidae) and possibly the asilids *Pamponerus germanicus* L. and *Dasypogon diadema*; the last is now presumed extinct and is discussed below.

As with other groups examples of relict fauna may be found on mountain tops, peat bogs, etc.

2. Climatic Factors

Changes in climate may affect distribution and may lead to invasion by some species and the extinction of others. *Musca vitripennis* Meigen (Muscidae) occured in southern England during the last century and its absence now is attributed to a cooling of the climate. Parallel cases on the continent are the Muscidae *Hydrotaea hirticeps* (Fallén) and *Phaonia scutellata* (Zetterstedt) which were both described from Sweden last century but do not occur there now and are restricted to Europe south of the Alps.

3. Geological Factors

Geology is an important and often overlooked factor affecting the distribution of a species, particularly if the species has a soil-dwelling larva or is associated with particular plants or hosts of limited distribution. The rhagionid *Sym-*

phoromyia immaculata F. is found chiefly on chalk or magnesium limestone (Stubbs, 1967).

The asilid *Leptarthrus* (= *Isopogon*) *brevirostris* (Meigen) is commoner in highland Britain, very local in lowland areas and in the south is possibly restricted to chalk downland (Stubbs, 1968) although some other factors may restrict it further; other Asilidae occur only in sandy areas. Stephenson and Knutson (1970) in a preliminary study of the British distribution of the family Sciomyzidae suggest that the aquatic members of the family seem to be more restricted to areas of basic rock substrata than do the terrestrial species. Of course, unless data labels on specimens include the grid reference, it is not possible to make correlations of this sort.

Other ecological factors may influence distribution. Some families of Diptera have aquatic immature stages and the presence or absence of certain species may serve as ecological indicators to the freshwater biologist. However even here the situation for the student of distribution may have taxonomic complications. For example, the aquatic larvae of Simuliidae are usually more readily identifiable than the adults, whereas in the aquatic larvae of Chironomidae the reverse is largely true and even the adults are not easy to identify.

FLUCTUATING POPULATIONS

Many species of wide distribution may show considerable fluctuations in population in a given area due to various ecological or behavioural factors. This may mean that the presence or absence of a species during survey may depend largely on being in the right place at the right time. An example of this is the occurrence of the rhagionid *Atrichops crassipes* in water-meadows in the south of England early this century. It has not been found since.

Many flies may form swarms or aggregations, such as Chironomidae and Trichoceridae and these phenomena are discussed by Oldroyd (1964) and Downes (1969). Often no behavioural gregarious factor is involved and swarms are formed by merely constituting an aggregation of individuals often over or under a fixed "marker". Other flies form swarms for the purpose of courtship and mating (e.g. Empididae). The females of *Atherix* (Rhagionidae) cluster on leaves in numbers for oviposition; others may swarm indoors for the purpose of hibernation, such as the cluster-fly *Pollenia rudis* (F.), or be carried indoors in large numbers through open windows by air currents and eddies, such as *Thaumatomyia notata* (Meigen) (Chloropidae) and other small Diptera.

The Sepsidae sometimes occur in vast numbers on vegetation, a phenomenon little understood. The larvae of some species such as Empididae (Smith, 1968) and leather-jackets (Tipulidae) may occur in vast numbers when old grassland is ploughed. In my own garden in north London, the normally rare stratiomyid fly *Xylomyia* (= *Solva*) *marginata* (Meigen) bred in hundreds under bark when some unsafe black poplar trees were felled and cut into logs.

Apart from such swarms or aggregations, individual species may suddenly become numerous in a relatively small area because of some local attractant such as carrion or excrement.

The otitid fly *Seioptera vibrans* (L.) may be rare in some areas, but suddenly becomes locally abundant when honeydew is secreted in large quantities by aphids that are unusually common. Russell (1959) reported on this in Yorkshire and Lancashire and I have seen the same in north London.

The platypezid genus *Microsania* has the strange habit of suddenly appearing in numbers in bonfire smoke then disappearing again. A family that occasionally becomes numerous enough even to gain notice by the popular press is the Coelopidae, the seaweed or kelp-flies. These breed in seaweed cast up on the shore and may occur in vast numbers, entering buildings, often inland, especially hospitals where they are attracted to the smell of chloroform, xylene and trichloroethylene (Oldroyd, 1954; Taylor, 1955).

DECLINING POPULATIONS AND SPECIES

1. Declining Populations of Widely Distributed Species

The specific constitution and population size of Diptera in towns and cities has varied greatly over the centuries. One can imagine the high incidence of house-flies and other Diptera in towns in times when horses provided the principal means of transport. There is a relationship between the number of horse-drawn vehicles and the incidence of the presumably house-fly borne summer diarrhoea, the principal cause of infant mortality around the turn of the century (Graham–Smith, 1929).

The number of horses in nineteenth-century London is indicated by the fact that in 1840 twelve cart-loads of dung were taken off Regent Street every day and later in the century the horse population became even greater. In those days vast quantities of this dung went back to fertilize the soil so that the population of dung-breeding flies must have been fairly evenly distributed. However, horse dung was not the only excrement and filth available. Human sanitation was almost non-existent and in large areas of London drinking water was taken from shallow wells and, to quote Sherwood Taylor (1942) this was "often grossly contaminated by seepage from cesspools and by carrion liquor which oozed from the crammed and ghastly burial grounds". From contemporary accounts we find that the graveyards were so overloaded that when graves were dug the ground steamed like a dunghill and the gravediggers were overcome by gases. To quote Sherwood Taylor again "no small addition to the filth was given by the slaughter-houses. Once a week thirty thousand sheep and oxen, which may be pictured as a procession four abreast, head to tail, and six miles long, were herded through London's streets to Smithfield . . . they were slaughtered in butchers' shops all over the town. Many of the slaughter-houses were underground, inches deep in carrion, and without any drainage. Nobody cared, and only now and then, when decomposing blood flowed down the gutters, was any comment made." The keeping of pigs, even in houses, was commonplace among the poorer slums in the 1840s and Sherwood Taylor remarks that "the Irish immigrants, whose low standard of living made that of the Englishmen still lower, lived on terms of peculiar intimacy with their pigs". I give these gory details simply to emphasize (in the absence of accurate numerical data) that apart from there being vastly greater numbers of common Diptera such as *Musca domestica* L., *Calliphora*, *Lucilia* and *Sarcophaga*, many other dung- and carrion-breeding Diptera such as Muscidae, Scatophagidae, Sphaeroceridae, Phoridae, *Eristalis* and *Helophilus* (Syrphidae) Piophilidae, Sepsidae and Ephydridae must have abounded. Only 60 years ago it was usual for London kitchens to buzz with flies and the relationship of the Diptera to public health and medical problems in those days is another story, and quite a story.

Similarly, outside towns reduction of ponds, marshes, forests, the use of insecticides and pollution have drastically reduced populations of many species. Though, as I shall mention later, some of these activities of man have increased the populations of a few species and may continue to do so.

2. Declining and Extinct Taxa

Many individual species may decline or become extinct in Britain due to a variety of ecological factors.

Blood-sucking species favouring declining animals show a reduction in range and numbers. The Hippoboscidae *Hippobosca equina* L. (on horses) and *Melophagus ovinus* (L.) (on sheep) were formerly commoner and more widespread than they are now.

The Tabanidae are blood-sucking flies with aquatic larvae which appeared in the late Tertiary period in association with the great evolution of ungulate mammals and appear to have had their maximum success in the past. With the rapid decline of their wild hosts and of suitable breeding habitats due to the effect of land drainage, cultivation, etc., the family may be expected to decline further.

Several families which share with Tabanidae the bog–marsh type of habitat for breeding purposes, such as Psychodidae (except in the vicinity of sewage works), Stratiomyiidae and Sciomyzidae, etc., may be expected to decline. The sad thing is that species are still being discovered in our fauna in these habitats. Some of these may be relict species of considerable interest such as the recently discovered Scatophagid *Coniosternum tinctinervis* Becker from Beanrig Moss, Roxburghshire (Nelson, 1972). It would be a pity if this element in our fauna was lost before being discovered, so to speak.

The family Sciomyzidae is represented in the British Isles by 63 species and of these the life histories of some 52 are known (Stephenson and Knutson, 1970). The larvae feed as parasites or predators on freshwater and terrestrial snails, slugs or orb and pea mussels (Sphaeriidae). As expected, the distribution follows that of their mollusc food and these flies are more common in lime-rich areas south of a line joining the Severn and Humber estuaries than in the more acidic, oligotrophic areas to the north. Forty species live in aquatic or semi-aquatic situations. Undoubtedly, the ecological tolerance of these flies and their hosts enabled many species to survive in glacial refuges in the south of England some 18 000 years ago. Until their recent drainage the fenlands of East Anglia were probably important centres of dispersal for the family, and with the general disappearance of fen or swamp-like habitats these flies must be on the decline now.

The bot-flies and others may also be expected to decline following the decline of their wild vertebrate hosts or the better husbandry of domesticated hosts. The bot-fly *Pharyngomyia picta* Meigen (Calliphoridae), a parasite of red deer (*Cervus elaphus* L.) is now regarded as extinct in Britain, but, as mentioned below, another species has been introduced with the reindeer. However, there are said to be more deer in England now than for centuries so *Pharyngomyia* may return again.

Others of parasitic habit whose particular hosts are declining may be expected to follow suit, e.g. some tachinid parasites of butterflies and moths.

The curious little family Thyreophoridae has been recorded from the dead bodies of various animals, including man, where the larvae develop. Strangely, none has been seen in Britain or Europe since the early 1900s. In the British Museum (Natural History) we have two specimens of *Centrophlebomyia furcata* F. found on a dead donkey at Mt Edgcumbe Park, Cornwall, in 1889 and a dozen (without habitat data) found at Porthcawl, Glamorgan, during 1903–06. It is generally assumed that improvements in hygiene regarding animal and human corpses is responsible for their decline and probable extinction, but it seems surprising that they did not reappear, even on the continent, during the two World Wars!

Many reputedly British species have been shown to be misidentifications by early authors, but others seem more credible. The fine Asilid, *Dasypogon* (= *Selidopogon*) *diadema* (F.) was recorded from Britain in the last century and although no authentic British specimen is known it was accurately illustrated by Stephens (1846) in his "Illustrations of British Entomology". It was stated to occur in sandy situations near Bristol and rarely near Swansea. Verrall (1909) states that R. C. Bradley once saw, but failed to catch, a large blackish asilid near Barmouth and some years ago G. S. Kloet told me that he had seen asilids definitely not on the British List among Welsh coastal dunes. Oldroyd (1969) suggested that "one possibility is that *D. diadema* was introduced from ships trading between Bordeaux and South Wales, flourished for a while then died". However, the species may yet be found to survive among dunes somewhere along the Welsh coast. On the continent it is common in scrubby areas from the Landes district of France to Asia Minor.

Another fine asilid, *Laphria gilva* L., was regarded as British by early authors but later relegated to the ranks of reputed British species. Then in 1938 it suddenly appeared again and became frequent for a few years in Berkshire, Sussex and Surrey, then disappeared again.

Another British fly illustrated by Stephens and probably now extinct is *Clitellaria ephippium* F. (Stratiomyiidae). Again, no authoritative British specimens exist but in the British Museum (Natural History) we have four from Stephens' Collection with data and one labelled "Locality unknown, possibly Darenth Woods, about 1850". The larvae of this genus are known to live in the nest of the ant *Lasius fuliginosus* (Latreille) and to overwinter in soil or debris. It seems unlikely that this handsome species would be over-looked and it must be presumed extinct.

INCREASING POPULATIONS AND SPECIES
1. Increasing Populations of Established Species

The population of a species in a particular area may increase quite rapidly especially if competition is suddenly reduced. Elton (1958) has referred to such occurrences as ecological explosions, but mainly with reference to invasions by species into areas where they previously did not occur (I shall deal with this aspect later). Species already established in an area may increase in a dramatic manner by natural means, but the most spectacular results are achieved by man's interference with the ecological balance.

The use of insecticides has the effect of increasing selection pressure and a good example of this among the Diptera is provided by the chemical control of Diptera in poultry houses (Conway, 1970). The nuisance species were

Musca domestica L., *Muscina stabulans* Fallén, *Fannia canicularis* L. (Muscidae) and various Sphaeroceridae and these were successfully controlled by the use of organo-phosphorus insecticides. On the disappearance of these species, another, normally rare, muscid, *Ophyra capensis* Wiedemann, appeared. A similar phenomenon has occurred in the U.S.A., where a related species, *Ophyra leucostoma* Wiedemann, has moved in as a larval predator of other muscid larvae in poultry houses. It is puzzling that O. *leucostoma* has not been involved here as it is a common British species. There are many examples in the history of chemical control of insects where one pest is eliminated only to be succeeded by another.

As long as man and some vegetation survive I regard as potentially successful families the following: Culicidae, Ceratopogonidae, Sciaridae, Syrphidae, Cecidomyiidae, Phoridae, Ephydridae, Sphaeroceridae, Sepsidae, Piophilidae, Agromyzidae, Tephritidae, Chloropidae, Drosophilidae, Muscidae, Sarcophagidae and Calliphoridae.

The Ephydridae are a very adaptive family and their larvae survive in the most hostile environments. In Iceland *Scatella thermarum* Collin breeds in geysers and the petrol-fly *Helaeomyia* (= *Psilopa*) *petrolei* (Coquillett) breeds in the pools of crude petroleum in the Californian oil fields. Species of this, and equally adaptable filth- and excrement-loving families such as Sphaeroceridae, will survive whatever man or natural agencies do to the environment.

2. Additions to the Fauna

Additions and expansions of our fauna can occur in several ways. These are now considered with examples, but these examples are few owing to lack of adequate documentation.

(a) *Genuine additions to the fauna by extension of range.* Probably one of the most spectacular additions to the British dipterous fauna in living memory is that of the syrphid *Volucella zonaria* Poda. This species occurs freely in suitable situations in west and south-west Europe and has been taken on a cross-channel boat. It is usually regarded as a migrant species but has not been seen actually migrating in large numbers, as have other Syrphidae. Verrall (1901) included *V. zonaria* among reputed British species on the basis of two old specimens in the collection of the Entomological Club, but these are now thought to have been taken in Jersey. Other nineteenth century specimens, probably from Kent and Hampshire lacked full documentation and the first fully authenticated specimen was taken by David Sharp in the New Forest in 1908 (Goffe, 1945). Since then single specimens were recorded in 1925, 1928, 1934, and every year from 1938 until 1945, the great migration years, when among many other previously rare insects *V. zonaria* invaded in numbers. Since 1945 specimens have been taken in greater or lesser numbers every year (with peaks in 1948, 1950, 1951, 1953, 1956 and 1960 (Fig. 3)). The species now appears to be established in England mainly on the south coast (Fig. 1). In common with some other species of the genus (e.g. *V. inanis* (L.), *V. pellucens* (L.)) the immature stages are spent as

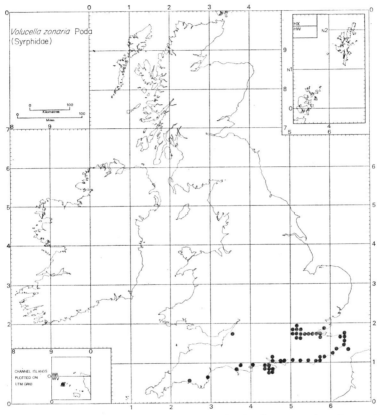

FIG. 1. Map of the recorded British distribution of *Volucella zonaria*, based on published and unpublished records.

scavengers in the nests of wasps and hornets (*Vespula* spp. and *Vespa crabro* L.) (d'Herculais, 1875). Fraser (1946) reared *Volucella zonaria* and *V. inanis* from the same nest of *Vespula vulgaris* L. in Bournemouth. In view of the biology of *V. zonaria* one might expect an association between numbers recorded and the fluctuations in populations of wasps. A relationship between the relative abundance of wasps and other British species of *Volucella* has been observed (Goffe,

1944; Edwards, 1944). It is well known that there are "wasp years" when high numbers are recorded, usually followed by a year of extreme paucity with intermediate years of varying intensity. Bierne (1944) and Fox-Wilson (1946) analysed these variations in wasp abundance and correlated them with climatic factors. An important factor affecting wasp population is the weather in

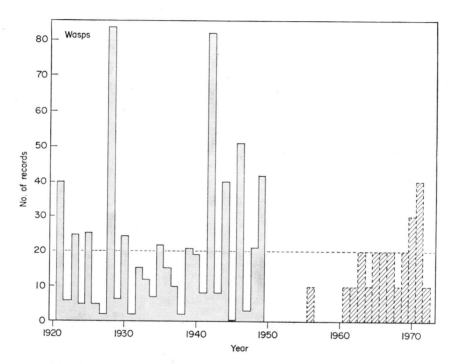

FIG. 2. Abundance or scarcity of wasps. 1921–49 (heavy) from numbers of nests collected per annum in a 100-acre plot at Wisley (Royal Horticultural Society); 1950 onwards (shaded) are approximate estimates of "good", "bad" and "average" years based on published records in the literature, records of Public Health Officers, etc. The horizontal dotted line is an average year (i.e. 20 nests at Wisley), and the estimated good and bad years are indicated as half of the difference above and below this figure.

April–June when the queens are establishing their nests. Rain or early warm weather followed by a wet or cold spell is harmful since queens emerge prematurely and are unable to establish themselves. Delayed emergence from prolonged cold weather does no harm but prolonged cold spells after nests are established may confine the foraging females to their nests, thus affecting the supply of food for the brood. A graph showing the abundance of wasps in given years is shown (Fig. 2).

It should be borne in mind that the available data for "wasps" include 5 species, *Vespula vulgaris* L., *V. germanica* F., *V. rufa* L., *V. norwegica* F. and *V. sylvestris* Scop., and that these species vary in abundance in relation to each

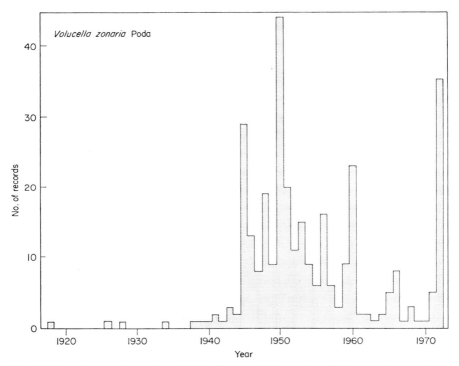

FIG. 3. Abundance and scarcity of *Volucella zonaria* from all available records, numbers per annum, both published and unpublished.

other in each season, but it does appear that in years of abundance more than one species is commoner than usual. Also, it does not always hold that a good year in one district is necessarily a good one elsewhere, although there are undoubtedly "wasp years" when they are abundant everywhere and on the continent. Nevertheless with the rough figures available there does appear to be a relationship in the relative abundance of wasps and *V. zonaria*, in particular the peak years 1945, 1950, and 1972 which seem to follow the "wasp years" and each is followed by a year of relative paucity.

The 1960 peak is probably very much indicative of a good "wasp year" also 1959; although I do not have precise wasp data for 1959, it was a long hot summer and a good early May start to summer usually yields many wasps.

There is some evidence (Rothschild, 1946; Stallwood, 1947) that the hornet was increasing in the period 1940–47, which is about the time when *V. zonaria* was establishing itself here. However, I believe the hornet is very much on the decline now and I think a link with wasps more likely.

(b) *Artificial introductions.* It is artificially introduced species that, once established, may give rise to the spectacular ecological explosions described by Elton (1958).

The Drosophilidae are well known to biologists as a very malleable group genetically, and as Oldroyd (1964) has observed, "there is a certain irony in the fact that, while the laboratory geneticist is showing his powers over nature by moulding his fruit-flies as he pleases, the flies themselves are staging a quiet demonstration of their own". Drosophilids are certainly a widely adaptive and successful family and because of their close association with man and his food and drink they get transported very easily. They readily fill any available ecological niche and as examples *Drosophila buscki* Coquillett, not normally a common species, and *D. funebris* (F.) along with the phorid *Paraspiniphora bergenstammi* (Mik), have become adapted to feeding on the curds left in empty milk-bottles where they pupate cemented to the glass sides. This parallels the sudden behavioural adaptation of blue-tits to feeding on the cream by perforating milk-bottle tops. *Drosophila repleta* Wollaston was first recorded in London in 1943 (Coe, 1943). The following year it was found to be a nuisance in canteens and a hospital in London where the immature stages were breeding in various decomposing vegetables and even in hospital bedpans. The species is now widespread and well established. As mentioned above, we have lost one bot-fly, *Pharyngomyia picta* (Meigen), but another bot-fly, *Cephanomyia trompe* Modeer has been introduced into Scotland with reindeer.

The yellow fever mosquito *Aedes aegypti* L. has been reared from a tree-hole in Epping Forest (Macgregor, 1919), but the species cannot become established here because of its temperature requirements. Escapes from laboratory cultures are always a possibility and a species may survive for a short while. In 1865 an outbreak of yellow fever in Swansea appears to have been associated with a single generation of mosquitoes which flew ashore from a ship. While discussing mosquitoes it is worth mentioning that we have British anopheline mosquitoes capable of carrying malaria and that the "ague" indigenous in Britain until about 1870 (Ross did not establish that mosquitoes transmit malaria until 1897) was this disease. The disease is no longer indigenous here (thanks to improvements in farming and rural housing and drainage reducing mosquito breeding sites) but after the two World Wars outbreaks occurred through our native mosquitoes transmitting the disease from returning infected soldiers (see Marshall, 1938; Busvine, 1966). Incidentally this resulted in the publication of one of the first (if not *the* first) detailed distribution maps of British Diptera (Lang, 1918).

Species of economic importance may be introduced with imported foodstuffs or ornamental plants. The Mediterranean fruit-fly *Ceratitis capitata* Weidemann breeds in various fruits and is often introduced in imported oranges, peaches, etc. and has been captured outdoors in a London park (Colyer and Hammond, 1968). It is doubtful if this species will establish itself in Britain due to its temperature requirements.

(c) *New, apparently endemic, species.* As stated above, apparently endemic species such as the asilid *Epitriptus cowini* Hobby may ultimately be shown to occur in continental Europe. Nevertheless, although many new British species prove to be synonymous (see below) with described continental forms, there may be a truly endemic element in the fauna, but if so it is probably very small.

(d) *Discovery of apparently long established species in well known groups common to the continent.* Here I include those apparently well established species that have only recently been discovered in Britain, presumably due to rarity, local discontinuous distribution or elusive habits.

The rather conspicuous syrphid fly, *Eriozona syrphoides* Fallén, was first recorded in Britain by Peter Crow (1969). His specimen was taken in Merionethshire but in fact Pennington had previously taken the species in Lancaster in 1957 though he did not record this until he found the species again, and not uncommonly, during September 1972 in Stirling (Pennington, in press). A further specimen has been recorded from Carmarthen (Cull, 1972). On the continent this species is not common and is said to be found locally on umbelliferous plants in mountainous limestone areas. In Britain it appears to be associated with high moorland areas.

It is difficult to believe that this conspicuous fly has escaped the attention of entomologists for so many years but this must be the case as there seems little reason to suppose that it has suddenly extended its range as did *Volucella zonaria*. Crow (*in litt.*) feels that the species may be associated with new forestry plantations of hardwoods but the life history is unknown, so no clues are available there. This rather looks like a case where the distribution of the species may be linked with the underlying geology. The empid fly *Syneches muscarius* F., the only west palaearctic species of a largely oriental genus was discovered in Britain in 1953 and remains known here from only one ditch in Dorset (see Collin, 1961).

Concentrated study of lesser worked groups in one area is bound to reveal unrecorded species. J. E. Collin resided at Newmarket and over a long period added many species to the British List from his garden and environs. Claude Morley similarly made many interesting additions from Suffolk, including the fine syrphid *Callicera spinolae* Rondani (see Blair, 1948) only recently re-discovered by Hammond (1973). Scotland, Ireland and the northern counties of England are still largely unexplored entomologically and many unrecorded and undescribed species undoubtedly occur there.

Similarly, concentrated study of a discrete ecological unit, such as dung, fungi, etc., yields unrecorded species.

(e) *Chance freak occurrences*. Examples of this type of occurrence are found in the family Hippoboscidae, ectoparasites introduced on migrating birds. *Icosta minor* Bigot, a common African and Mediterranean species, was taken from a little bittern in this country and *Olfersia spinifera* Leach has been taken from a frigate-bird which was driven ashore on our western coast. Such accidental occurrences may also turn up on common birds such as the common white-throat which yielded *Ornithophila metallica* Schiner, a tropical species, obviously acquired on migration between Africa and Britain.

It is very doubtful if these freak occurrences could ever become established.

(f) *Re-instated species*. As discussed under extinct species, some species thought to be extinct may reappear and thus be reinstated.

Eustalomyia hilaris (Fallén) (Anthomyiidae) was included on the British List by Curtis (1829) but no British specimens were known and it was assumed that it had been misidentified, or had become extinct. In 1967 I found this species on a stump in my London garden and have found it sparingly each year since (Smith, 1971), but so far this is the only known British locality.

However, a species recently reinstated on the British list may in fact be extinct. Again, to quote an example from personal experience, the empid *Chersodromia cursitans* Zetterstedt was recorded by Walker (1851) as a rare British species inhabiting the sea coast, but no specimens were known and it became a reputed British species. However, in the collections at the Hope Department of Entomology in Oxford I found two specimens taken at Studland in 1904 and thus re-established this species on the British List (Smith, 1964). But I have since searched for the species without success and I suspect that it may now be extinct in Britain since other members of the genus occupy its ecological niche on the sea shore, and on the continent the species is said to occur on the shores of streams and inland lakes.

The flies discussed as added to the British fauna are but a small fraction on the number of species added to the British List, and this brings me to the purely taxonomic aspects of the changes in our British fauna.

TABLE I. The British Lists, as indicated in check-lists and comprehensive works

Author	Date	Title	Genera included	Species included
Forster	1770	"Catalogue of British Insects"	10	114
Harris	1792	"Exposition of English Insects"		299
Samouelle	1819	"A Nomenclature of British Entomology"	74	75
Curtis	1829	"Guide to an arrangement of British Insects"	224	1980[a]
Curtis	1837	"Guide to an arrangement of British Insects"	293	2134[a]
Stephens	1829	"Systematic Catalogue of British Insects"	219	1673[a]
White	1853	"List of animals in the British Museum" (No Nematocera)	198	1557
Walker	1851–53	"Insecta Britannica Diptera"		2074
Morris	1865	"Catalogue of British Insects"	367	2344
Verrall	1888	"List of British Diptera"	603	3527 ("843 require confirmation")
Verrall	1901	"List of British Diptera"	665	2881 ("303 require confirmation")
Kloet and Hincks	1945	"Check List of British Insects"	1133	5219
Present estimate	—		1166	5728

[a] Include many *nomina nuda*.

THE BRITISH LIST

As I said at the beginning, only after comprehensive taxonomic works are available for the whole order can we hope to understand our fly fauna completely.

Our knowledge of the British Diptera is still at a stage where the major changes in the British List are due to the activities of taxonomists rather than of the flies themselves.

The size of the British dipterous fauna as reflected in "check-lists" is shown in Table I.

Taxonomic changes can arise in a variety of ways: (1) as a result of a revision when all available material, including types, is studied. This is obviously the most satisfactory way of effecting changes and the following categories are merely divisions of this; (2) synonymizing supposed endemic species (dealt with above) by recognition of conspecificity with described continental taxa; (3) correction of misidentifications of species with closely allied continental taxa.

When a neglected family is studied a great change in the British List is to be expected. Of the 507 additions to the British List since Kloet and Hincks' "Check List" of 1945, no less than 223 were Agromyzidae (leaf-miners) added largely by the efforts of two workers, Spencer and Griffiths.

Species "readily identifiable on sight" are rarely checked in keys, thus the presence of two superficially very closely related species may remain undetected. In the family Conopidae, the difficult genus *Myopa* has one or two more distinct and readily identifiable species. One such species, *Myopa polystigma* Rondani, has a more spotted wing and has been more readily separated from its fellows. However, this was recently shown (Smith, 1970) to be two distinct species and the true *polystigma* was shown to be the rarer, the one usually called *polystigma* being in fact *M. tessellatipennis* Motschulsky, a species described from Russia. From modern collections it now appears that *M. tessellatipennis* is commoner in northern Europe and *M. polystigma* in southern Europe.

The previous example of misidentification resulted in an addition of one species to our List; the following example results in a *status quo*.

The handsome fly *Myennis octopunctata* Coquillett (Otitidae) was included on early British lists (Stephens, 1829; Curtis, 1837; Verrall, 1909) as *Ortalis* (or *Loxodesma*) *lacustris* Meigen. It was found in Suffolk in some numbers in 1944, when its correct identity was established.

ACKNOWLEDGEMENTS

For unpublished information I am grateful to P. N. Crow and H. Pennington (*Eriozona*); R. Edwards and K. M. Harris (wasp abundance); and G. R. Else and C. O. Hammond (*Volucella zonaria*).

I thank the following colleagues at the British Museum (Natural History) for useful information and discussion: B. H. Cogan, A. M. Hutson, H. Oldroyd and A. C. Pont.

REFERENCES

BIERNE, B. P. (1944). The causes of the occasional abundance or scarcity of wasps (*Vespula* spp.) (Hym., Vespidae). *Entomologist's mon. Mag.* **80**, 121–124.

BIERNE, B. P. (1952). "The Origin and History of the British Fauna." Methuen, London.

BLAIR, K. G. (1948). Some recent additions to the British insect fauna. *Entomologist's mon. Mag.* **84**, 51–52.

BUSVINE, J. R. (1966). "Insects and Hygiene", 2nd edition. Methuen, London.

CARPENTER, G. H. (1930). A review of our present knowledge of the geological history of insects. *Psyche, Camb.* **37**, 15–34.

COE, R. L. (1943). *Drosophila repleta* Wollaston (Dipt., Drosophilidae), new to Britain. with notes on the species and some account of its breeding habits. *Entomologist's mon. Mag.* **79**, 204–207.

COLLIN, J. E. (1920). Some records of British Diptera. *Entomologist's mon. Mag.* **56**, 137–138.

COLLIN, J. E. (1961). "British Flies, 6. Empididae." Cambridge University Press.

COLYER, C. N. and HAMMOND, C. O. (1968). "Flies of the British Isles", 2nd edition. Warne, London.

CONWAY, J. A. (1970). *Ophyra capensis* Wiedemann (Dipt., Muscidae)—a new ecological niche for this species in Britain? *Entomologist's mon. Mag.* **106**, 15.

CROW, P. N. (1969). *Eriozona syrphoides* Fallén (Diptera, Syrphidae) in North Wales—a new British species and genus. *Entomologist's Rec. J. Var.* **81**, 237–238.

CULL, S. B. (1972). A second British record of *Eriozona syrphoides* Fallén (Dipt., Syrphidae) from Carmarthenshire. *Entomologist's mon. Mag.* **107** [1971], 173.

CURRAN, C. H. (1920). Notes on some Syrphidae (Diptera) collected in England and France during 1917–18. *Can. Ent.* **52**, 35–37.

CURTIS, J. (1829–31). "A Guide to an Arrangement of British Insects." Westley and Davis, London.

CURTIS, J. (1837). "A Guide to an Arrangement of British Insects", 2nd edition. Pigot, London.

DOWNES, J. A. (1969). The swarming and mating flight of Diptera. *A. Rev. Ent.* **14**, 271–298.

EDWARDS, J. (1944). Relative abundance of *Volucella* spp. (Dipt., Syrphidae) and wasps (Hym., Vespidae). *Entomologist's mon. Mag.* **80**, 293.

ELTON, C. (1958). "The Ecology of Invasions by Animals and Plants." Methuen, London.

EMDEN, F. I. VAN (1954). Diptera Cyclorrhapha Calyptrata 1. (a) Tachinidae and Calliphoridae. *Handbk Ident. Br. Insects* **10**(4a), 1–133.

FORSTER, J. R. (1770). "A Catalogue of British Insects." Privately published, Warrington.

FOX-WILSON, G. (1946). Factors affecting populations of social wasps, *Vespula* species, in England (Hymenoptera). *Proc. R. ent. Soc. Lond.* (A), **21**, 17–26.

FRASER, F. C. (1946a). *Volucella* (Dipt., Syrphidae) larvae sp., breeding in a nest of *Vespula vulgaris* (Hym., Vespidae). *Entomologist's mon. Mag.* **82**, 55–57.

FRASER, F. C. (1946b). A final report on the breeding of *Volucella* (Dipt., Syrphidae) larvae in a nest of *Vespula vulgaris* L. (Hym., Vespidae). *Entomologist's mon. Mag.* **82**, 158.

GOFFE, E. R. (1944). The scarcity of wasps (Hym., Vespidae). *Entomologist's mon. Mag.* **80**, 186.

GOFFE, E. R. (1945a). Migration in the Syrphidae (Diptera). *Entomologist's mon. Mag.* **81**, 61–62.

GOFFE, E. R. (1945b). *Volucella zonaria* (Poda, 1761) (Dipt., Syrphidae) in Britain. *Entomologist's mon. Mag.* **81**, 159–162.

GRAHAM-SMITH, G. S. (1929). The relation of the decline in the number of horse-drawn vehicles, and consequently of the urban breeding grounds of flies, to the fall of the summer diarrhoea death rate. *J. Hyg.* **29**, 132–138.

HAMMOND, C. O. (1973). *Callicera spinolae* Rondani—extended range. *Entomologist's Rec. J. Var.* 85, 22–24.

HARRIS, M. (1782). "An Exposition of English Insects." Privately published, London.

D'HERCULAIS, J. K. (1875). "Recherches sur l'organisation et le développement des *Volucelles* insectes diptères de la famille des Syrphides." Académie de médecine, Paris.

HOBBY, B. M. (1946). What is the present status of *Volucella zonaria* Poda (Dipt., Syrphidae) in Britain? *Proc. R. ent. Soc. Lond.* (A), **21**, 1–2.

JOHNSON, C. G. (1961). Syrphid (Dipt.) migration on the Norfolk coast in August 1960. *Entomologist's mon. Mag.* **96**, 196–197.

JOHNSON, C. G. (1969). "Migration and Dispersal of Insects by Flight." Methuen, London.

KLOET, G. S. and HINCKS, W. D. (1945). "A Check List of British Insects." Privately published, Stockport.

LANG, W. D. (1918). "A map showing the known distribution in England and Wales of the Anopheline mosquitoes, with explanatory text and notes." British Museum (Natural History), London.

MACGREGOR, M. E. (1919). On the occurrence of *Stegomyia fasciata* in a hole in a beech tree in Epping Forest. *Bull. ent. Res.* **10**, 91.

MARSHALL, J. F. (1938). "The British Mosquitoes." British Museum (Natural History), London.

MORRIS, F. A. (1865 [–67]). "A Catalogue of British Insects, in all the Orders." Longman, London.

NELSON, M. (1972). *Coniosternum tinctinervis* Becker, a Scatophagid fly new to Britain (Diptera). *Entomologist's Gaz.* **23**, 247.

OLDROYD, H. (1954). The seaweed-fly nuisance. *Discovery, Lond.* **15**, 198–202.

OLDROYD, H. (1964). "The Natural History of Flies." Weidenfeld and Nicolson, London.

OLDROYD, H. (1969). Diptera Brachycera (a) Tabanoidea and Asiloidea. *Handbk Ident. Br. Insects* **9**(4), 1–132.

OLDROYD, H. (1970). Diptera 1. Introduction and key to families (3rd edn). *Handbk Ident. Br. Insects* **9**(1), 1–105.

PENNINGTON, H. (1974). *Eriozona syrphoides* Fallén (Dipt., Syrphidae) new to England and Scotland. *Entomologist's mon. Mag.* (in press).

ROTHSCHILD, M. (1946). A note on the hornet *Vespa crabro* L. at Ashton Wold. *Entomologist* **79**, 175–176.

RUSSELL, H. M. (1959). *Seioptera vibrans* L. (Dipt., Otitidae) feeding on honeydew. *Entomologist's mon. Mag.* **95**, 129.

SAMOUELLE, G. (1819). "A Nomenclature of British Entomology." Boys, London.

SMITH, K. G. V. (1964). *Chersodromia cursitans* Zetterstedt (Dipt., Empididae) reinstated as a British species. *Entomologist's mon. Mag.* **99**, 127–128.

SMITH, K. G. V. (1968). The immature stages of *Rhamphomyia sulcata* Meigen (Diptera: Empididae) and their occurrence in large numbers in pasture soil. *Entomologist's mon. Mag.* **104**, 65–68.

SMITH, K. G. V. (1969). Diptera Conopidae. *Handbk Ident. Br. Insects* **10**(3a), 1–19.

SMITH, K. G. V. (1970). The identity of *Myopa polystigma* Rondani, and an additional British and continental species of the genus (Diptera). *Entomologist* **103**, 186–189.

SMITH, K. G. V. (1971). *Eustalomyia hilaris* Fallén (Diptera, Anthomyiidae) confirmed as British, with notes on other species of the genus. *Entomologist's Gaz.* **22**, 55–60.

STALLWOOD, B. R. (1947). The increase of the hornet (*Vespa crabro*). *Entomologist* **80**, 119.

STEPHENS, J. F. (1829). "A Systematic Catalogue of British Insects", vol. 2. Baldwin and Cradock, London.

STEPHENS, J. F. (1846). "Illustrations of British Entomology, Supplement." Bohn, London.

STEPHENSON, J. W. and KNUTSON, L. V. (1970). The distribution and snail-killing flies (Dipt., Sciomyzidae) in the British Isles. *Entomologist's mon. Mag.* **106**, 16–21.

STUBBS, A. E. (1967–68). Geology as an ecological factor in the distribution of an insect. The fly *Symphoromyia immaculata* F. (Dipt., Rhagionidae). *Entomologist's Rec. J. Var.* **79**, 292–293, 313–316, **80**; 22–59, 80–83.

TAYLOR, E. (1955). Seaweed flies (Diptera, Coelopidae) at Oxford. *Entomologist's mon. Mag.* **91**, 97.

TAYOR, F. SHERWOOD (1942). "The Century of Science." Heinemann, London.

TUTT, J. W. (1898–1902). Migration and dispersal of insects. *Entomologist's Rec. J. Var.* **10–14** [many parts].

VERRALL, G. H. (1888). "A List of British Diptera", 3 vols. Pratt, London.

VERRALL, G. H. (1901a). "A List of British Diptera", 3 vols. Cambridge University Press.

VERRALL, G. H. (1901b). "British flies, 8. Platypezidae, Pipunculidae, and Syrphidae of Great Britain." Gurney and Jackson, London.

VERRALL, G. H. (1909). "British flies, 5. Stratiomyidae and succeeding Families of the Diptera Brachycera of Great Britain." Gurney and Jackson, London.

WAINWRIGHT, C. J. (1944). Migratory Diptera. *Entomologist's mon. Mag.* **80**, 225–226.

WALKER, F. (1851). "Insecta Brittannica. Diptera", 3 vols. Lovell Reeve, London.

WHITE, A. (1853). "List of the specimens of British animals in the collection of the British Museum. XV. nomenclature of Diptera." British Museum (Natural History), London.

WINGATE, W. J. (1906). A preliminary list of Durham Diptera, with analytical tables. *Trans. nat. Hist. Soc. Northumb.* **2**, 1–416.

21 | Arthropod Ectoparasites of Man

K. MELLANBY

Monks Wood Experimental Station, The Nature Conservancy,
Abbots Ripton, Huntingdon

Abstract: Man supports four arthropod ectoparasites on or in his skin. These are the Body Louse, Head Louse, Crab Louse and Itch Mite. Social factors and methods of treatment have changed, and there have been changes in the incidences of the parasites. Body Lice are nearly extinct in Britain, but the others are commoner in 1973 than they were in 1945.

This short paper may be thought of as an appendix to that of K. G. V. Smith on the Diptera (Chapter 20) which dealt with the problems of blood-sucking insects which attack man. We also have a group of arthropods which not only live on man's blood or tissue fluids as a diet, but which are only found living on his person. They cannot exist and breed successfully on any other animal. They are also interesting because they have no obvious natural enemies, except man himself, to control their numbers, and, although their habitat is so specialized and restricted, the different species do not appear to compete with one another. The numbers of these creatures have fluctuated in recent years, and these fluctuations illustrate a series of interesting ecological problems.

The animals concerned are:

Pediculus humanus humanus L. (Body Louse)
Pediculus humanus capitis De G. (Head Louse)
Phthirus pubis L. (Crab or Pubic Louse)
Sarcoptes scabiei L. (Itch Mite)

Body lice (Buxton, 1947) live normally on the inner sides of clothes, and the eggs are attached to the fabric. The insects feed by sucking blood, which they normally do many times a day. When not feeding they retreat onto the garments. Under favourable conditions, when the clothes are worn throughout the

---

Systematics Association Special Volume No. 6, "The Changing Flora and Fauna of Britain", edited by D. L. Hawksworth, 1974, pp. 393–397. Academic Press, London and New York.

24 h, the eggs hatch in under a week and the life cycle takes about a fortnight. If the clothes are removed at night the lice and the eggs are cooled, the insects are starved, and the life cycle is considerably prolonged. The insects sometimes come onto the outside of the garments, and may thus infest a new host, but he will not develop a flourishing culture of parasites unless he picks up a fertilized female or specimens of both sexes, and he keeps the same clothes on all the time for several weeks. Louse populations were common on the very poor in Britain in the nineteenth century, and in wartime on soldiers unable to wash or change their garments. Today the body louse is almost extinct in Britain. A few old tramps harboured the only specimens until recently, when reports of young "hippy-type dropouts" with lice have been received. Nevertheless, the body louse is likely to disappear from the British fauna quite soon, unless efforts to conserve it are made. As I formerly reared colonies of this insect in pillboxes with gauze tops kept in my socks for many years, I do not favour their conservation. I think we should allow Pediculus humanus humanus to disappear.

The head louse is a subspecies with very different habits. It lives only on the head, and sticks its eggs on to hairs near to the scalp. Head lice do sometimes crawl onto hats and onto the outside of the coiffure, whence transmission to another person takes place, but they usually remain near the scalp, and have a generation time of about two weeks. Head lice are still common in Britain, and the numbers of infestations seem to be on the increase (Maunder, 1971). Head lice were commonest on children, particularly girls, whose longer hair provided a more favourable habitat. In 1940, an outcry arose when so many town children who were evacuated to rural areas were found to be lousy, in spite of official statistics from the Board of Education's Medical Service which suggested that this parasite was comparatively rare. I therefore investigated the situation (Mellanby, 1941) and found that, in Liverpool for instance, over 50% of school girls were infected whilst boys were about half as lousy. The picture in other cities was similar though not quite so serious, while rural children were much less commonly infested. Lousiness was commonest in large families, where transmission was easiest and care most difficult and could be correlated with various social groupings.

Soon after my report was published, the louse population fell, and in the immediate post-war years many thought that Pediculus humanus capitis would soon be extinct. The reduction occurred because the problem was tackled more vigorously. Many previously undetected infestations were recognized and treated. Also we now had medicaments which were more effective. Previously it had not been difficult to kill the adults and larvae on the scalp, but the eggs usually survived unless they succumbed to the lengthy and difficult process of combing with a "louse comb". We first used lauryl thiocyanate and then DDT, substances with sufficient persistance to render the hair louse-proof for a period. From the point of view of the insect, "environmental pollution" was now a limiting factor!

Unfortunately we have had, in recent years, a considerable increase in the rate of infestation, reaching as high as 10% of children in some areas. The reasons for this are twofold. First, we have seen the appearance of head lice which are resistant to DDT—natural selection has thus produced a change in the insect in response to this new factor in the environment. Secondly, the effort to detect and control infestation has relaxed, and modern trends in fashion, decreeing that both boys and girls should grow their hair longer, has provided the louse with a more favourable habitat. A head of long, unkempt hair is as favourable a breeding place as one with an elaborate style kept undisturbed day and night in a hair net.

The crab louse, *Phthirus pubis*, is a completely different species. Its life history is similar to the head louse, but it breeds only among the pubic hair and consequently only affects individuals over the age of pubity. The anatomy of the insect makes it best able to exist among hair which has a broad cross-section which fits its claws, and where the hairs are comparatively widely spaced. Thus the head is an unsuitable habitat, though the insect has been found inhabiting the eyebrows, eyelashes and even, in a few cases, the beard, although breeding probably only occurs in the pubic regions. Crab lice are usually transmitted venereally, though garments and bedding cannot be entirely ruled out as means of infestation. Crab lice were quite common among soldiers during the war, and here again the rate of infestation fell rapidly after 1945. It seems probable that numbers are again rising. Insecticide resistance does not seem to be an important factor here, and growing promiscuity seems to be the most likely cause, as the insects are found throughout groups of young people who enjoy what would formerly have been an unusual degree of shared intimacy.

The fourth ectoparasite is *Sarcoptes scabiei*, the itch mite, which causes the disease known as scabies by burrowing in the horny layer of the human epidermis (Mellanby, 1972). This disease was rare in the 1920s, it rose slowly in the 1930s, and appeared to be epidemic during the early years of the 1939–45 War. The reasons for these fluctuations were unknown, though blame was (wrongly) attributed to "wartime conditions". In 1941 at least 2% of the population of Britain was affected. When I investigated this situation, I found that, contrary to common belief, blankets and clothing ("fomites") were seldom the source of infection. The mite usually remained in or on the human skin, and close personal contact was necessary for transmission commonly to occur. Thus children in overcrowded houses passed the parasites to and fro, and sharing a bed with a patient was found to be the normal means of picking up the disease. Young soldiers frequently obtained the infection venereally, but non-venereal transmission was most common. An important finding was that an individual never previously infected showed no symptoms for perhaps six weeks, during which time a substantial mite population developed in his epidermis, and during which period he was also a potential source of infection.

TABLE I. The present state of the populations of the arthropod ectoparasites of man in Britain and factors responsible for the population changes

Species	State of the population	Probable reasons for population changes
Pediculus humanus humanus (Body Louse)	Almost extinct	Improved social conditions and cleanliness
Pediculus humanus capitis (Head Louse)	Increasing	(a) Resistance to insecticides (b) Social (longer and dirtier hair)
Phthirus pubis (Crab Louse)	Probably increasing	Social (greater sexual promiscuity)
Sarcoptes scabiei (Itch Mite)	Increasing	(a) Infestations not always properly diagnosed and treated (b) Populations now more susceptible to infestation

On the other hand when a cured patient was reinfected he started to itch at once, reinfection was much more difficult, and mite populations if they developed at all were small. So spread was clearly much easier in a community where scabies was a new factor.

During and after the 1939–45 War *Sarcoptes scabiei* was vigorously attacked, and the incidence reduced so much that many medical students between 1945 and 1965 never saw a single case during their training. Thus when the disease started to reappear in the 1960s, cases escaped recognition and were not treated. Spread was also easy in a population most of whose younger members had never had the disease. These two factors are, in my opinion, sufficient to explain the spread which is still continuing.

The situation for these four ectoparasites is summarized in Table I. It will be seen that the factors controlling the population differ for each species, though they all exist in the same, specialized, habitat.

REFERENCES

BUXTON, P. A. (1947). "The Louse." Arnold, London.
MAUNDER, J. W. (1971). The use of malathion in the treatment of lousy children. *Community Medicine* **126**(10), 145–147.
MELLANBY, K. (1941). The incidence of head lice in England. *Medical Officer* **65**, 39–45.
MELLANBY, K. (1972). "Scabies", 2nd edition. Classey, Middlesex.

22 | Some Comments on the Aculeate Fauna

J. C. FELTON

Shell Research Ltd, Woodstock Laboratory
Sittingbourne, Kent

Abstract: The number of species of aculeate Hymenoptera, excluding Bethyloidea and Chrysidoidea, recognized as British has increased steadily at about one species per year during the present century. In 1972 it stood at 483, a 28% increase on the total in 1896. An examination of the net gain of 25 species since 1943 reveals that 79% of the change represents an increase in our knowledge of the existing fauna and only 21% (6 species) *may* represent actual newly established species. No estimate of possible extinctions has been attempted.

Detailed consideration of the Kent fauna (384 species) suggests that 75% are stable and that the remainder are about equally divided between species captured before 1920 but not since (15%) and species captured since 1920 but not before (10%).

Some examples of the extreme persistence of aculeate species, even at very low population density, are given, together with some examples of increases and possible decreases.

An examination of the distribution patterns of aculeates in Britain suggest possible explanations for the observed facts and also the types of changes that might be expected.

Four main reasons are advanced for the desirability of studying population trends in aculeates. Suggestions for such studies compatible with conservation interests include a more specialized handling of the British list, particular recording methods, and relevant localities and species.

INTRODUCTION

The Aculeata, the ants, bees and wasps, are marked off from the rest of the Hymenoptera more by behavioural than structural characters. They are primarily specialized predators in which the adult is the raptor and provides invertebrate food for the larva, for which it also expends more or less effort to protect within a nest. The adult itself obtains its energy source from plants although in some cases juices are also sucked from the animal prey intended for the larva. The most advanced aculeate group, the bees (Apoidea), have returned

Systematics Association Special Volume No. 6, "The Changing Flora and Fauna of Britain", edited by D. L. Hawksworth, 1974, pp. 399–418. Academic Press, London and New York.

to complete direct dependence on plant food by using pollen as the primary protein source for the larva.

Aculeates are small to medium sized, mobile, rare animals, with specialized requirements for food and nesting sites, often aggregated populations, difficult to catch and to identify. This has led to a corresponding rarity of Hymenopterists and this again to the major difficulty in discussing changes in their distribution and abundance: the extent to which apparent changes relate to the actual creatures or merely to our knowledge of them. This will first be examined by considering the British List.

For the purpose of the present paper, the two most primitive groups of the Aculeata, the Bethyloidea and Chrysidoidea, have not been considered for a combination of reasons: neither group was included in Edward Saunders' book (1896); our knowledge of the Bethyloidea is not at a stage when their overall distribution can be realistically assessed; within the Chrysidoidea it is known that *Chrysis ignita* in fact represents a complex of seven or eight species and this question has not yet been resolved.

CONSIDERATION OF THE BRITISH LIST

The British Aculeata were first comprehensively reviewed by Saunders (1896). In 1937, the relevant part of the "Generic Names of British Insects" appeared, including a check list of species by Richards. An annotated check list by Yarrow

TABLE I. Summary of the British list of Aculeata

| Group | Number of species recognized in | | | |
	1896 (Saunders)	1937 (Richards)	1943 (Yarrow)	1972
Scolioidea	8	8	8	8
Formicoidea	21 (+7)[a]	37	38	43
Pompiloidea	30	38	40	39
Vespoidea	23	27	27	29
Sphecoidea	91 (+1)[a]	103	104	114
Apoidea	203 (+2)	240	241	250
Total	376 (+10)[a]	453	458	483
No. of species added	77 (67)[a]	5	25	
No. of years	41	6	29	
No. of species added per year	1·9 (1·6)[a]	0·8	0·9	

[a] Ten named forms mentioned by Saunders are now recognized as species.

appeared in the "Hymenopterists' Handbook" (1943). These three lists have been collated, and the additions made since 1943 included, in an Appendix.* A summary of the number of species recognized is given in Table I.

The data in Table I show that during this century the number of species of Hymenoptera recognized as British has been increasing steadily. Over the first 40 years, the rate was about 1·5 species per year. Over the last 35 years the rate has remained steady at very close to one species per year.

TABLE II. Summary of changes to the British list of Aculeata since 1943

Category of change	Number of species
Species deleted as identifications or provenance of records unacceptable	2
Previously recognized forms raised to species level	1
Previously recognized forms now actually named	1
Newly recognized species for which early dated captures are known	19
Newly recognized species for which no early dated captures are known	6

To see what types of change this increase involves, the net gain of 25 species since 1943 has been examined further in Table II. These data show that, of the 29 changes, only 6 (21%) may reflect an actual increase in the fauna rather than in our knowledge of it.

However, this method of analysis does not locate extinctions. Indeed, the very rarity of many aculeates renders this extremely difficult. Let us attempt to do so by considering a county list in more detail.

CONSIDERATION OF THE KENT LIST

Kent is a county rich in aculeates and also comparatively well collected. The species currently known from Kent are also indicated on the list in the Appendix.* They have been collated in a somewhat different way from the British list. Rather than indicating which species were recognized as Kentish at a

---

* This Appendix is not reproduced here. Copies are deposited in the Biological Records Centre of The Nature Conservancy, Monks Wood Experimental Station, Abbots Ripton, Huntingdon, and in the Maidstone Museum, Maidstone, Kent.

particular time, the capture data have been examined and lists of the species actually taken in Kent in five time periods, last century and the four twenty-year periods this century, have been compiled. The totals are summarized in Table III. In Table IV a further collation into two time periods is given. The totals for each time period in Table III reflect the amount of collecting in the county, intense up to about 1920 (Felton, 1963; Frisby, 1928) and again since

TABLE III. Summary of the Kent list of Aculeata

Group	Number of species captured in Kent					
	pre-1899	1900–19	1920–39	1940–59	1960–72	Ever
Scolioidea	6	3	2	3	6	7
Formicoidea	21	12	9	21	29	29
Pompiloidea	27	13	13	4	19	31
Vespoidea	19	15	12	12	17	22
Sphecoidea	69	46	20	48	69	91
Apoidea	175	162	69	107	156	203
Total	317	251	125	195	298	384

TABLE IV. Stability of the Kent aculeate fauna

Group	Number of species captured		
	Pre-1920 only	Post-1920 only	Both periods
Scolioidea	1	1	5
Formicoidea	0	6	23
Pompiloidea	8	1	22
Vespoidea	1	3	18
Sphecoidea	15	15	61
Apoidea	34	12	157
Total	57	39	288

1950 (Leclerq, 1968; Felton, many notes in *Bull. Kent Fd Club*). The data in Tables III and IV also give a good indication of personal collecting preferences. The recent emphasis on collecting ants (Felton, 1967a, 1969) is reflected in the fact that all the indigenous ant species known from Kent have been taken in the county since 1960. Similarly, the Sphecoidea, of particular interest to Leclerq

(1968) and myself, are better represented in recent time periods than the Apoidea, favourites with Elgar (Felton, 1963) and, of course, Sladen (1908).

The overall impression left by the data in Table IV is one of stability. 75% of the species occurred in both the main time periods. In fact 62 species (16% of the total) occurred in all of the five time periods of Table III, a very high number considering that only 125 species in total were captured between 1920 and 1939. Beneath this body of stable species, there is considerable flux among the remaining 25% of species. However, there is no marked trend, 15% of the species being recorded only from the earlier time period and 10% only from the later.

EXAMPLES OF STABILITY AND CHANGE

Several examples of the persistence or stability of ant populations have been quoted by Collingwood (1971). Perhaps two examples are sufficient to illustrate the point for other groups. In 1900, Elgar captured the rare pompilid *Aporus unicolor* at Upper Halling (Felton, 1963). This is at first sight an unlikely capture on the chalk of a species that is normally coastal or associated with the southern heathlands. Yet in 1967, Eric Philp took a specimen of the same species probably within 2–3 km of the original locality (Felton, 1968c). An even more extreme example is that of the small sphecid wasp *Crossocerus exiguus* (Fig. 1). Three specimens are known for the country, from Barming in 1896 (Felton, 1963), Bedgebury in 1935 (Richards, 1935) and from Pluckley in 1967 (Felton, 1968a). These localities are only about 20 km apart and clearly suggest a persisting population.

Among the examples of species that are known to have extended their range quoted by Richards (1964) one aculeate was included, the eumenid wasp *Microdynerus exilis*. To this list, the sphecid wasp *Ectemnius dives* can probably be added. Apart from one early specimen from Chester, the species was first recorded in numbers from S.E. London in 1926 (Nixon), it reached Glamorgan in 1935 (Hallett), but the spread eastwards was slower, West Kent by 1939 (Yarrow and Guichard, 1941) and East Kent by 1966 (Felton, 1967b). One more example of a spread is the cuckoo bee *Coelioxys conoidea*. As its synonym *vectis* suggests, this species is long known from the Isle of Wight through central England to S.W. Yorkshire (Saunders, 1896). Another species of *Coelioxys*, *C. mandibularis*, is known from the same host, *Megachile maritima*, on the Glamorgan, Lancashire and Cheshire sand-dunes. The absence of *C. conoidea* from Glamorgan is particularly remarked upon by Hallett (1928). In 1972, of 13 specimens of *Coelioxys* taken at Methyrmawr and Kenfig, Glamorgan, two from each locality were *C. conoidea* (Felton, in preparation). It is also only

o

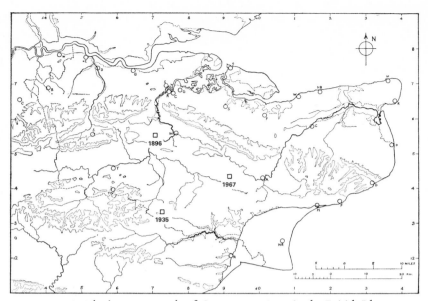

FIG. 1. The known records of *Crossocerus exiguus* in the British Isles.

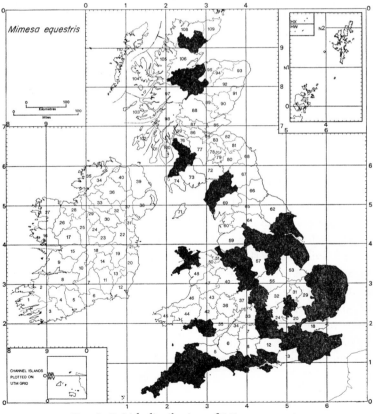

FIG. 2. British distribution of *Mimesa equestris*.

recently that *C. conoidea* has been added to the Kent list, from Kemsing in 1952 (Leclerq, 1968), and Deal in 1964 (Chalmers-Hunt, 1965). *C. mandibularis* is also now known from this latter area (Felton, 1959; Chalmers-Hunt, 1965).

As already stated, decreases are much more difficult to establish in aculeates than increases. Several species of infrequent occurrence which have not been recaptured recently are mentioned by Richards (1964), notably *Philanthus triangulum* and *Bombus pomorum*. These species are little more than accidentals and scarcely differ in status from *Pollistes gallicus* and *Xylocopa violacea*, both of which have bred in this country and yet are not accepted on the British list (Richards, 1964). Two species that have been actively sought for and yet not recaptured over recent years are *Andrena polita* in West Kent, known up to 1933 (Frisby, 1934) but not found recently (Guichard, *in litt.*), and *Bombus cullumanus*, not refound on the Berkshire Downs (Yarrow, *in litt.*).

TABLE V. Occurrence of two *Ancistrocerus* species in Kent

Species	Number of specimens recorded				
	pre-1899	1900–19	1920–39	1940–59	1960–72
parietum	12	7	1	2	0
gazella	7	9	3	10	20
% *gazella*	37%	56%	75%	83%	100%

A particular case combining the preceding two types of change is that of replacement. One example illustrates the point. *Ancistrocerus parietum* and *A. gazella* are closely similar species, the latter only recently recognized as British (Yarrow, 1954a). There is some indication that the former preponderates in early collections, the latter in more recent ones. The data from Kent are summarized in Table V and certainly agree with this suggestion.

However, these data are merely suggestive and it is an assumption that the changes are real and that they are directly related.

DISTRIBUTION PATTERNS

We may better detect changes in distribution if we first examine typical distribution patterns. A general account of these is given by Richards (1964) and, for ants in particular, by Collingwood (1965, 1971).

Many species are found throughout the British Isles. The distribution of *Mimesa equestris* (Fig.2) is typical except that this species is not found in Ireland. *Psenulus atratus* (Fig. 3) does not extend far north but is found in the south of

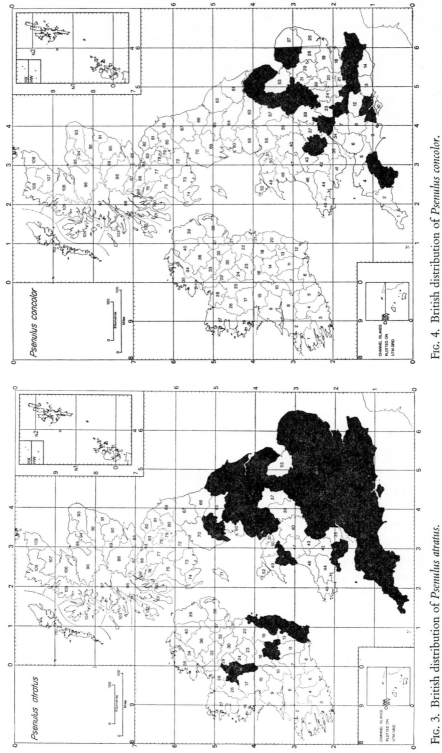

Fig. 3. British distribution of *Psenulus atratus.*

Fig. 4. British distribution of *Psenulus concolor.*

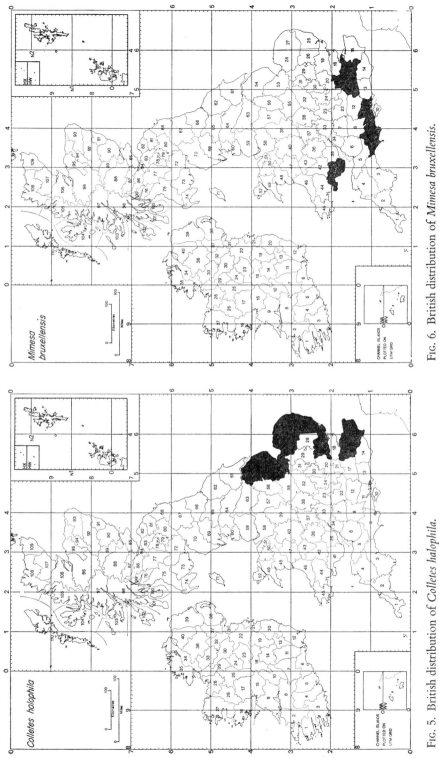

FIG. 5. British distribution of *Colletes halophila*.

FIG. 6. British distribution of *Mimesa bruxellensis*.

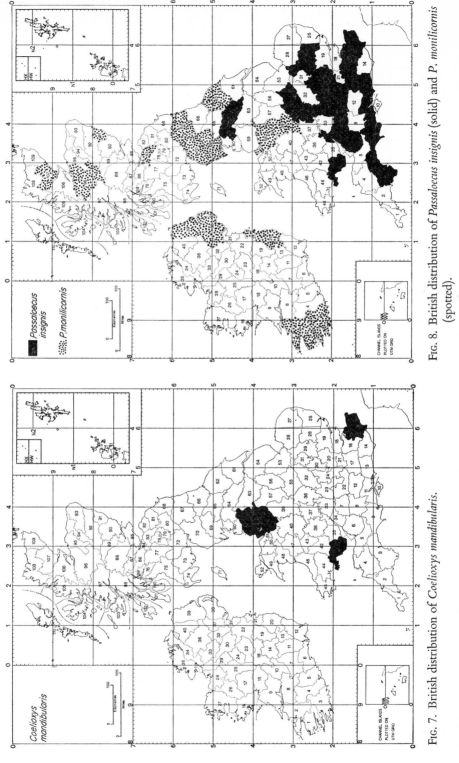

FIG. 8. British distribution of *Passaloecus insignis* (solid) and *P. monilicornis* (spotted).

FIG. 7. British distribution of *Coelioxys mandibularis*.

Ireland. This is not a common pattern as the usual generalization is "Aculeates found in Ireland also occur in S.W. Scotland". The distribution of *Psenulus concolor* (Fig. 4) is typical of a number of species, the northern limit being distinctly angled north-east to south-west, in this case following the Humber–Severn line.

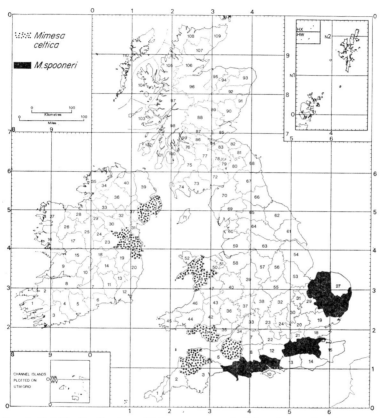

FIG. 9. British distribution of *Mimesa spooneri* (solid) and *M. celtica* (spotted).

Colletes halophila (Fig. 5) represents a peculiar eastern fringing distribution, a "Doggerland" species. The natural habitat of this species, salt marsh with adjacent gravel banks, is not nearly so restricted. A southern fringing distribution is exemplified by *Mimesa bruxellensis* (Fig. 6) and in even more extreme form by *Coelioxys mandibularis* (Fig. 7).

Figure 8 illustrates the distribution of a north-west to south-east species pair, in this case *Passaloecus monilicornis* and *P. insignis*, although several other examples could have been chosen. An example of west to east species pair with a southern fringing distribution is given in Fig. 9, *Mimesa celtica* and *M. spooneri* (the name proposed by Richards (1947) for British specimens previously called *M. unicolor*).

Some ant distribution patterns from Collingwood (1971) are illustrated in Fig. 10. Northern limits of distribution running both east to west and north-east to south-west are found. The rare southern *Formica rufibarbis* falls in the latter pattern, yet the even more restricted *Myrmica specioides* represents the north-west extremity of a predominantly south-east European distribution. Perhaps this species has indeed spread north-westwards as has the Collared Dove (*Streptopelia decaocto*).

The possible factors causing these distribution patterns can now be discussed.

FACTORS DETERMINING RECENT DISTRIBUTION

1. Climatic Factors

Aculeate Hymenoptera are sun and warmth loving and this is generally taken as a dominant factor in their distribution (Richards, 1964; Haeseler, 1972). The richness of the fauna of the Dorset, Hampshire and Surrey heathlands is explained in this way.

The distribution patterns of *Psenulus atratus* (Fig. 3) and *Stenamma westwoodii* (Fig. 10) suggest that these shade-loving species may be limited by mean summer temperature alone. The 59°F daily mean isotherm for July is an approximation to the northern limit of both species in the British Isles.

However, the fringing distributions of *Mimesa bruxellensis* (Fig. 6), *Coelioxys mandibularis* (Fig. 7), and *Mimesa celtica* and *M. spooneri* (Fig. 9) suggest the importance of insolation in these open-ground nesters. Perhaps the case cited above of *Aporus unicolor* occurring on the chalk is also indicative. The form of chalk slopes is well studied (Clark, 1965) and the two commonest scarp slope angles, 26° and 33°, span the optimum angle for maximum incident energy at the summer solstice at 50° N which is 27° (Geiger, 1966).

These fringing distributions also suggest an association with "growing season" as Fig. 11 (from Coppock, 1964) shows. The significance of this has not yet been investigated.

Rainfall is another important climatological variable. It is clearly implicated in determining the abundance of social wasps (Fox-Wilson, 1946). It is tempting to surmise that the Humber–Severn line so often observed as the north-west extent of a species in England (Figs 4, 8) is the result of a balance between the higher rainfall in the west and the higher temperature in the south during the summer. It seems that rainfall itself may affect seasonal abundance rather than actual occurrence.

Wind is a factor particularly important in exposed habitats. It has been found to affect the micro-distribution of ant species on the Dorset heathlands (Brian, personal communication). It is also evident that certain bumble bee

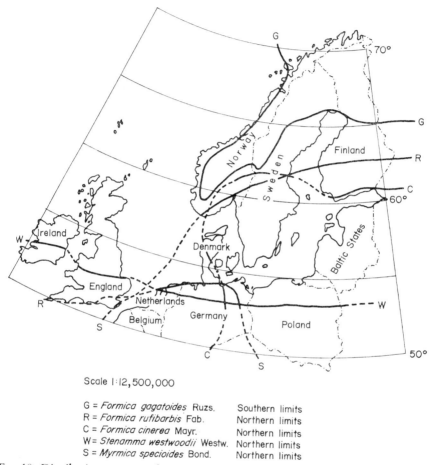

Scale 1:12,500,000

G = *Formica gagatoides* Ruzs.   Southern limits
R = *Formica rufibarbis* Fab.   Northern limits
C = *Formica cinerea* Mayr.   Northern limits
W = *Stenamma westwoodii* Westw.   Northern limits
S = *Myrmica specioides* Bond.   Northern limits

FIG. 10. Distribution patterns of some ants (Formicidae) in north-western Europe (after Collingwood, 1971).

species, notably *Bombus muscorum*, are better adapted to windy conditions than others (Yarrow, *in litt.*).

2. Habitat Requirements

Many aculeate species have special requirements for nesting. This is well illustrated by the study of Danks (1971) which clearly indicates that the presence

of suitable cut ends of bramble stems can limit the population size of stem
nesting wasps and bees.

Predacious species often have very specific prey requirements, and certain
bees are also very specialized in their flower requirements. For example, the

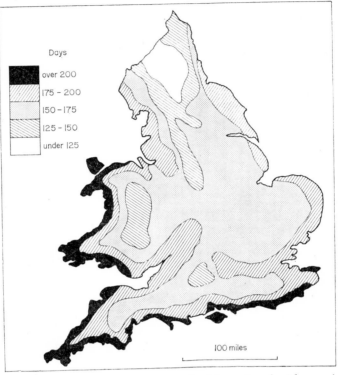

FIG. 11. Length of the growing season in England (reproduced with permission from
Coppock, 1964).

newly added species *Melitta dimidiata* depends upon *Onobrychis viciifolia*, but is
much more limited in its distribution than is the plant (Baker, 1965; Yarrow,
1968).

It seems that this group of factors has more influence on abundance than on
occurrence.

3. Behavioural Factors

In general, aculeate Hymenoptera are strong fliers and at first sight it seems
unlikely that their distribution within these Islands could be limited by their
ability to reach an area. However, there do appear to be behavioural limits

imposed within the mechanical ones. Migratory type movements of bumble bees are now well documented (Rothschild, 1965) and *Bombus pomorum* is said to have crossed the Channel (Richards, 1964), yet no specimens of the white-tailed Continental form of *Bombus terrestris* has ever been recorded from this country although it is common along the coasts of north-west France and Belgium. This does suggest a behavioural barrier.

The study of Butler (1965) clearly demonstrates the importance of a behaviour pattern in limiting distribution. Females of the colonial nesting bee *Andrena flavipes* will only accept a mate and dig a new nesting burrow within the confines of the colony from which they were reared. This is obviously an important factor in limiting the rate of spread of such species. It could well be operating in two of the cases already quoted, *Colletes halophila* and *Melitta dimidiata*.

Perhaps the most interesting and important behavioural factor is that in interspecific competition. The existence of species pairs, such as the two *Passaloecus* spp. (Fig. 8), and of species replacement (Table V), suggest that this factor may often be dominant. The mechanism(s) through which it functions have scarcely been investigated.

4. Human Activities

The recent extensive study of Haeseler (1972) in Germany shows that man-made secondary habitats, such as waste ground and municipal parks, can prove advantageous refuges for aculeates between cities and intense agricultural areas. In this country, the aculeate fauna of Hampstead Heath, Middlesex, has been surveyed by Guichard and Yarrow (1948) and by Yarrow (1954b). A conclusion from this study is that a number of species, particularly bees, have been lost from the Heath over the past 100 years. The Bushy Park area, also in Middlesex, has been surveyed by Yeo (1957) and Felton (1968a). The fauna proved similar to that recorded this century at Hampstead, except that it was poorer in bees. No historical data exist in the case of Bushy Park to establish long-term trends.

Certainly, many wood nesting species must have benefited from the use of spile fencing and the presence of pruned garden shrubs. It is symptomatic that *Psenulus concolor*, considered a very rare species by Saunders, is now known to have a wide distribution (Fig. 4) and *Psenulus schencki* has also extended its range from a limited area of Surrey in the 1920s to reach East Kent by 1970 (Felton, 1971).

Cities are on average warmer by 1°C or more than their surrounding country-side (Ashby, 1971: 41). This level of increase should be sufficient to extend the range of a species northwards, but no examples are as yet known within the aculeates. As the temperature difference is most notable in the winter, it may in fact have little relevance to aculeates.

Few aculeates prefer damp habitats, and so the extensive draining of wet lands, which is a feature leading to change in many groups, may not be detectable with aculeates. Indeed, the two notable wet land species have both been captured in new localities recently, *Passaloecus clypealis* previously restricted to Wicken Fen at Benfleet, Essex, by P. J. Chandler, and *Prosopis pectoralis* in South Hampshire by G. Else.

FURTHER WORK

1. The Desirability of Further Work

It is clear that our knowledge of the changes in aculeate populations, and their causes, is fragmentary and we must first ask whether it is worth while to have a better knowledge in this area. For four main reasons, the answer to this question must be in the affirmative: (a) Aculeates form an important part of the "girder system" (Elton, 1966, p. 377) within ecosystems. (b) Study of their distribution can lead to an increase in our knowledge of the mechanisms by which distribution is determined. (c) Bees are both ecologically and economically important through their function in pollination. (d) Solitary wasps stand in the same relation to their prey as do the birds of prey to theirs. Sphecids could therefore prove very sensitive indicators to changes occurring in ecosystems.

2. Suggested Studies

Bearing in mind that aculeates are in general rare animals, the following types of study are suggested as being likely to yield the desired information and yet be compatible with conservation interests. The objectives must be to obtain maximum information from every capture.

(a) *Handling of the British list.* The British list is significant for two main reasons: only species on the list will be looked for by the majority of workers, and if a species is once accepted on the list it will be considered extinct if not recaptured reasonably frequently. Introduced and accidental species present problems, some examples of which have already been cited. A method of listing which allows of several categories is needed to meet all requirements. An accepted authority, perhaps based on the British Museum (Natural History), the Royal Entomological Society of London, or the Biological Records Centre, would have to be established to operate such a system.

(b) *Recording by vice-county.* While the exact locality of each capture should be known, at least down to the 1 km square, the vice-county unit seems to be the right one for summarizing distribution and frequency of occurrence. It makes optimum use of published data and the often vague locality labels of early (and even some recent) collections. It has been used in Figs 2–9 quite deliberately to emphasize this point. In particular it requires the capture of fewer specimens to establish an initial pattern, desirable both to the conservationists and to the few (over-worked) Hymenopterists.

A complete vice-county survey of the country is an essential pre-requisite to further studies.

(c) *Studies of particular areas.* If the objective is to detect change, and bearing in mind known distribution patterns, some areas of the country are more critical than others. Because good early data exist, it is relevant to continue to monitor populations in Glamorgan, Dorset, Kent and Bedfordshire. Surrey and Middlesex should also be included as being particularly exposed to human pressures. Bushy Park and Hampstead Heath are particular localities. Further north, the triangle marked out by the Wash–Severn–Humber is particularly important as it includes the northern or southern limits of several species in England and merits study. The Solway coast is another area of significance. Crowson (1967) has discussed certain warmth-loving species that are found there and study of the aculeates is obviously merited.

Continued study in relatively few vice-counties, taken in conjunction with the initial vice-county survey suggested above, would materially increase our knowledge.

(d) *Studies of particular groups and species.* The two aculeate groups for which national recording schemes are already operating, the ants and bumble bees, are excellent choices, the former being almost as easy to study as plants, and the latter representing the important pollination aspect. Study of the Sphecidae has already been suggested.

Particular interest centres around species pairs. If change is taking place, one would expect the boundary between two similar, mutually exclusive, species to be particularly sensitive to such change. An example would be the metallic species of *Halictus* (now to be referred to *Lasioglossum*) *morio, smeathmanellus* and *scoticum* (the name now published by Ebmer (1970) for the "sp. nr *smeathmanellus*" of British authors).

CONCLUSIONS

(1) In general the aculeate fauna of the British Isles has been reasonably stable during the present century.

(2) The changes that have been detected are nicely balanced between increases and decreases.

(3) Man's activities have probably played a part in both types of change.

(4) A fuller knowledge of changes in aculeate populations, and their causes, is desirable.

(5) Some acceptable and accepted organization will be required if such knowledge is to be acquired.

REFERENCES

ASHBY, E. (1971). "Royal Commission on Environmental Pollution. First Report." H.M.S.O., London.

ASHBY, E. (1971). *Royal Commission on Environmental Pollution. First Report.* H.M.S.O., London.

BAKER, D. B. (1965). Two bees new to Britain (Hym., Apoidea). *Entomologist's mon. Mag.* **100** (*1964*), 279–286.

BUTLER, C. G. (1965). Sex attraction in *Andrena flavipes* Panzer (Hymenoptera: Apidae), with some observations on nest-site restriction. *Proc. R. ent. Soc. Lond.*, A, **40**, 77–80.

CHALMERS-HUNT, J. M. (1965). Meeting Note. *Proc. S. Lond. ent. nat. Hist. Soc.* **1964**, 35.

CLARK, M. J. (1965). The form of chalk slopes. *Southampton Research Series in Geography* **2**, 3–34.

COLLINGWOOD, C. A. (1965). The distribution of ants (Hymenoptera) of the *Formica fusca* species group in Europe. *Proc. XIIth Int. Cong. Ent.*, 446.

COLLINGWOOD, C. A. (1971). A synopsis of the Formicidae of North Europe. *Entomologist* **104**, 105–176.

COPPOCK, J. T. (1964). "An Agricultural Atlas of England and Wales." Faber and Faber, London.

CROWSON, R. A. (1967). Refuges for warmth-loving species along the Scottish south coast. *Entomologist's mon. Mag.* **102** (*1966*), 245–246.

DANKS, H. V. (1971). Populations and nesting-sites of some aculeate Hymenoptera nesting in *Rubus. J. Anim. Ecol.* **40**, 63–77.

EBMER, A. W. (1970). Die bienen des genus *Halictus* Latr. s.1. im Grossraum von Linz (Hymenoptera, Apidae). *Naturkundliches Jahrbuch der Stadt Linz*, **1970**, 19–82.

ELTON, C. S. (1966). "The Pattern of Animal Communities." Methuen, London.

FELTON, J. C. (1959). Two Aculeate Hymenoptera new to East Kent. *Entomologist's mon. Mag.* **95**, 71.

FELTON, J. C. (1963). The Hymenoptera in the Maidstone Museum Collection. *Trans. Kent Fd Club* **1**, 171–190.

FELTON, J. C. (1967a). A preliminary account of the ants (Hymenoptera, Formicidae) of Kent. *Trans. Kent Fd Club* **3**, 95–121.

FELTON, J. C. (1967b). Reports of recorders: Hymenoptera. *Bull. Kent Fd Club* **12**, 6–7.

FELTON, J. C. (1968a). Notes on the Hymenoptera of the Bushy Park area, Middlesex. *Lond. Nat.* **46**, 105–109.

FELTON, J. C. (1968b). Reports of recorders: Hymenoptera. *Bull. Kent Fd Club* **13**, 6–7.

FELTON, J. C. (1968c). Meeting Note: Upper Halling. *Bull. Kent Fd Club* **13**, 25–26.

FELTON, J. C. (1969). Further records of ants (Hymenoptera, Formicidae) from Kent. *Trans. Kent Fd Club* **3**, 202–210.

FELTON, J. C. (1971). Sphecidae (Hym.) from East Kent. *Entomologist's mon. Mag.* **106** (*1970*), 213.

FOX-WILSON, G. (1946). Factors affecting populations of social wasps, *Vespula* species, in England. *Proc. R. ent. Soc. Lond.*, A, **21**, 17–27.

FRISBY, G. E. (1928). The Hymenoptera of the Rochester District. *Rochester Naturalist* **6**, (131), 90–101.

FRISBY, G. E. (1934). *Andrena polita* Smith ♂ at Halling, Kent. *Entomologist's mon. Mag.* **70**, 136.

GEIGER, R. (1966). "The Climate near the Ground." Harvard University Press, Cambridge, Mass.

GUICHARD, K. M. and YARROW, I. H. H. (1948). The Hymenoptera Aculeata of Hampstead Heath and the surrounding district, 1832–1947. *Lond. Nat.* **27**, 81–111.

HAESELER, V. (1972). Anthropogene Biotope (Kahlschlag, Kiesgrube, Stadtgärten) als Refugien für Insekten, untersucht am Beispiel der Hymenoptera Aculeata. *Zool. Jb. Syst. Bd.* **99**, 133–212.

HALLETT, H. M. (1928). The Hymenoptera Aculeata of Glamorgan. *Trans. Cardiff Nat. Soc.* **60**, 33–67.

HALLETT, H. M. (1935). *Ectemnius dives* Lep. in Glamorgan. *Entomologist's mon. Mag.* **71**, 185–186.

LECLERCQ, J. (1968). Solitary bees and wasps (Hymenoptera Aculeata) in Kent, in the summer. *Entomologist's mon. Mag.* **104**, 30–42.

NIXON, G. E. J. (1935). A crabronid (Hymenoptera) new to the British List. *Entomologist's mon. Mag.* **71**, 57–58.

RICHARDS, O. W. (1935). A new British wasp, *Crossocerus exiguus* (van der Linden), (Hym., Sphecoidea). *Entomologist's mon. Mag.* **71**, 176–177.

RICHARDS, O. W. (1937). A check list of the British Hymenoptera Aculeata. *In* "The generic names of the British Hymenoptera Aculeata with a check list of British species" (R. B. Benson, C. Ferrière and O. W. Richards, eds), *The Generic Names of British Insects*, Part 5, 94–116. Royal Entomological Society of London, London.

RICHARDS, O. W. (1947). *Mimesa unicolor* of British authors (Hym, Sphecidae) is an undescribed species. *Ann. Mag. Nat. Hist.*, ser. 2, **14**, 871–876.

RICHARDS, O.W. (1964). The entomological fauna of Southern England with special reference to the country around London. *Trans. Soc. Br. Ent.* **16**, 1–48.

ROTHSCHILD, M. (1956). Notes on insect migration on the North Coast of France. *Entomologist's mon. Mag.* **92**, 375–376.

SAUNDERS, E. (1896). "The Hymenoptera Aculeata of the British Islands." Reeve, London.

SLADEN, F. W. L. (1908). *In* "Victoria History of the County of Kent" (W. Page, ed.), **1**, 108–122. London.

YARROW, I. H. H. (1943). Collecting bees and wasps. *In* Hymenopterist's Handbook. *Amat. Ent.* **7**, 55–81.

YARROW, I. H. H. (1954a). *Ancistrocerus gazella* (Panzer) (=*A. pictipes* Thomson), an abundant but hitherto undetected Eumenine wasp in Britain. *J. Soc. Br. Ent.* **5**, 78–82.

YARROW, I. H. H. (1954b). Records of Aculeate Hymenoptera at Hampstead. *Entomologist* **87**, 168–169.

YARROW, I. H. H. (1968). Recent additions to the British bee-fauna, with comments and corrections. *Entomologist's mon. Mag.* **104**, 60–64.

YARROW, I. H. H. and GUICHARD, K. M. (1941). Some rare Hymenoptera Aculeata, with two species new to Britain. *Entomologist's mon. Mag.* **77**, 2–13.

YEO, P. F. (1954). *Halictus smeathmanellus* (Kirby) (Hym., Apidea) and its allies in the midlands. *Entomologist's mon. Mag.* **90**, 221.

YEO, P. F. (1957). Bees and wasps in Bushy Park and at Hampton Hill, Middlesex. *Lond. Nat.* **36**, 16–24.

23 | Summing Up

K. MELLANBY

Monks Wood Experimental Station, The Nature Conservancy,
Abbots Ripton, Huntingdon

At the meeting of the Linnean Society of London held on 5 December 1935 on the topic "Changes in the British Fauna and Flora in the past fifty years", Dr K. G. Blair, who dealt with the insects, concluded his paper by saying:

> "... the need for records was stressed, records not only of the occasional occurrence of rarities, but of the annual fluctuation in numbers of some of the commoner species, records best accumulated by local societies throughout the country."

The papers reported here record not only changes in the numbers and distribution of many of our animals and plants, but also they show how we have proceeded to make our records in just the ways advocated by Dr Blair.

As Director of Monks Wood Experimental Station, I have been very gratified to note the contribution being made to systematic recording by our Biological Records Centre. However I, and the staff of that Centre, fully realize that any success depends on the accuracy of the material which reaches it, and I would like to express our gratitude to the many observers and recorders, both professional scientists and amateur naturalists, and particularly co-operating scientific societies, who provide the material which the Centre uses. I believe that this collaboration by those who are actively concerned with recording augurs well for the future, and may be one of the ways in which we can ensure that the scientific staff of the Nature Conservancy (who are being reorganized as the Institute of Terrestrial Ecology within the Natural Environment Research Council) will continue to direct much of their research into the field of wildlife conservation rather than into other and perhaps more academic branches of terrestrial ecology.

Systematics Association Special Volume No. 6, "The Changing Flora and Fauna of Britain", edited by D. L. Hawksworth, 1974, pp. 419–423. Academic Press, London and New York.

Our recording has vastly improved, but it is still very imperfect. Thus for the flowering plants we have a good picture of the range of most species, and very detailed records, perhaps the actual sites of every single plant, of a few of our greatest rarities. For birds we have good records of breeding success of many species, and here again every nest of the rarest breeds is identified, — though for both birds and rare flowering plants the actual locations may have to be kept secret to thwart sabotage by unscrupulous collectors. A number of speakers has already referred to the drastic effects of overcollecting rare species (Chapter 4). At the other end of the scale, we have organisms like some micro-fungi, whose spores are probably ubiquitous and so could be recorded, if they could be recognized, at almost every site; here we are recording the existence of the conditions which make the fungi manifest themselves (e.g. as an outbreak of a disease such as potato blight) in a form that can be easily recognized.

Dr Blair recognized the danger of devoting too great a proportion of our efforts to the study of rarities, and this danger remains. Here the label of "elitist" can justifiably be applied to many "conservationists". However, there has been a great improvement. The Common Bird Census of the British Trust for Ornithology is now providing invaluable information about species which were previously thought to be too widely distributed and too numerous to warrant our attention. Thus we have in the past found it difficult to quantify the effects of pesticides on Sparrow Hawks, which were previously so numerous, though we have been able to make good use of the excellent records for the previously much rarer Peregrine.

However, elitist or not, the study of rarities has a great value. These organisms may be very delicate indicators of environmental change, as well as worthy of conservation in their own right. Without them, our land would be seriously impoverished. All too few data are available on the factors limiting the occurrence of our rarest species and more research in this area is necessary. An accurate knowledge of a species' requirements is prerequisite to serious discussions on how to preserve it.

Throughout the meeting, speaker after speaker stressed the importance of conserving the habitat as crucial to conserving the various species of plants and animals. However, there was some disagreement about what "conserving the habitat" meant. Instances are known of ornithologists, wishing to persuade a species of bird now extinct in Britain and reduced to the status of an occasional visitor to remain and nest on his reserve, who would introduce exotic plants to that end. The whole problem of reserve management and conservation and the intentional or unintentional spread of animals and plants still remains unsolved.

A few speakers referred to the interactions between different organisms that occur in nature. It is clear that we still know very little of this aspect of ecology and the effects of the decline of an organism of one group on those of others. An enormous amount of co-operation between specialist systematists in different groups will be needed to remedy this situation. At the moment biologists should be concerned if any species is becoming endangered as a result of the activities of man simply because we do not know what effects its decline will have on other organisms, i.e. the ecosystem of which it is a part.

It can be argued, therefore, that at the present time systematists should be concerned with the conservation of ecosystems and not only rarities. From the papers presented here perhaps the most threatened single ecosystem in Britain at the present time is that of ancient deciduous woodland. The felling and disturbance of woodland has had the greatest effect on the species with the poorest means of dispersal to other woodland sites (which now are often likely to be many miles away). The loss of ancient woodlands has affected bryophytes (Chapter 3), lichens (Chapter 4), a few mammals (Chapter 11), terrestrial molluscs (Chapter 15), butterflies and moths (Chapter 16), spiders (Chapter 17), and beetles (Chapter 19). Some of the most sensitive indicators of ancient woodlands known are the lichens which can provide an indication of their degree of disturbance (Chapter 4). Ancient deciduous woodlands, such as the New Forest, are of a type now almost unknown elsewhere in Europe so their conservation (by leaving them unmanaged and undisturbed) is of more than national importance. Other habitats in which changes have led to the loss of species from diverse groups include heathlands, calcareous grasslands, hedgerows, fens and marshes. Few authors referred to any montane species being endangered on a national scale and it will be interesting to see how these fare in the future as public pressure in upland areas increases.

Speakers were all concerned with the "native" flora and fauna of Britain, but they did not agree in their definition of the term "native". The ornithologist welcomes any newcomer which arrives in Britain by its own efforts, and much work is devoted to recording "new" species which have been reliably sighted (Chapters 12, 13). In the nineteenth century these birds were shot and their skins treasured, and a disreputable trade of bogus importations flourished. The eggs of rare birds were also avidly collected. Today shooting rarities is considered wrong, and the nests of invaders are guarded with para-military efficiency. Any species introduced and released by man is not admitted to the records, and there is a fear that in the near future ornithological work will be jeopardized by the increasing number of species which will escape or be released from captivity. Entomologists, whose subjects are also mobile, and, to some extent those who are concerned with fungi with mobile spores, share the point of view of the ornithologists.

The views of botanists are somewhat different. They have had to live for the last 200 years with the situation now feared by the ornithologists. Thousands of exotic plants have been introduced by gardeners and agriculturalists, and many of these have escaped and become widespread. In the case of freshwater invertebrates, the main attention of the speaker was to such importations (Chapter 9). Britain has several new mammals which have been introduced and are now widespread (Chapter 11). Some accept the rabbit and the fallow deer, introduced many hundred years ago, as part of the native fauna, though they may hesitate to give this recognition to the grey squirrel which has only been with us for a hundred years, or the mink which has been present for a quarter of that time. But all agreed that it was important to keep accurate records of all these unintentional or intentional importations.

Introduced species have featured strongly in many of the papers. In some cases such species appear to have occupied a niche previously vacant, or one from which our native species have been eliminated by man-made factors, but in others their spread has been at the expense of native species. The subject of introduced species merits a symposium in its own right.

A great deal of attention was given to the understanding of the causes of changes in distribution and in the sizes of the populations. It was made clear that lichens (Chapter 4) and bryophytes (Chapter 3) are accurate indicators of air pollution, and that fish (Chapter 10), invertebrates (Chapter 9) and algae (Chapter 8) can "monitor" the state of rivers and lakes. This has produced a welcome interest in the commonest as well as the rarer species; the study of both can contribute to a better understanding of the environment.

Most of the changes that have taken place arise from the activities of man. In addition to the destruction and disturbance of "natural" and "semi-natural" habitats which have led to declines in some species, improved hygiene has, perhaps fortunately, affected some plant pathogenic fungi (Chapter 6), arthropod ectoparasites of man (Chapter 21), flies (Chapter 20), beetles (Chapter 19) and orthopterans (Chapter 18). Man has also created new habitats into which "native" species have been able to expand, and aided the spread of introduced and "native" species. This is seen in the extension of ranges of some vascular plants (Chapter 2), bryophytes (Chapter 3), lichens (Chapter 4), freshwater algae (Chapter 8), freshwater invertebrates (Chapter 9), freshwater fish (Chapter 10), terrestrial molluscs (Chapter 15) spiders (Chapter 17) and beetles (Chapter 19) in particular. In most groups, however, more species appear to have declined than to have been favoured by man. While the success of synanthropic species seems to be assured that of species not able to tolerate man's activities is less certain. Careful planning of new developments is needed to ensure that species in the latter category will remain a part of our fauna and flora.

Climatic changes in the last 100 years or so have affected some organisms much more than others. In the case of microfungi (Chapter 6) and macrofungi (Chapter 5) seasonal variations from year to year are important, while longer term trends related to climatic fluctuations are seen in the case of birds (Chapters 12, 13), seaweeds (Chapter 7), butterflies and moths (Chapter 16), terrestrial molluscs (Chapter 15) and aculeate Hymenoptera (Chapter 22). The less mobile bryophytes (Chapter 3), lichens (Chapter 4) and beetles (Chapter 19), in contrast, seem to be little affected by the climatic changes Britain has experienced in this period.

It is clear that neither animals nor plants remain exactly the same over long periods of time, and that evolution is still an active process. There are the records of the sudden spread of a previously rare species, now become "aggressive". There is the selection by pollution or other means of a strain which is resistant to a new condition. Isolated populations of Natterjack toads or of beetles or of flowering plants soon develop their own identity. Many conservationists oppose the transport of a species from its remaining site to others where it has become extinct on genetic grounds—"preserving the purity of the gene pool". Others think that species are so labile that the pool is already impure.

The general impression left by the symposium was one of surprising optimism Few recent extinctions, in any groups, were reported, and many new records of birds which have once more returned to Britain or of insects thought extinct which have been rediscovered were described. Most of the British mammals, for instance, are holding their own. Nevertheless the dangers of pollution and human pressure were not minimized, and the need for the active conservation of endangered species and habitats was obvious.

Vice-Counties of Great Britain and Ireland

In the foregoing chapters several authors have employed vice-county names and(or) numbers. The Watsonian vice-county system was originally proposed by Watson (1852) and has subsequently been extensively used for the recording of distributions of many groups of plants and animals, although it is now being superseded by 10 km square grid distribution maps, as will be evident from this volume.

The vice-county boundaries of Great Britain and Ireland are shown in Fig. 1, and the names to which the vice-county numbers correspond are listed below. In the past difficulties arose over defining the precise lines of division between adjacent vice-counties because of changes in the boundaries of administrative counties since Watson's time. The Systematics Association formed a Sub-committee on Maps and Censuses in 1947–48 to determine the precise boundaries of the vice-counties in the British Isles and their decisions have been incorporated into a definitive account of the vice-counties of Great Britain which has been published by Dandy (1969). Dandy's work is accompanied by two Ordnance Survey maps at a scale of 1:625 000 (about ten miles to one inch) on which the vice-county boundaries are clearly marked. A map at this same scale showing the boundaries of vice-counties in Ireland, as defined by Praeger (1896), was issued by the Ordnance Survey of Ireland in 1949.

Vice-county numbers and corresponding vice-counties

ENGLAND AND WALES

1. West Cornwall (with Isles of Scilly)	7. North Wiltshire	14. East Sussex
2. East Cornwall	8. South Wiltshire	15. East Kent
3. South Devon	9. Dorset	16. West Kent
4. North Devon	10. Isle of Wight	17. Surrey
5. South Somerset	11. South Hampshire	18. South Essex
6. North Somerset	12. North Hampshire	19. North Essex
	13. West Sussex	20. Hertfordshire

ENGLAND AND WALES—*contd.*

21. Middlesex
22. Berkshire
23. Oxfordshire
24. Buckinghamshire
25. East Suffolk
26. West Suffolk
27. East Norfolk
28. West Norfolk
29. Cambridgeshire
30. Bedfordshire
31. Huntingdonshire
32. Northamptonshire
33. East Gloucestershire
34. West Gloucestershire
35. Monmouthshire
36. Herefordshire
37. Worcestershire
38. Warwickshire
39. Staffordshire

40. Shropshire (Salop)
41. Glamorgan
42. Breconshire (Brecknockshire)
43. Radnorshire
44. Carmarthenshire
45. Pembrokeshire
46. Cardiganshire
47. Montgomeryshire
48. Merionethshire
49. Caernarvonshire
50. Denbighshire
51. Flintshire
52. Anglesey
53. South Lincolnshire
54. North Lincolnshire
55. Leicestershire (with Rutland)

56. Nottinghamshire
57. Derbyshire
58. Cheshire
59. South Lancashire
60. West Lancashire
61. South-east Yorkshire
62. North-east Yorkshire
63. South-west Yorkshire
64. Mid-west Yorkshire
65. North-west Yorkshire
66. Durham
67. South Northumberland
68. North Northumberland (Cheviotland)
69. Westmorland (with North Lancashire)
70. Cumberland
71. Isle of Man

SCOTLAND

72. Dumfrieshire
73. Kirkcudbrightshire
74. Wigtownshire
75. Ayrshire
76. Renfrewshire
77. Lanarkshire
78. Peebleshire
79. Selkirkshire
80. Roxburghshire
81. Berwickshire
82. East Lothian (Haddington)
83. Midlothian (Edinburgh)
84. West Lothian (Linlithgow)
85. Fifeshire (with Kinross-shire)

86. Stirlingshire
87. West Perthshire (with Clackmannanshire)
88. Mid Perthshire
89. East Perthshire
90. Angus (Forfar)
91. Kincardineshire
92. South Aberdeenshire
93. North Aberdeenshire
94. Banffshire
95. Morayshire (Elgin)
96. East Inverness-shire (with Nairn) (Easterness)
97. West Inverness-shire (Westerness)

98. Main Argyllshire
99. Dunbartonshire
100. Clyde Isles (Buteshire)
101. Kintyre
102. South Ebudes
103. Mid Ebudes
104. North Ebudes
105. West Ross
106. East Ross
107. East Sutherland
108. West Sutherland
109. Caithness
110. Outer Hebrides
111. Orkney Islands
112. Shetland Islands (Zetland)

IRELAND

H.1. South Kerry
H.2. North Kerry
H.3. West Cork

H.4. Mid Cork
H.5. East Cork
H.6. Waterford

H.7. South Tipperary
H.8. Limerick
H.9. Clare

IRELAND—*contd.*

H.10. North Tipperary	H.19. Kildare	H.30. Cavan
H.11. Kilkenny	H.20. Wicklow	H.31. Louth
H.12. Wexford	H.21. Dublin	H.32. Monaghan
H.13. Carlow	H.22. Meath	H.33. Fermanagh
H.14. Leix (Queen's County)	H.23. West Meath	H.34. East Donegal
	H.24. Longford	H.35. West Donegal
H.15. South-east Galway	H.25. Roscommon	H.36. Tyrone
H.16. West Galway	H.26. East Mayo	H.37. Armagh
H.17. North-east Galway	H.27. West Mayo	H.38. Down
H.18. Offaly (King's County)	H.28. Sligo	H.39. Antrim
	H.29. Leitrim	H.40. Londonderry

NOTE: The Channel Islands may be numbered "113" or "C" in some publications, and a few authors treat the Isles of Scilly as a separate unit 'S' and do not include them in vice-county 1. Dandy (1969) retains Watson's original, often shortened names for vice-counties (with a few minor exceptions) some of which differ from their full names presented above.

REFERENCES

DANDY, J. E. (1969). "Watsonian Vice-counties of Great Britain." Ray Society, London.
PRAEGER, R. L. (1896). On the botanical subdivision of Ireland. *Ir. Nat.* **5**, 29–38.
WATSON, H. C. (1852). "Cybele Britannica," vol. 3. Longman, London.

D. L. HAWKSWORTH

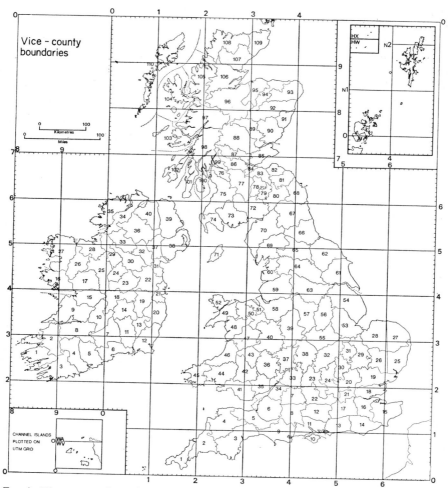

F<small>IG</small>. 1. Vice-county boundaries of the British Isles (reproduced from a map kindly supplied by the Biological Records Centre of the Nature Conservancy).

Species Index

Pages including distribution maps of species are placed in **bold** type.
Species are listed by both common and scientific names where appropriate whether one or both of these names appears on a particular page.

A

Abdera triguttata, 344
Abida secale, 257
Abramis brama, 158, 163, 165
Acanthinula lamellata, 267
Acanthis flammea, 211
Acanthophyma gowerensis, 297
Acanthoxyla prasina, 318–319
Acarospora fuscata, 69
    heppii, 69
Accipiter nisus, 206, 420
Acer, 55
Achatina fulica, 264
Acheta domesticus, 308, 316–318
Acicula fusca, 256, 266
Acilius canaliculatus, 145
Acipenser sturio, 176
Acme fusca, see *Acicula fusca*
Acrida bingleii, 311
Acritus homoeopathicus, 345
Acrocephalus arundinaceus, 224
    palustris, 210
Acronicta strigosa, 278–279
Acrulia inflata, 344
Adalia bipunctata, 343
    decempuncta, 361
Adder (*Vipera berus*), 229, 240-242
Adiantum capillus-veneris, 23
Adonis Blue (*Lysandra bellargus*), 280, 282
Aedes aegypti, 384
Agabus striolatus, 341
    wasastjernae, 330

Agile Frog (*Rana dalmatina*), 251
Agonum dorsale, 361
    ruficorne, 361
    sahlbergi, 340
Agriolimax agrestis, 259-260
    carnuanae, 259–260
    laevis, 258, 261
    reticulatus, 257, 260, 264, 266, 268, 270–271
Agrostemma githago, 19
Agrostis, 250
Ahasverus advena, 345
Albugo tragopogonis, 94
Alburnus alburnus, 159, 165
Alces alces, 181
Alectoria capillaris, 61
    fuscescens, 53, 66
    nitidula, 50–51
    ochroleuca, 72
    sarmentosa subsp. *sarmentosa*, 51
    trichodes agg., 50
Aleochara, 333
Allis Shad (*Alosa alosa*), 158, 170
Allomengea scopigera, 299
    warburtoni, 299
Aloina brevirostris, 34, 37
    rigida, 34, 37
Alopercus myosuroides, 19
Alosa alosa, 158, 170
    fallax, 158, 170
    sapidissima, 170
Alphitobius diaperinus, 345
Alphitophagus bifasciatus, 345

Subject Index

Names of suprageneric taxa occurring within the major groups treated in this volume are omitted.

A

Acari, 75

Aculeates, 309, 379, 381–384, 399–418, 423

Agricultural chemicals, 5–6, 19–21, 35, 39, 58–60, 85, 194, 200, 270, 276, 342, 365, 379–380, 422

Agriculture, 2, 5–6, 19, 35, 196, 250, 255, 269–272, 276, 278, 280, 341, 360, 363, 384, 413

Air pollution, 3–4, 34–38, 40–43, 51–58, 66, 73, 94, 342–343, 365, 422

Airports, 313

Algae, 58–59, 97–141, 348, 376, 422–423

Algal blooms, 60, 128

Algicides, 118

Ammonium-nitrogen, 60

Amphibians and reptiles, 229–254, 423

Amphipoda, 150

Anglers, 102, 132, 161, 175

Ants, *see* Aculeates

Arthropods, 293–305, 333, 350, 373, 393–397, 421–422

B

Bacteria, 115

Barnacles, 102

Bats, 179, 181, 190–191, 200

Beetles, 143–145, 166, 271, 323–369, 422–423

Birds, 5, 16–17, 66, 130, 172–173, 199, 203–227, 333, 373, 386, 420–421, 423

Bogs, 18, 33–35, 37–38, 297

Brickworks, 3

Bryophytes, 3, 27–46, 71, 115, 298, 421–423

Buildings, 2, 18, 66, 271, 294, 303–304, 310, 316–318, 341–342, 345, 347, 353, 384

Butterflies and moths, 275–292, 378, 421, 423

C

Canals, 23, 132, 148–150, 152, 160–161, 164–166, 175

Carbon dioxide, 2
monoxide, 37

Cars, 2–4, 185, 270

Cattle, 70, 183

Caves, 190, 304, 373

Centrarchids, 160–162

Climate, 1–2, 72–73, 84, 92–94, 104–105, 109, 118, 158, 215, 224–226, 239, 249, 255, 265–266, 269, 271, 275–276, 280, 282, 284, 286, 291–292, 333, 338, 340, 343, 346, 365, 374, 382, 410–411, 413, 423

Climbing, 3, 70

Coal mines, 304
tar products, 89

Coccidiosis, 187

Coleoptera, 143–145, 166, 271, 323–269, 422–423

Collecting, 18, 40, 71–72, 213, 246–248, 150–151, 263, 280, 420

Commercial exploitation, 40, 213

Coniferous forests and trees, 5, 18, 39, 85, 187, 195, 214, 251, 276, 286, 341, 344, 349, 353, 359